Organic Chemistry II FOR DUMMIES®

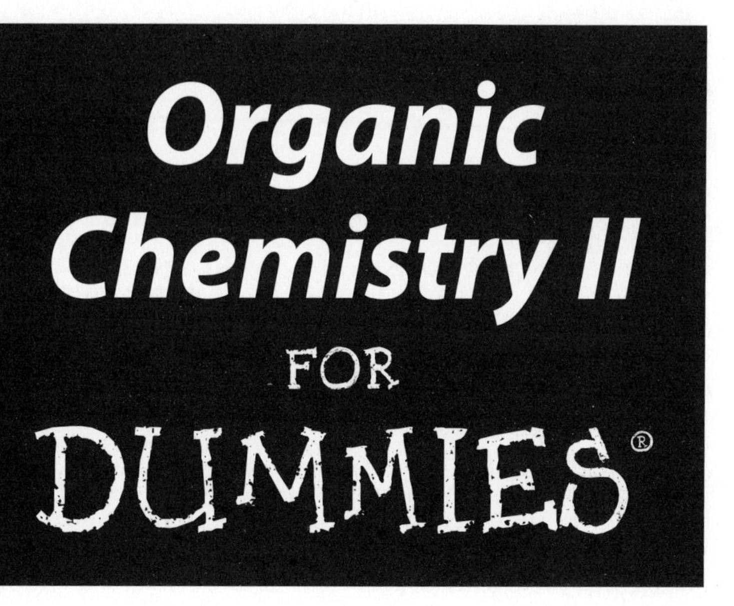

by John T. Moore, EdD, and
Richard H. Langley, PhD

Wiley Publishing, Inc.

Organic Chemistry II For Dummies®

Published by
Wiley Publishing, Inc.
111 River St.
Hoboken, NJ 07030-5774
www.wiley.com

For general information on our other products and services, please contact our Customer Care Department within the U.S. at 877-762-2974, outside the U.S. at 317-572-3993, or fax 317-572-4002.

For technical support, please visit www.wiley.com/techsupport.

Wiley also publishes its books in a variety of electronic formats. Some content that appears in print may not be available in electronic books.

Library of Congress Control Number: 2010926849

ISBN: 978-0-470-17815-7

Manufactured in the United States of America

10 9 8 7 6 5 4 3 2 1

About the Authors

John T. Moore, EdD, grew up in the foothills of western North Carolina. He attended the University of North Carolina–Asheville where he received his bachelor's degree in chemistry. He earned his master's degree in chemistry from Furman University in Greenville, South Carolina. After a stint in the United States Army, he decided to try his hand at teaching. In 1971 he joined the chemistry faculty of Stephen F. Austin State University in Nacogdoches, Texas, where he still teaches chemistry. In 1985 he went back to school part time and in 1991 received his doctorate in education from Texas A&M University. For the past several years he has been the co-editor (along with one of his former students) of the "Chemistry for Kids" feature of *The Journal of Chemical Education.* In 2003 his first book, *Chemistry For Dummies,* was published by Wiley, soon to be followed by *Chemistry Made Simple* (Broadway) and *Chemistry Essentials For Dummies* (Wiley). John enjoys cooking and making custom knife handles from exotic woods.

Richard H. Langley, PhD, grew up in southwestern Ohio. He attended Miami University in Oxford, Ohio, where he received bachelor's degrees in chemistry and in mineralogy and a master's degree in chemistry. His next stop was the University of Nebraska in Lincoln, Nebraska, where he received his doctorate in chemistry. Afterwards he took a postdoctoral position at Arizona State University in Tempe, Arizona, followed by a visiting assistant professor position at the University of Wisconsin–River Falls. In 1982 he moved to Stephen F. Austin State University. For the past several years he and John have been graders for the free-response portion of the AP Chemistry Exam. He and John have collaborated on several writing projects, including *5 Steps to a Five AP Chemistry* and *Chemistry for the Utterly Confused* (both published by McGraw-Hill). Rich enjoys jewelry making and science fiction.

Dedication

John: I dedicate this book to my wife, Robin; sons, Matthew and Jason; my wonderful daughter-in-law, Sara; and the two most wonderful grandkids in the world, Zane and Sadie. I love you guys.

Rich: I dedicate this book to my mother.

Authors' Acknowledgments

We would not have had the opportunity to write this book without the encouragement of our agent Grace Freedson. We would also like to thank Chrissy Guthrie for her support and assistance in the early portion of this project and to Sarah Faulkner who helped us complete it. We would also like to thank our copy editor, Caitie Copple, and our technical editors, Susan Klein and Joe Burnell.

Many thanks to our colleagues Russell Franks and Jim Garrett who helped with suggestions and ideas. Rich would also like to acknowledge Danica Dizon for her suggestions, ideas, and inspiration. Thanks to all of the people at Wiley publishing who help bring this project from concept to publication.

Publisher's Acknowledgments

We're proud of this book; please send us your comments at http://dummies.custhelp.com. For other comments, please contact our Customer Care Department within the U.S. at 877-762-2974, outside the U.S. at 317-572-3993, or fax 317-572-4002.

Some of the people who helped bring this book to market include the following:

Acquisitions, Editorial, and Media Development

Project Editors: Sarah Faulkner, Christina Guthrie

Senior Acquisitions Editor: Lindsay Sandman Lefevere

Copy Editor: Caitlin Copple

Assistant Editor: Erin Calligan Mooney

Senior Editorial Assistant: David Lutton

Technical Editors: Susan J. Klein, PhD, Joe C. Burnell, PhD

Editorial Manager: Christine Meloy Beck

Editorial Assistants: Jennette ElNaggar, Rachelle S. Amick

Cover Photos: © Haywiremedia | Dreamstime.com/ © iStock

Cartoons: Rich Tennant (www.the5thwave.com)

Composition Services

Project Coordinator: Patrick Redmond

Layout and Graphics: Nikki Gately

Proofreaders: Laura Albert, Sossity R. Smith

Indexer: Sharon Shock

Special Help Jennifer Tebbe

Publishing and Editorial for Consumer Dummies

Diane Graves Steele, Vice President and Publisher, Consumer Dummies

Kristin Ferguson-Wagstaffe, Product Development Director, Consumer Dummies

Ensley Eikenburg, Associate Publisher, Travel

Kelly Regan, Editorial Director, Travel

Publishing for Technology Dummies

Andy Cummings, Vice President and Publisher, Dummies Technology/General User

Composition Services

Debbie Stailey, Director of Composition Services

Contents at a Glance

Table of Contents

Introduction

Welcome to *Organic Chemistry II For Dummies.* We're certainly happy you decided to delve further into the fascinating world of organic chemistry. It's a complex area of chemistry, but understanding organic chemistry isn't really that difficult. It simply takes hard work, attention to detail, some imagination, and the desire to know. Organic chemistry, like any area of chemistry, is not a spectator sport. You need to interact with the material, try different study techniques, and ask yourself why things happen the way they do.

Organic Chemistry II is a more intricate course than the typical freshman introductory chemistry course, and you may find that it's also more involved than Organic I. You may actually need to use those things you learned (and study habits you developed) in Organic I to be successful in Organic II. But if you work hard, you can get through your Organic II course. More importantly, you may grow to appreciate the myriad chemical reactions that take place in the diverse world of organic chemistry.

About This Book

Organic Chemistry II For Dummies is an overview of the material covered in the second half of a typical college-level organic chemistry course. We have made every attempt to keep the material as current as possible, but the field of chemistry is changing ever so quickly as new reactions are developed and the fields of biochemistry and biotechnology inspire new avenues of research. The basics, however, stay the same, and they are where we concentrate our attention.

As you flip through this book, you see a lot of chemical structures and reactions. Much of organic chemistry involves knowing the structures of the molecules involved in organic reactions. If you're in an Organic Chemistry II course, you made it through the first semester of organic chemistry, so you recognize many of the structures, or at least the functional groups, from your previous semester's study.

If you bought this book just to gain general knowledge about a fascinating subject, try not to get bogged down in the details. Skim the chapters. If you find a topic that interests you, stop and dive in. Have fun learning something new.

If you're taking an organic chemistry course, you can use this rather inexpensive book to supplement that very expensive organic textbook.

Conventions Used in This Book

We have organized this book in a logical progression of topics; your second semester organic chemistry course may progress similarly. In addition, we set up the following conventions to make navigating this book easier:

- *Italics* introduce new terms that you need to know.
- **Bold** text highlights keywords within a bulleted list.
- We make extensive use of illustrations of structures and reactions. While reading, try to follow along in the associated figures, whether they be structures or reactions.

What You're Not to Read

You don't have a whole lot of money invested in this book, so don't feel obliged to read what you don't need. Concentrate on the topic(s) in which you need help. Feel free to skip over any text in a gray shaded box (which we refer to as sidebars). Although interesting, they aren't required reading.

Foolish Assumptions

We assume — and we all know about the perils of assumptions — that you are one of the following:

- A student taking a college-level organic chemistry course.
- A student reviewing organic chemistry for some type of standardized exam (the MCAT, for example).
- An individual who just wants to know something about organic chemistry.

If you fall into a different category, you're special and we hope you enjoy this book anyway.

How This Book Is Organized

The topics in this book are divided into six parts. Use the following descriptions and the table of contents to map out your strategy of study.

Part I: Brushing Up on Important Organic Chemistry I Concepts

Part I is really a rapid review of many of the concepts found in an Organic Chemistry I course. It's designed to review the topics that you need in Organic II. We set the stage by giving you an overview of Organic Chemistry II, and then review mechanisms. Next we cover alcohols and ethers, their properties, synthesis, and reactions; followed by an overview of conjugated unsaturated systems. We end this review section with a discussion of spectroscopy, including IR, UV-visible, mass spec, and, of course, NMR. A whirlwind tour of Organic I!

Part II: Discovering Aromatic (And Not So Aromatic) Compounds

In Part II we concentrate on aromatic systems, starting with the basics of structure and properties of benzene and then moving on to related aromatic compounds. We even throw in a section of spectroscopy of aromatic compounds. Chapters 7 and 8 finish up this part by going into detail about substitution reactions of aromatic compounds. You find out all you ever wanted to know (and maybe more) about electrophilic and nucleophilic substitutions, along with a little about elimination reactions.

Part III: Carbonyls: Good Alcohols Gone Bad

In Part III we cover that broad category of organic compounds called the carbonyls. First we give you an overview of carbonyl basics, including structure, reactivity, and spectroscopy. Then we go into more detail on aldehydes and ketones, enols and enolates, and carboxylic acids and their derivatives.

Part IV: Advanced Topics (Every Student's Nightmare)

In Part IV we start by taking a closer look at nitrogen compounds and their structure, reactivity, and reactions. Then we move on to organometallic compounds, where we meet the infamous Grignard reaction. We then finish up this part by addressing some more-involved reactions of the carbonyls and biomolecules. You pick up some good hints for synthesis and roadmaps here.

Part V: Pulling It All Together

In Part V we show you how to pull all the previous information together and use it to develop strategies for designing synthesis reactions. We talk about both one-step and multistep synthesis as well as retrosynthetic analysis. Then we tackle the dreaded organic roadmaps. (We all wish we had an organic chemistry GPS here.)

Part VI: The Part of Tens

In this final part of the book we discuss ten surefire ways to flunk your organic chemistry class (so you know what to avoid) along with ten ways to increase your grade on those organic chemistry exams.

Icons Used in This Book

If you have ever read other *For Dummies* books (such as the wonderful *Chemistry For Dummies* or *Biochemistry For Dummies,* written by yours truly and published by Wiley), you recognize the icons used in this book. The following four icons can guide you to certain kinds of information:

This icon is a flag for those really important things that you shouldn't forget as you go deeper into the world of organic chemistry.

We use this icon to alert you to a tip on the easiest or quickest way to learn a concept. Between the two of us, we have almost 70 years of teaching experience. We've learned a few tricks along the way and we don't mind sharing.

The warning icon points to a procedure or potential outcome that can be dangerous. We call it our Don't-Try-This-At-Home icon.

We try to avoid getting too technical throughout this book (believe it or not), but every now and then we can't help but throw something in that is a little more in-depth than you might need. You won't hurt your education by skipping it.

Where to Go from Here

The answer to this question really depends of your prior knowledge and goals. As with all *For Dummies* books, this one attempts to make all the chapters independent so that you can dive right into the material that's causing you trouble without having to read other chapters first. If you feel comfortable with the topics covered in Organic Chemistry I, feel free to skip Part I. If you want a general overview of organic chemistry, skim the remainder of the book. Take a deeper plunge into a chapter when you find a topic that interests you or one in which you really need help.

And for all of you, no matter who you are or why you're reading this book, we hope you have fun reading it and that it helps you to understand and appreciate organic chemistry.

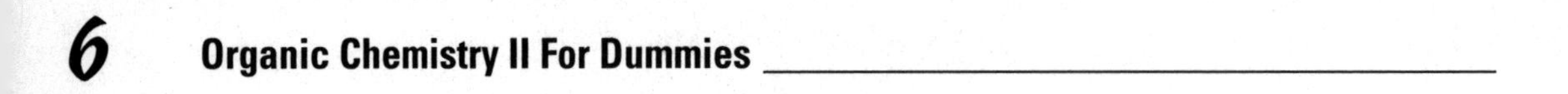

Part I

Brushing Up on Important Organic Chemistry I Concepts

In this part . . .

Part I is a review of some general chemistry and Organic Chemistry I topics you need a firm grounding in before moving on to Organic Chemistry II. Different books and different instructors break Organic I and Organic II material at different places. We use the most common break, but some Part I material may, in fact, be new to you. Even if you covered these concepts last semester, some of them have a high vapor pressure and may have escaped between semesters.

We begin by bringing you up to speed on mechanisms and reminding you how to push electrons around with those curved arrows. We jog your memory with a discussion of substitution and elimination reactions and their mechanisms, in addition to free radical reactions. Next you review the structure, nomenclature, synthesis, and reactions of alcohols and ethers, and then you get to tackle conjugated unsaturated systems. Finally, we remind you of spectroscopic techniques, from the IR fingerprints to NMR shifts. The review in this part moves at a pretty fast pace, but we're sure you can keep up.

Chapter 1

Organic Chemistry II: Here We Go Again!

In This Chapter

- Reviewing the material you learned in Organic I
- Previewing what you find out in Organic II

If you're looking at this chapter, it's probably because you're getting ready to take the second half of organic chemistry, are in the midst of Organic II, or you're trying to figure out what Organic II covers in time to change your major from pre-med to art history. In any respect, you probably successfully completed Organic Chemistry I. Many of the study techniques (and coping mechanisms) you learned that helped you do well in Organic I are helpful in Organic II. The two primary things to remember are

- *Never* get behind.
- Carbon has four bonds.

In this book we use larger, more complex molecules than you may have encountered in Organic I. We chose to do this because, firstly, that's the nature of Organic II — larger and more complex molecules. Secondly, many of you will be taking biochemistry at some point, and to succeed in that subject you need to become comfortable with large, involved molecules. (If you do take biochemistry, be sure to check out *Biochemistry For Dummies* by John T. Moore and Richard H. Langley [Wiley]. We understand the authors are really great guys.)

To get you started, this chapter does a quick review of the topics commonly found in Organic I, and then gives an overview of what we cover in Organic II.

Recapping Organic Chemistry I

In Organic I you learned that organic chemistry is the study of carbon compounds. Until the mid-1800s, people believed that all carbon compounds were the result of biological processes requiring a living organism. This was called the *vital force theory*. The synthesis (or formation) of urea from inorganic materials showed that other paths to the production of carbon compounds are possible. Many millions of organic compounds exist because carbon atoms form stable bonds to other carbon atoms. The process of one type of atom bonding to identical atoms is *catenation*. Many elements can catenate, but carbon is the most effective, with apparently no limit to how many carbon atoms can link together. These linkages may be in chains, branched chains, or rings, providing a vast combination of compounds.

Carbon is also capable of forming stable bonds to a number of other elements, including the biochemically important elements hydrogen, nitrogen, oxygen, and sulfur. The latter three elements form the foundation of many of the functional groups you studied in Organic I.

Intermolecular forces

You also learned about intermolecular forces in Organic I. Intermolecular forces (forces between chemical species) are extremely important in explaining the interaction between molecules. Intermolecular forces that you saw in Organic I and see again in Organic II include dipole-dipole interactions, London, hydrogen bonding, and sometimes ionic interactions.

Dipole-dipole forces exist between polar regions of different molecules. The presence of a dipole means that the molecule has a partially positive ($\delta+$) end and a partially negative ($\delta-$) end. Opposite partial charges attract each other, whereas like partial charges repel.

Hydrogen bonding, as the name implies, involves hydrogen. This hydrogen atom must be bonded to either an oxygen atom or a nitrogen atom. (In nonbiological situations, hydrogen bonding also occurs when a hydrogen atom bonds to a fluorine atom.) Hydrogen bonding is significantly stronger than a normal dipole-dipole force, and is stronger than London dispersion forces, the forces between nonpolar molecules due to the fluctuations of the electron clouds of atoms or molecules. The hydrogen bonded to either a nitrogen or oxygen atom is strongly attracted to a different nitrogen or oxygen atom. Hydrogen bonding may be either intramolecular or intermolecular.

In organic reactions, ionic interactions may serve as intermolecular or intramolecular forces. In some cases, these may involve metal cations, such as Na^+, or anions, such as Cl^-. Cations may include an ammonium ion from an amino group, such as RNH_3^+. The anion may be from a carboxylic acid, such as $RCOO^-$. The oppositely charged ions attract each other very strongly.

Functional groups

Carbon is an extremely versatile element because it can form many different compounds. Most of the compounds have one or more *functional groups,* which contain atoms other than carbon and hydrogen and/or double or triple bonds, and define the reactivity of the organic molecule.

In Organic I you probably started with the hydrocarbons, compounds of carbon and hydrogen, including the alkenes and alkynes that contained double and single bonds, respectively. Then you probably touched on some of the more common functional groups, such as alcohols and maybe even some aromatic compounds.

Reactions

You encountered a lot of reactions in Organic I. Every time you encountered a different functional group, you had a slew of reactions to learn. Reactions that told how the functional group could be formed, common reactions that the functional group underwent — reactions, reactions, and more reactions.

Two of the most important ones you learned were substitution and elimination reactions: S_N1, S_N2, E1, and E2. We hope you learned them well, because you'll be seeing them again quite often.

Spectroscopy

In Organic I you probably learned a lot about the different types of spectroscopy and how they're used in structure determinations. You discovered how mass spectroscopy can give you an idea about molar mass and what fragments may be present in the molecule. You found out that infrared spectroscopy can be used to identify functional groups, and you learned to look at the fingerprint region. Then finally you progressed to nuclear magnetic resonance (NMR) spectroscopy, one of the main tools of organic chemists, which can be used to interpret chemical shifts and splitting patterns to give you more clues about structure.

Isomerism and optical activity

During Organic I you were exposed to the concepts associated with isomerism and optical activity. You need to be familiar with these concepts in Organic II, so we take a few minutes here for a brief review.

Isomers are compounds with the same molecular formula but different structural formulas. Some organic and biochemical compounds may exist in different isomeric forms, and these different isomers have different properties. The two most common types of isomers in organic systems are *cis-trans* isomers and isomerism due to the presence of a chiral carbon.

Cis-trans isomers

The presence of carbon-carbon double bonds leads to the possibility of isomers. Double bonds are rather restrictive and limit molecular movement. Groups on the same side of the double bond tend to remain in that position *(cis)*, while groups on opposite sides tend to remain across the bond from each other *(trans)*. You can see an example of each in Figure 1-1. However, if the two groups attached to either of the carbon atoms of the double bond are the same, *cis-trans* isomers are not possible. *Cis* isomers are the normal form of fatty acids, but processing tends to convert some of the *cis* isomers to the *trans* isomers.

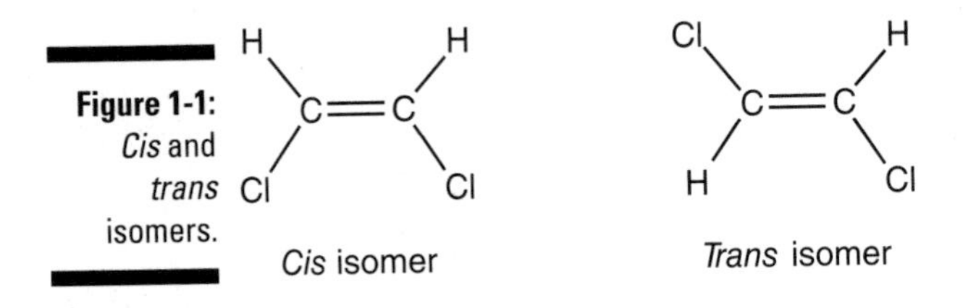

Figure 1-1: *Cis* and *trans* isomers.

Cis-trans isomers are also possible in cyclic systems. The *cis* form has similar groups on the same side of the ring, while the *trans* form has similar groups above and below the ring.

Chiral compounds

A carbon atom with four different groups attached is *chiral.* A chiral carbon rotates plane-polarized light, light whose waves are all in the same plane, and has an *enantiomer* (non-superimposable mirror image). Rotation, which may be either to the right (dextrorotatory) or to the left (levorotatory), leads to one optical isomer being *d* and the other being *l.* Specific rotation (represented by $[\alpha]_D^T$, where α = observed rotation, T = temperature, and D = sodium D line) is a measure of the ability of a compound to rotate light. The specific rotation comes from the observed rotation (α) divided by the product of the concentration of the solution and the length of the container. Other than optical activity, the physical properties of enantiomers are the same.

A *racemic mixture* is a 50:50 mixture of the enantiomers.

A *meso compound* is a compound with chiral centers and a plane of symmetry. The plane of symmetry leads to the optical rotation of one chiral carbon cancelling the optical rotation of another.

Diastereomers are stereoisomers that aren't enantiomers.

R-S notation is a means of designating the geometry around the chiral center. This method requires the groups attached to the chiral center to be prioritized in order of decreasing atomic weight. To assign the center, place the lowest priority group (the group with the lowest atomic weight) on the far side and count the remaining groups as 1, 2, and 3. Counting to the right is *R* and counting to the left is *S*. Any similarity between *d* and *l* and *R* and *S* is coincidental.

Some important organic compounds have more than one chiral center. Multiple chiral centers indicate the presence of multiple stereoisomers. The maximum number of stereoisomers is 2^n where n is the number of non-identical chiral centers. Figure 1-2 shows the four stereoisomers present in a molecule with two chiral centers. Non-superimposable mirror images are enantiomers, while the other species in the figure are diastereomers. Unlike enantiomers, diastereomers have different physical properties.

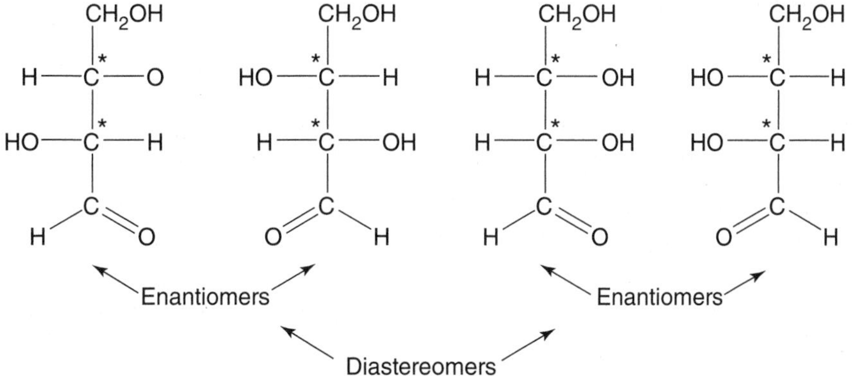

Figure 1-2: Representations of a molecule with two chiral centers.

Emil Fischer developed a method of drawing a compound to illustrate which stereoisomer is present. Drawings of this type, called *Fischer projection formulas,* are very useful in biochemistry. In a projection formula, a chiral carbon is placed in the center of a + pattern. The vertical lines (bonds) point away from the viewer, and the horizontal lines point towards the viewer. Fischer used the *D* designation if the most important group was to the right of the carbon, and the *L* designation if the most important group was to the left of the carbon. (See Figure 1-3.)

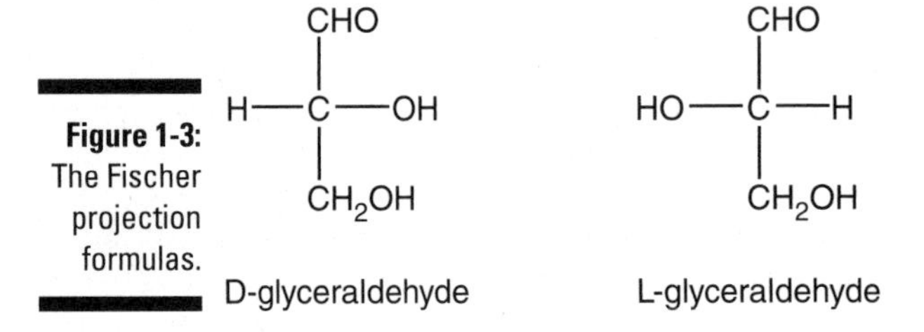

Figure 1-3: The Fischer projection formulas.

The use of D and L is gradually being replaced by the R and S system of designating isomers, which is particularly useful when more than one chiral carbon atom is present.

Looking Ahead to Organic Chemistry II

One of the keys to Organic II is *mechanisms,* the specific way in which a reaction proceeds. Recall from Organic I that this involves pushing around electrons, showing where they're going with curved arrows. We give you a good review of these concepts in Chapter 2, along with some basic reaction moves.

In Chapter 3 we go into some depth about alcohols and ethers. Like Organic I, when we encounter a new functional group we examine the structure, nomenclature, properties, synthesis, and reactions. In some courses and textbooks, alcohols are covered in the first semester, but for those readers who haven't gotten to them yet, we include them in this book. If you're already comfortable with that material, please feel free to skip that chapter and go on to another.

Conjugated unsaturated systems are an important part of organic chemistry, so in Chapter 4 we spend a little time talking about those systems, setting the stage for our discussion of aromatic compounds that you can find in Chapter 6.

To bring you up to speed on spectroscopy, we cover the basics in Chapter 5. We give you the executive summary on infrared (IR), ultraviolet-visible (UV-vis), mass spectrometry (mass spec), and nuclear magnetic resonance (NMR). In addition, many of the chapters in this book have a spectroscopy section at the end where we simply cover the essentials concerning the specific compounds that you study in that chapter.

Aromatic compounds and their reactions are a big part of any Organic II course. We introduce you to the aromatic family, including the heterocyclic branch, in Chapter 6. (You may want to brush up on the concept of resonance beforehand.) Then in Chapters 7 and 8, you find out more than you ever wanted to know about aromatic substitution reactions, starring electrophiles and nucleophiles.

Another important part of Organic II is carbonyl chemistry. We look at the basics of the carbonyls in Chapter 9. It's like a family reunion where I (John, one of your authors) grew up in North Carolina — everybody is related. You meet aldehydes, ketones, carboxylic acids, acyl chlorides, esters, amides, and on and on. It's a quick peek, because later we go back and examine many of these in detail. For example, in Chapter 10 you study aldehydes and ketones, along with some of the amines, while in Chapter 11 we introduce you to other carbonyl compounds, enols and enolates, along with nitroalkanes and nitriles.

Carboxylic acids and their derivatives are also an important part of Organic II. We spend quite a few pages looking at the structure, nomenclature, synthesis, reactions, and spectroscopy of carboxylic acids. While on this topic in Chapter 12, we use a lot of acid-base chemistry, most of which you were exposed to in your introductory chemistry course. (For a quick review, look over a copy of *Chemistry For Dummies* or *Chemistry Essentials For Dummies,* both written by John T. Moore and published by Wiley.)

Carbon compounds that also contain nitrogen, such as the amines, play a significant part of any Organic II course. You encounter more acid-base chemistry with the amines, along with some more reactions. We hit this topic in Chapter 13 and give you some tips for multistep synthesis.

You probably haven't considered the fact that some organic compounds may contain a metal, so we give you an opportunity to become familiar with the organometallics in Chapter 14. In this chapter you meet the Grignard reaction. It's a very important organic reaction that you may have the opportunity to run in organic lab.

You just can't get away from those carbonyls, so you get another taste of these reactions, many of them named reactions, in Chapter 15. You may be able to avoid biomolecules if your course doesn't cover them, but if it does, Chapter 16 is there for you.

Finally, what's a good organic course without multistep and retrosynthesis along with roadmaps? We hope that our tips can ease your pain at this point. Roadmaps are the bane of most organic chemistry students, but just hang in there. There is life after organic chemistry, and you may just find in the end that you actually enjoyed organic. And for those of you who missed the chemical calculations, there's always quantitative analysis and physical chemistry.

Chapter 2

Remembering How We Do It: Mechanisms

In This Chapter

- Analyzing arrows
- Breaking down basic moves
- Contemplating combining basic moves
- Mastering free-radical mechanisms

Mechanisms are the key to organic chemistry. Understanding the mechanism allows organic chemists to control the reaction and to avoid unwanted side reactions. Understanding the mechanism many times allows chemists to increase the yield of product.

In this chapter you review the basics of mechanisms and their conventions and look at some of the more common ways that electrons shift during a reaction. You also see how these individual steps can fit together in the overall reaction mechanism and apply some of these techniques in free-radical mechanisms.

Duck — Here Come the Arrows

Many types of arrows are used in organic chemistry, and each of them conveys information about the particular reaction. These arrows include the resonance arrow, equilibrium arrow, reaction arrow, double-headed arrow, and single-headed arrow.

The *resonance arrow,* a single line with arrow heads at both ends (see Figure 2-1), separates different resonance structures. The actual structure is a weighted average of all resonance forms. More resonance forms usually indicate a more stable structure. One or more of the resonance structures may be useful in predicting what will happen during a reaction (a mechanism).

Figure 2-1: The resonance arrow.

The *equilibrium arrow,* which has two lines pointing in opposite directions (see Figure 2-2), separates materials that are in equilibrium. Materials on each side of the arrow are present. Unlike a resonance arrow, the materials actually exist and aren't a hybrid. If one of the arrows is longer than the other is, it indicates that one side of the equilibrium predominates over the other.

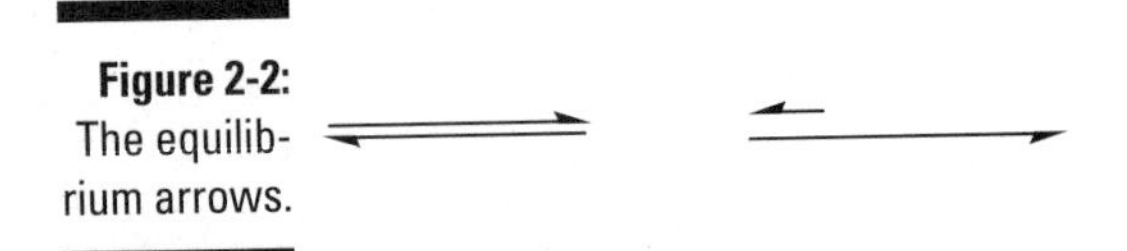

Figure 2-2: The equilibrium arrows.

A *reaction arrow,* a single arrow pointing in one direction (see Figure 2-3), simply separates the reactants from the products. The use of this arrow usually indicates that the reaction proceeds in only one direction (unlike the equilibrium arrow).

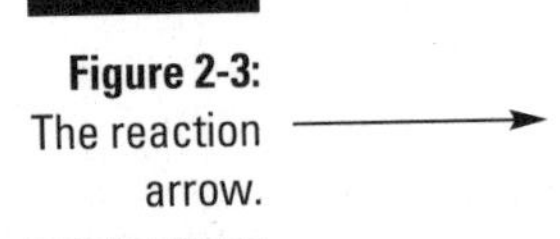

Figure 2-3: The reaction arrow.

Double- and single-headed curved arrows indicate the movement of electrons. *Double-headed curved arrows* (shown in Figure 2-4a) show the movement of two electrons, whereas *single-headed curved arrows* (Figure 2-4b) indicate the movement of one electron. The electrons always move in the direction indicated by the arrow. The head (point) of the arrow is where the electron is going, and the tail is the electron's source.

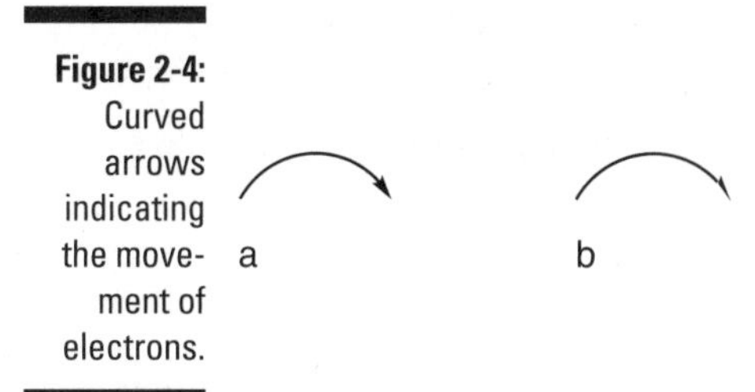

Figure 2-4: Curved arrows indicating the movement of electrons.

Coming Around to Curved Arrows

Mechanisms, like resonance structures, utilize curved arrows. (*Resonance structures* are ways of illustrating the various resonance forms that contribute to the resonance hybrid. If you need more review, refer to *Organic Chemistry I For Dummies.*) Many of the same rules apply to both; however, there are some important differences:

- In resonance, the electrons don't actually move, whereas in mechanisms there is an actual movement of electrons.
- In resonance, you should never, ever break a single bond; however, many mechanisms involve the breaking of a single bond. Nonetheless, you should never, ever exceed an octet of electrons for any atom in the second period.

A mechanism provides a means toward understanding why a reaction occurred. When you understand why a reaction occurred, you're much closer to understanding organic chemistry. Reactions involve the breaking and the forming of bonds. The mechanism shows how the electrons move (flow) to break old bonds and to form new bonds. Curved arrows indicate the flow of the electrons from the nucleophile (electron donor) to the electrophile (electron acceptor).

To be successful in organic chemistry, you must know the mechanism for the reaction you're studying.

Mechanisms in this book are, in general, advanced examples of mechanisms appearing in *Organic Chemistry I For Dummies.* A college organic chemistry course presents very few completely new mechanisms. Perfecting a few mechanisms goes a long way toward understanding all reaction mechanisms, and, therefore, all organic reactions. Although many students feel that memorization is important, understanding the mechanism is what's necessary to comprehend organic chemistry. If you simply memorize mechanisms, you'll become hopelessly confused by even minor changes; however, if you understand a mechanism thoroughly, you can accommodate any changes.

Keep two things in mind when drawing curved arrows: The tail of the arrow needs to be in the right place, and the head of the arrow needs to be in the right place. (Simple, right?) Don't forget that electrons occupy orbitals. Other than radicals, the electrons in the orbitals are either bonding pairs or lone pairs. This means that the tail of the curved arrow must be at a lone pair or a bonding pair. (A radical may have the tail of the curved half arrow originating at the unpaired electron.) The head of the curved arrow indicates where a lone pair is going or where a bond will form.

Getting Ready for Some Basic Moves

The tail of a curved arrow has two possible positions, and the head of a curved arrow has two possible positions. This means that, in theory, four combinations are possible. These combinations are

- Bond → lone pair
- Bond → bond
- Lone pair → bond
- Lone pair → lone pair

The last combination doesn't work, at least not in a single step, because it tends to force an atom to exceed an octet. This leaves only three important types.

The basic idea behind these reactions is the same: An electron-rich atom with a lone pair (a nucleophile) donates that lone pair to an electron-poor atom (an electrophile).

Bond → lone pair

An example of the bond–to–lone pair combination is shown in Figure 2-5. In this example, the tail of the curved arrow begins at the bonding pair. The head of the curved arrow is at the chlorine atom where it forms a lone pair. The overall charge doesn't change. The original compound was neutral (0 charge), the products are +1 – 1 = 0.

Figure 2-5: Bond–to–lone pair movement.

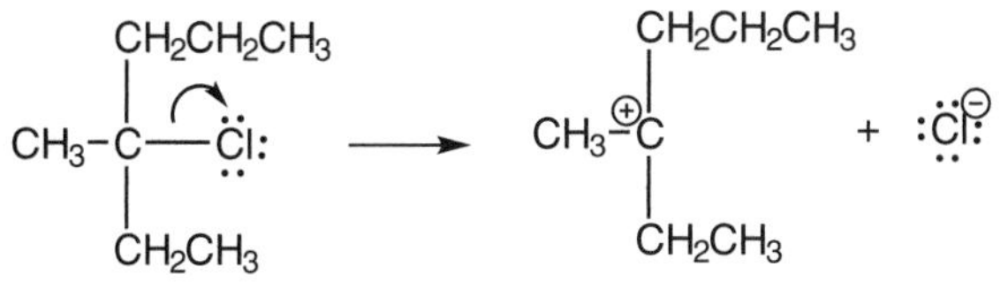

Bond → bond

An example of a bond-to-bond step is shown in Figure 2-6. The tail of the curved arrow begins at one of the bonding pairs of the double bond (the π-bond), while the head points to where the new π-bond will form.

Figure 2-6: Bond-to-bond movement.

A more common example of this process involves two arrows and the shifting of two electron pairs. An example of this process is shown in Figure 2-7. The tail of the curved arrow again begins at one of the bonding pairs of the π-bond, while the head points to where the new bond will form. This movement forces the bonding pair between the hydrogen and oxygen to move to the oxygen atom to create a lone pair on the oxygen atom.

Figure 2-7: Two electron-pair movements.

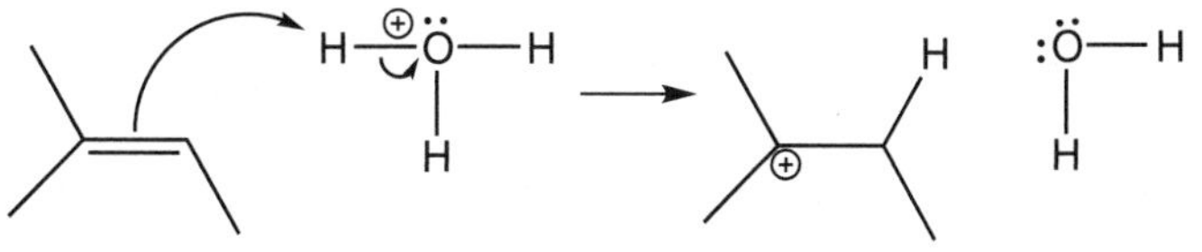

When more than one curved arrow is present, they should all point in the same general direction and never toward each other or away from each other. However, curved single-headed arrows do not necessarily follow this rule.

Lone pair → bond

An example of the lone pair–to–bond step is shown in Figure 2-8. In this step, the tail of the curved arrow begins at the lone pair. The head of the curved arrow is going to form the C-N bond. Notice that there's conservation of the positive charge. In any mechanism, the overall charge must remain the same.

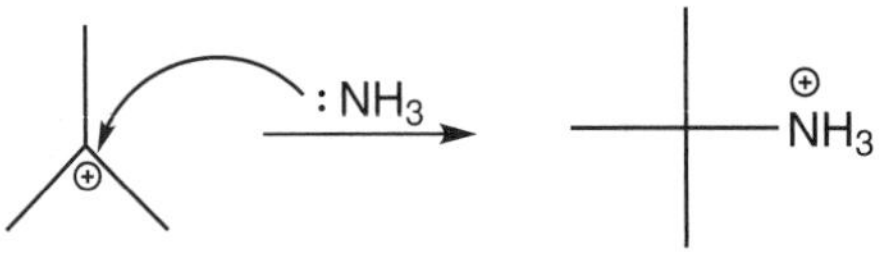

Figure 2-8: Lone pair–to–bond movement.

$$(CH_3)_3C^{\oplus} + :NH_3 \longrightarrow (CH_3)_3C\text{-}\overset{\oplus}{N}H_3$$

Combining the Basic Moves

A common error in a mechanism is to attempt to do too much in a single mechanism step, and a sure sign that you're trying to do too much is having arrows pointing in opposite directions in the step. You can have arrows pointing in one direction in a step and in the opposite direction in the next step, but resist the temptation to combine these two steps.

The best way to see how these steps work together is with an example. Begin by examining the conversion of t-butyl alcohol to t-butyl chloride. This process has a 96 percent yield. (This is a good thing!) The overall reaction is shown in Figure 2-9.

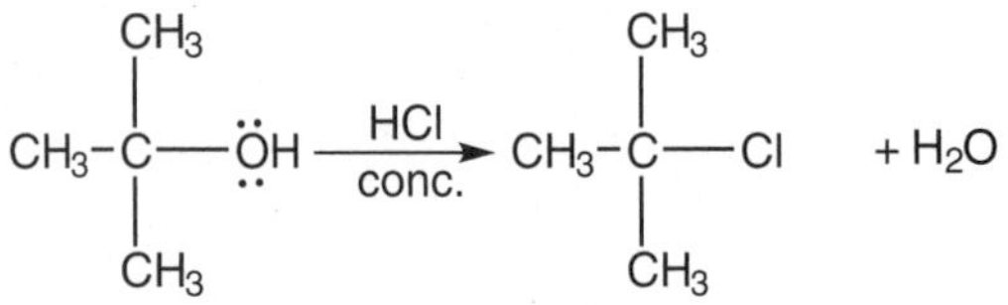

Figure 2-9: Conversion of t-butyl alcohol to t-butyl chloride.

$$(CH_3)_3C\text{—}\ddot{O}H \xrightarrow[\text{conc.}]{HCl} (CH_3)_3C\text{—}Cl + H_2O$$

Step 1: This reaction, like most reactions involving an acid, begins with protonation. In this case, a lone pair from the oxygen forms a bond to the hydrogen from the hydrochloric acid. (See Figure 2-10.) The movement of the oxygen lone pair to the hydrogen "pushes" the bonding pair from the H-Cl bond onto the chlorine. This is a lone pair–to-bond transfer, which induces a bond–to–lone pair transfer.

Protonation is almost always the first step in mechanisms involving an acid.

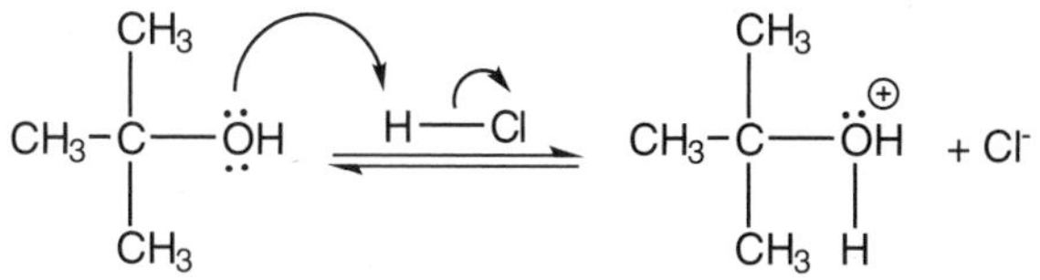

Figure 2-10: Step 1: Lone pair–to–bond movement.

$$(CH_3)_3C\text{–}\ddot{O}H + H\text{–}Cl \rightleftharpoons (CH_3)_3C\text{–}\ddot{O}H_2^{\oplus} + Cl^-$$

Step 2: The presence of a positive charge on the oxygen atom is unstable because the oxygen has such a high electronegativity. The bonding pair from the C-O bond moves to the positive charge on the oxygen to become a lone pair. (See Figure 2-11.) In this case, this is the rate-controlling step, which is why this is an example of an S_N1 mechanism. The water molecule formed is a good leaving group, which facilitates this reaction. (The OH group is not a good leaving group.) This is a bond–to–lone pair transfer.

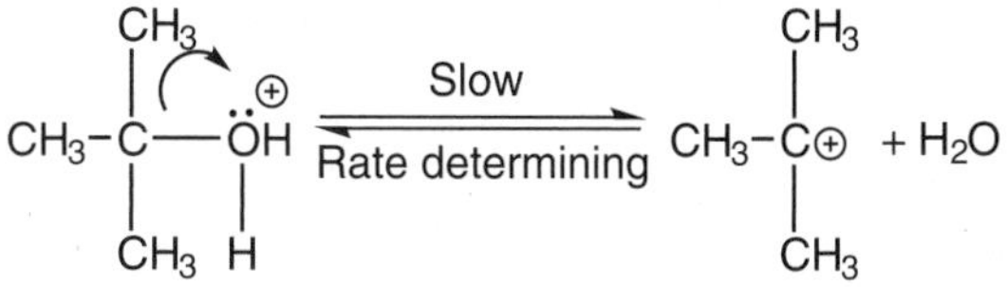

Figure 2-11: Step 2: Bond–to–lone pair movement.

$$(CH_3)_3C\text{–}\ddot{O}H_2^{\oplus} \underset{\text{Rate determining}}{\overset{\text{Slow}}{\rightleftharpoons}} (CH_3)_3C^{\oplus} + H_2O$$

Step 3: Forming a carbocation is difficult; however, tertiary carbocations, such as this one, can form as *intermediates,* or species that exist for a short time during the reaction. (See Figure 2-12.) The positive charge on the carbon makes this a strong electrophile that seeks a lone pair. In the final step of this mechanism, the carbocation accepts a lone pair from the chloride ion generated in the first step. The transfer is lone pair to bond.

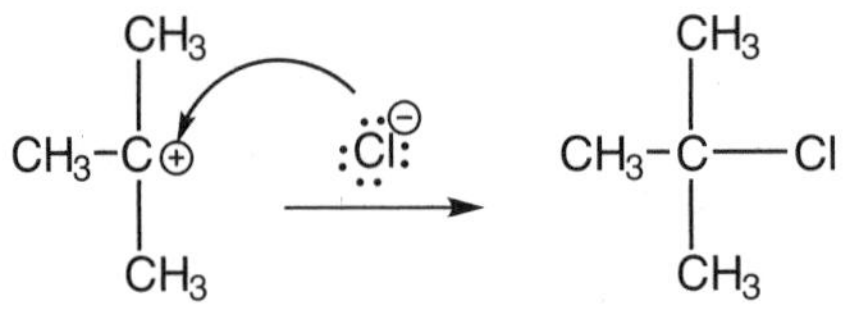

Figure 2-12: Step 3: Lone pair–to–bond movement.

Each step includes a conservation of charge. Conservation of charge is an important part of all mechanisms.

Intermediates

In the preceding mechanism, the carbocation was an intermediate (a species that exists for a short time during the reaction). The form of the intermediate is often essential to understanding the mechanism. The curved arrows help you in drawing the intermediate. Because you can use curved arrows in only three ways (bond to lone pair, bond to bond, and lone pair to bond), you have limited options for drawing intermediates.

In the next example, a nucleophile attacks a double bond. (See Figure 2-13.) In this case, the nucleophile is the hydroxide ion. The process begins with the hydroxide ion attacking the carbon atom at one end of the carbon-carbon bond. This is a lone pair–to–bond step. Next, a pair from the π-bond shifts to form another π-bond on the other side of the carbon atom. This is a bond-to-bond transfer. Finally, a bond–to–lone pair transfer takes place.

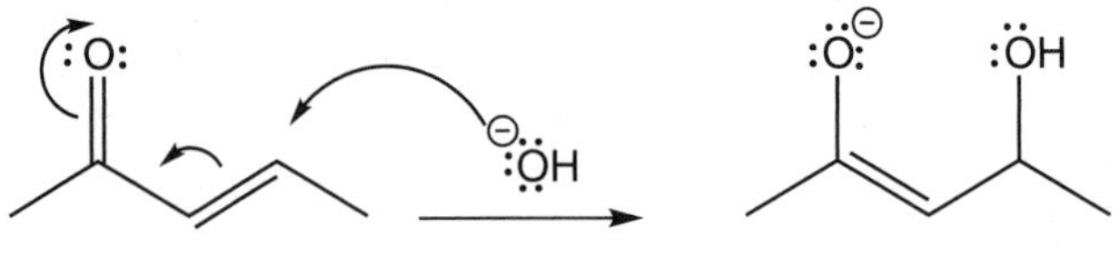

Figure 2-13: Nucleophilic attack of a double bond.

You need to be very careful to keep the formal charges correct. It may help to remember that charge will be conserved.

Don't forget: The nucleophile is at the tail of the arrow and the electrophile is at the head of the arrow.

Some materials may behave as either a base or a nucleophile. The hydroxide ion is an example. When the nucleophile attacks and removes a hydrogen ion, it is behaving as a base. When the nucleophile is attacking at some other point than a hydrogen atom, it is acting like a nucleophile. For example, both the methoxide ion (CH_3O^-) and the t-butoxide ion ($(CH_3)_3O^-$) are strong bases, but only the methoxide ion is a strong nucleophile. The t-butoxide ion is too big and bulky to attack efficiently. The effect of the bulky nature of the t-butoxide ion on its reactivity is an example of *steric hindrance,* which was discussed in your Organic I course (and, naturally, in *Organic I For Dummies*).

A molecule with a lone pair of electrons to donate can behave as a nucleophile. The strength of the nucleophile (the nucleophilicity) is often related to basicity. A strong nucleophile is usually a strong base and vice versa. But nucleophilicity and basicity aren't the same. Basicity refers to the ability of a molecule to accept (bond with) an H^+. The base strength is shown by its equilibrium constant. On the other hand, nucleophilicity refers to the ability of a lone pair of electrons to attack a carbon on an electrophile.

When working with nucleophiles, keep a few additional points in mind:

- Nucleophiles that are negatively charged are stronger nucleophiles than neutral ones.
- Generally, nucleophilicity increases as you go down the periodic table.
- Nucleophilicity is decreased by steric hindrance.

Keys to substitution and elimination mechanisms

Four types of mechanisms are inherent to Organic Chemistry I. These are substitution reaction mechanisms (S_N1 and S_N2) and elimination reaction mechanisms (E1 and E2). The principles of these four types apply to Organic Chemistry II, and no review would be complete without a few reminders about these processes.

The S_N refers to a nucleophilic substitution process where some nucleophile attacks an electrophile and substitutes for some part of the electrophile. The E refers to an elimination process where the nucleophile attacks an electrophile and causes the elimination of something. The 1 and 2 refer to the order of the reaction. A 1 (first order) means only one molecule determines the rate of the reaction, whereas a 2 (second order) means that a combination of two molecules determines the rate of the reaction. In many cases, two or more of these mechanisms are competing and more than one product may result.

Increasing the strength of the nucleophile increases the likelihood of a substitution occurring instead of elimination. Increasing substitution on the electrophile tends to increase the likelihood of a first-order process over a second-order process. This means a tertiary (3°) carbon is more reactive than a secondary (2°) carbon atom, which, in turn, is more reactive than a primary (1°) carbon atom.

All nucleophilic substitution reactions require a good leaving group. Ions like OH^-, RO^- (alkoxide), and NH_2^- are terrible leaving groups and don't normally form. The more likely leaving groups, in these cases, are H_2O, ROH, and NH_3, respectively.

The following four lists summarize the main features of each of these mechanisms. Remember the following when working with an S_N1 mechanism (Me = methyl):

- Reactivity increases in the order Me < 1° << 2° < 3°.
- A racemic mixture results if the attack is on a stereogenic center.
- It's a two-step process.
- It requires a good nucleophile.
- It relies on a carbocation intermediate.
- The rate depends entirely on the concentration of the electrophile.
- The intermediate carbocation can undergo rearrangement.
- Polar solvents, especially those with hydrogen bonding, promote this type of reaction.

Good S_N1 substrates make stable carbocation intermediates. Also, solvents that can supply an H^+ (protic solvents) will stabilize carbocations in S_N1 reactions.

The following are the main features of an S_N2 mechanism:

- Reactivity increases in the order 3° << 2° < 1° < Me.
- Inversion of configuration results if attack is at the stereogenic center.
- It's a one-step process.
- It requires a strong nucleophile.
- The rate depends on the concentration of both the nucleophile and the electrophile.
- Backside attack occurs.
- Rearrangement is not possible.
- Polar aprotic solvents promote this type of reaction.

S_N2 reactions lead to an inversion of stereochemistry. Nucleophilicity is decreased by protic solvents in S_N2 reactions. The presence of a polar aprotic solvent is a clue that the mechanism is S_N2.

The main features of an E1 mechanism are as follows:

- Reactivity increases in the order Me < 1° << 2° < 3°.
- The major product is the most substituted alkene.
- It's a two-step process.
- It requires a weak base.
- The rate depends entirely on the concentration of the electrophile.
- It uses a carbocation intermediate.
- Polar (hydrogen bonding) solvents promote this type of reaction.
- It's promoted by high temperatures.

The following list describes the main features of an E2 mechanism:

- Reactivity increases in the order Me < 1° < 2° < 3°.
- The major product is the most substituted alkene.
- It's a one-step process.
- The mechanism requires a strong base.
- The rate depends on the concentration of both the base and the electrophile.
- The intermediate is periplanar.
- It's promoted by high temperatures.

Revisiting Free-Radical Mechanisms

Free-radical mechanisms obviously involve free radicals. A *free radical* is a species with an unpaired electron. In these mechanisms, single-headed curved arrows are the norm. In Organic Chemistry I, these free radicals first appear when examining the chlorination of an alkane such as methane. The process begins with an initiation step as shown in Figure 2-14. (All initiation steps increase the number of free radicals.)

The initiation step is homolytic bond cleavage where each of the chlorine atoms receives one of the two electrons originally present in the bond and two chlorine free radicals form. The chlorine free radicals, like all free radicals, are very reactive.

Figure 2-14: Initiation step of a free-radical mechanism.

$$Cl{-}Cl \longrightarrow Cl\cdot + Cl\cdot$$

A chlorine free radical attacks an alkane molecule like methane to form hydrogen chloride and a methyl radical (see Figure 2-15).

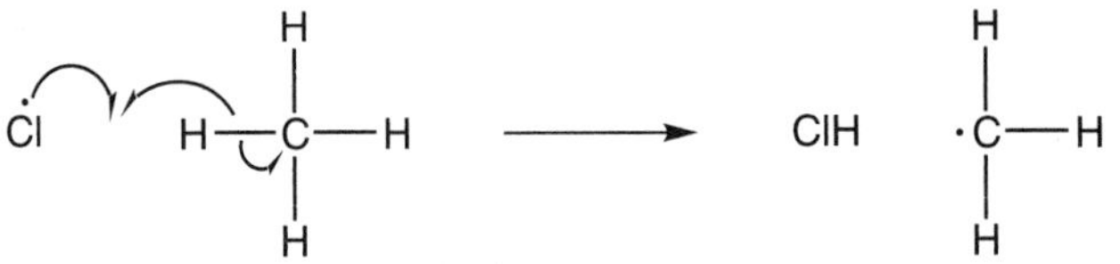

Figure 2-15: Free-radical attack of an alkane.

$$Cl\cdot + H{-}CH_3 \longrightarrow ClH + \cdot CH_3$$

The methyl radical, once formed, is also a very reactive species, which attacks other species to continue the reaction through a series of propagation steps. (All propagation steps maintain the number of free radicals.) Finally, the reaction of a free radical with another free radical gives rise to a termination step. (All termination steps decrease the number of free radicals.)

In a free-radical mechanism, the reaction of a free radical with a molecule results in a free radical, and a free radical reacts with a free radical to produce a molecule that isn't a free radical.

There are two important considerations concerning free-radical mechanisms. One of these factors is the identity of the halogen and the other is the stability of the alkyl free radical. The more substituted the carbon atom is, the more stable the free radical. The stability of the alkyl free radicals increase in the following order: Me < 1° < 2° < 3°. The relative stabilities rarely lead to rearrangement of the free radical.

The formation of a more-stable free radical increases the selectivity of the reaction. For this reason, the replacement of a particular hydrogen atom by a halogen isn't simply a matter of probability. In propane, replacement of one of the hydrogen atoms on the central carbon should occur one-fourth ($\frac{2}{8}$) of the time. (You may want to draw this reaction to see why this is true.) However, chlorination shows a distribution where replacement occurs at the second carbon about three-fourths of the time, and for bromination, the replacement is almost exclusively on the central carbon atom. Table 2-1 indicates the relative selectivity of chlorine and bromine.

Table 2-1 Selectivity of Chlorine and Bromine in Free-Radical Halogenation of Alkanes

	1° *RCH3*	*2°* *R2CH2*	*3°* *R3CH*
Chlorine (Cl_2)	1	3.9	5.3
Bromine (Br_2)	1	82.0	1640

Living with mechanisms

Mechanisms are very important in the understanding of organic reactions. Many mechanisms are presented in any organic chemistry course, and beginning students can get into trouble in a number of places. If you keep the following items in mind while studying and working mechanisms, life will be easier.

- You probably won't use all the reactants in each step of the mechanism.
- Be careful not to mistake a multistep synthesis problem with a mechanism. Both involve a number of steps, but curved arrows only appear in mechanisms.
- The solvent isn't a reactant. It may promote a particular type of mechanism; however, that doesn't make it a reactant.
- Materials may be added to prevent a build-up of undesired products. For example, a base may be present to trap released acid. These compounds aren't part of the mechanism.
- Draw out all the atoms in the vicinity of the reaction center, especially if there are charges or lone pairs.
- Don't try to do too much in one step.
- In many cases, you should draw the possible resonance structures, especially for intermediates. (Not everything will have a resonance structure.)
- Keep your goal in mind. It's easy to go off on a tangent.
- Don't attempt to overanalyze the process. Pick the best reaction from the ones you studied to get from point A to point B.
- Ions such as sodium, Na^+, potassium, K^+, and lithium, Li^+, are usually spectator ions and therefore aren't part of the reaction mechanism.
- Before going to the next step, make sure the step you just finished is reasonable. For example, are the electrons moving in the same direction? Are the intermediates reasonably stable? Do you have like charges close together? Is there conservation of charge?
- Acidic conditions may yield cationic or neutral products. Basic conditions may yield anionic or neutral products.
- Watch out for those pesky five-bonded carbon atoms.
- When you finish the mechanism, go over each step and check your assumptions. Make sure none of your intermediates are unstable.

Chapter 3

Alcohols and Ethers: Not Just for Drinking and Sleeping

In This Chapter

- Checking out alcohols: How they look and what they're called
- Taking a look at the synthesis and reactions of alcohols
- Tackling the basics of ethers
- Finding out what ethers do
- Clarifying the spectra of alcohols and ethers

Two types of organic compounds contain single-bonded oxygen atoms: the alcohols and the ethers. In the alcohols, an oxygen atom is between a carbon atom and a hydrogen atom, whereas in an ether an oxygen atom is between two carbon atoms. The generic representation of an alcohol is ROH and the generic representation of an ether is ROR'. If the carbon atom attached to the OH group is part of an aromatic ring, the compound is a phenol, which, unlike the alcohols, is an organic acid.

In this chapter you investigate the properties, synthesis, and reactions of alcohols and ethers. So drink up, and let's go.

Getting Acquainted with Alcohols

The alcohols are an important group of organic compounds. Even though an OH group is present, these are not basic compounds but are neutral to weakly acidic materials. The OH (hydroxyl) group is the source of hydrogen bonding, which increases the melting and boiling points of the alcohols relative to those of comparably sized alkanes. Hydrogen bonding also makes alcohols more soluble in water than less polar materials.

Structure and nomenclature of alcohols

In this section we take a look at how you can classify alcohols and the nomenclature of alcohols (no, just calling them bourbon, gin, and Scotch won't work).

Classifying alcohols

There are three general categories of alcohols:

- Primary (1°)
- Secondary (2°)
- Tertiary (3°)

The categories depend upon the number of carbon atom attached to the alcohol carbon atom. Figure 3-1 illustrates the three types of alcohols. Note the number of carbon atoms attached to the bold-faced carbon atoms.

$CH_3-CH_2-CH_2-\mathbf{C}(H)(H)-OH$ — 1°

$CH_3-CH_2-\mathbf{C}(H)(OH)-CH_3$ — 2°

$CH_3-\mathbf{C}(CH_3)(OH)-CH_3$ — 3°

Figure 3-1: Primary, secondary, and tertiary alcohols.

To distinguish between the alcohols in the figure, simply count the carbon atoms attached to the carbon atom in boldface. Primary alcohols have one carbon atom attached to the central carbon, secondary alcohols have two, and tertiary alcohols have three.

Naming alcohols

The nomenclature of the alcohols is an extension of the rules for the naming of other organic compounds. The general changes in the rules for alkanes are

- The parent chain contains the OH.
- The carbon with the OH gets the smaller number.
- Drop the final *-e* and add *-ol.*

The common names of the alcohols consist of the name of the alkyl group and the word *alcohol.* For example, CH_3OH is methyl alcohol and CH_3CH_2OH is ethyl alcohol. Some examples of naming alcohols are shown in Figure 3-2.

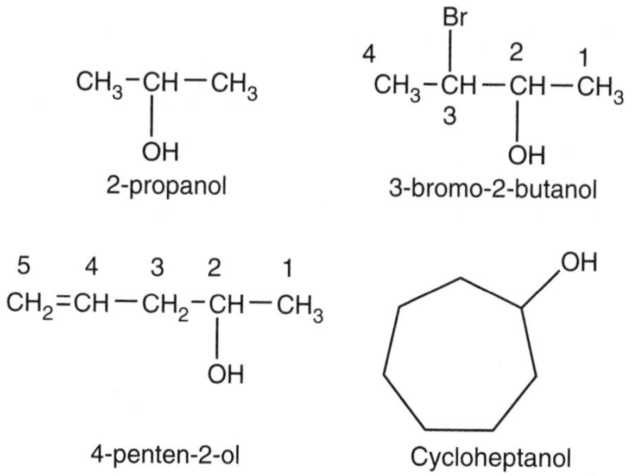

$CH_3-CH(OH)-CH_3$

2-propanol

$CH_3-CH(Br)-CH(OH)-CH_3$ (carbons numbered 4, 3, 2, 1)

3-bromo-2-butanol

$CH_2=CH-CH_2-CH(OH)-CH_3$ (carbons numbered 5, 4, 3, 2, 1)

4-penten-2-ol

Cycloheptanol

Figure 3-2: Examples of naming alcohols.

Under extreme conditions, alcohols may behave as acids and lose an H^+ to leave an anion with the general formula RO^-. These are alkoxides. Alkoxides are important in organic synthesis because they are very strong bases and may be good nucleophiles. Figure 3-3 illustrates two common alkoxides.

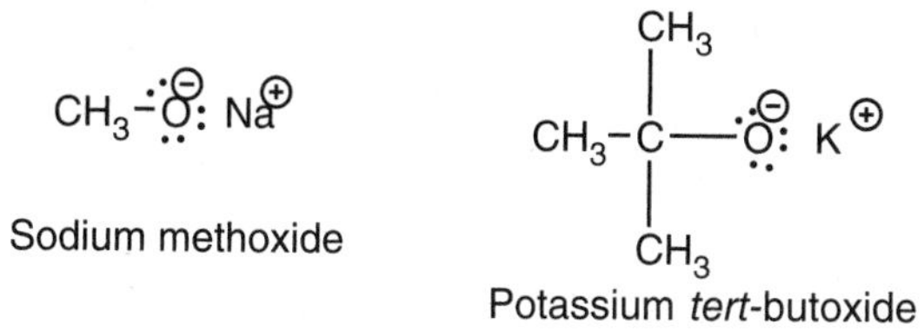

$CH_3-O^{\ominus}\ Na^{\oplus}$

Sodium methoxide

$(CH_3)_3C-O^{\ominus}\ K^{\oplus}$

Potassium *tert*-butoxide

Figure 3-3: Two common alkoxides.

Physical properties of alcohols

The important physical properties of organic compounds include melting and boiling points and solubility and density. Whenever you compare physical properties of different compounds, stick to compounds with similar molecular weights.

Melting and boiling points

The presence of hydrogen bonding causes alcohols to have significantly higher melting and boiling points than alkanes. Figure 3-4 illustrates the formation of hydrogen bonds between alcohol molecules.

Impurities increase the boiling point and reduce the freezing points of materials.

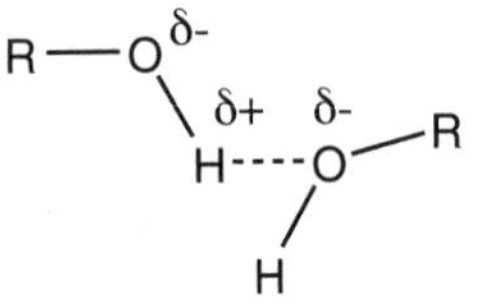

Figure 3-4: Hydrogen bonding between alcohol molecules.

Solubility and density

Alcohols with three or fewer carbons atoms are miscible in water in all proportions. The solubility of alcohols in water decreases with increasing number of carbon atoms, so that alcohols with more than six carbon atoms are nearly insoluble. As the number of carbon atoms increases, the solubility in nonpolar solvents increases.

Alcohols, like most organic compounds containing oxygen or nitrogen, are soluble in concentrated sulfuric acid because the acid protonates the oxygen atom, as illustrated in Figure 3-5.

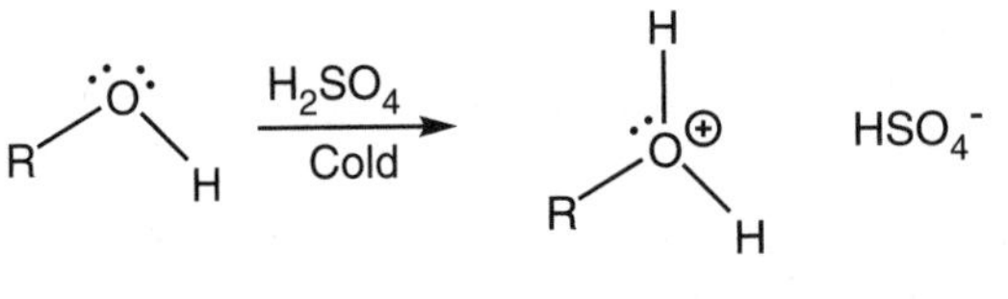

Figure 3-5: Protonation of an alcohol by sulfuric acid.

Alcohols tend to be denser than hydrocarbons with comparable carbon content.

Making moonshine: Synthesis of alcohols

There are a number of methods for synthesizing alcohols. Some of the methods are suitable only for the preparation of small quantities of alcohol, while other methods are industrially important for the synthesis of thousands of gallons of alcohol.

Hydration of alkenes

The hydration of alkenes is one important method of synthesizing alcohols. Industrially, sulfuric acid is used as a catalyst, while small-scale preparations often utilize toxic mercury compounds. (Definitely not the kind of stuff you want to drink.)

Catalytic hydration of alkenes

The general reaction for the catalytic hydration of an alkene to produce an alcohol is shown in Figure 3-6, and the mechanism is in Figure 3-7. This process is an example of a Markovnikov addition (as seen in Organic Chemistry I).

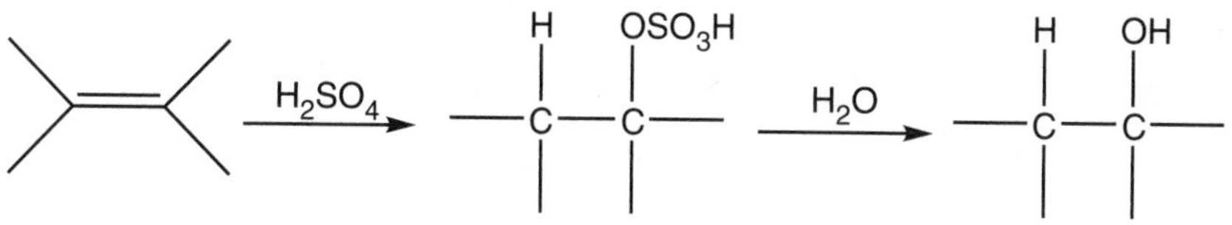

Figure 3-6: Catalytic hydration of an alkene to produce an alcohol.

H_2SO_4 H OSO_3H C C H_2O H OH C C

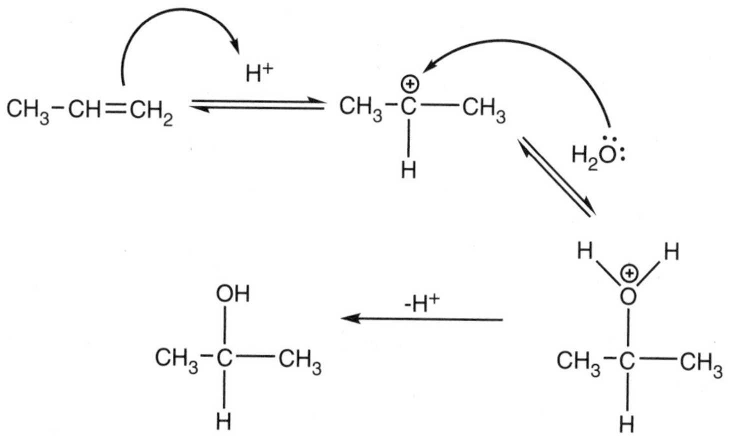

Figure 3-7: Mechanism of the Markovnikov addition of water to an alkene to yield an alcohol.

$CH_3-CH{=}CH_2$ H^+ $CH_3-\overset{\oplus}{C}-CH_3$ H $H_2\ddot{O}:$ H H $\overset{\oplus}{O}$ CH_3-C-CH_3 H $-H^+$ OH CH_3-C-CH_3 H

Oxymercuration-demercuration reactions with alkenes

Oxymercuration-demercuration is a useful laboratory method for the synthesis of small quantities of alcohol. Like the catalytic hydration reaction, this process is an example of Markovnikov addition. It's a useful procedure because it tends to result in high yields and rearrangements rarely occur.

The general reaction is shown in Figure 3-8, and you can see a specific example in Figure 3-9.

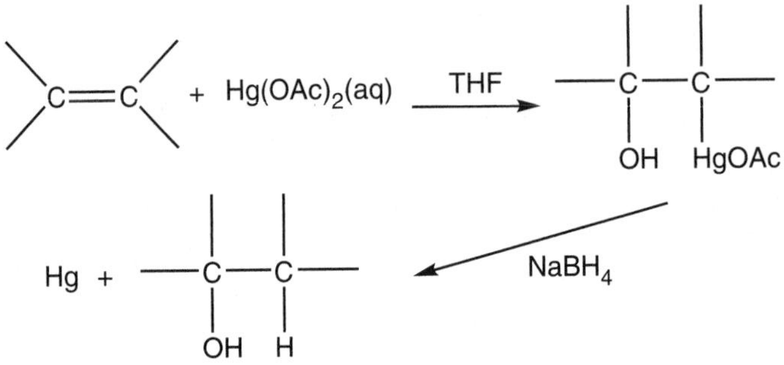

$$\text{>C=C<} + Hg(OAc)_2(aq) \xrightarrow{THF} \text{—C(OH)—C(HgOAc)—} \xrightarrow{NaBH_4} Hg + \text{—C(OH)—C(H)—}$$

Figure 3-8: Oxymercuration-demercuration of an alkene to yield an alcohol.

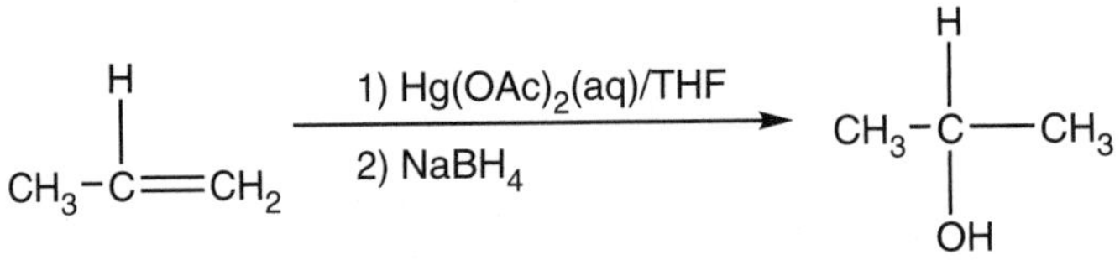

$$CH_3\text{–}CH\text{=}CH_2 \xrightarrow[\text{2) } NaBH_4]{\text{1) } Hg(OAc)_2(aq)/THF} CH_3\text{–}CH(OH)\text{–}CH_3$$

Figure 3-9: Synthesis of 2-propanol by oxymercuration-demercuration.

The numbering of the steps shown in Figure 3-9 is essential. The numbers indicate that Step 1 must take place before and be separate from Step 2.

Hydroboration-oxidation reactions with alkenes

Hydroboration-oxidation is a useful method when the desired product is the anti-Markovnikov alcohol. The borane-containing reactant is normally diborane (B_2H_6) or the borane tetrahydrofuran complex (BH_3•THF). No matter what hydroboration agent is used, it's usually simplified to BH_3 in the mechanism. BH_3 is a useful reactant because it's a good Lewis acid.

An example illustrating the general mechanism is shown in Figure 3-10.

The attack in Figure 3-10 is on the terminal carbon for two reasons:

- Steric (more important) — the outer carbon is more accessible.
- Electronic — the 2° carbocation is more stable.

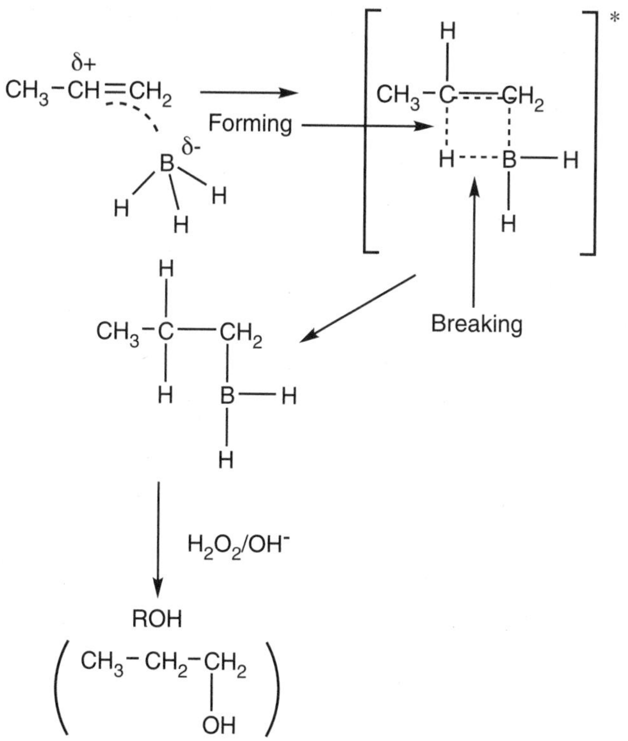

Figure 3-10: Synthesis of 1-propanol from propene by hydroboration-oxidation.

The transition state is four-centered, which means there is *syn addition:* both groups add to the same face of the double bond. (*Anti addition* is where the two groups add to opposite faces of a double bond.)

When doing a reaction, you can usually call it quits at this point. In reality, the process continues until the replacement of all B-H with B-R to give BR_3 (as long as the alkene is small).

A question that may appear on some Organic Chemistry exams is: What will be the product of the reaction sequence given in Figure 3-11?

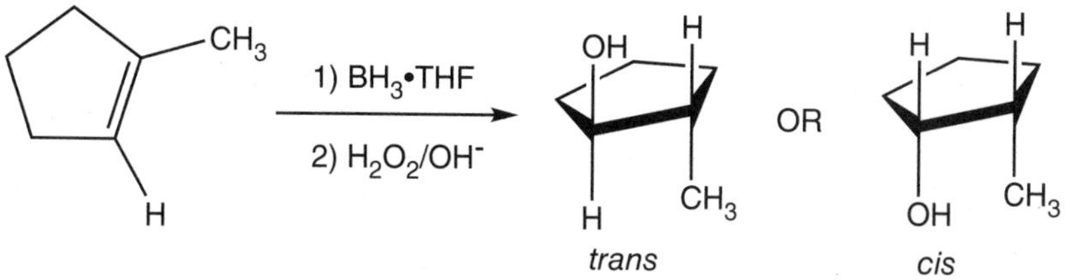

Figure 3-11: Reaction of methylcyclopentene by hydroboration-oxidation.

The BH_3 undergoes syn addition so the *trans* isomer will form. However, the *cis* isomer can be produced by heating the mixture to 160 degrees Celsius. Heating to 160 degrees initiates a series of eliminations and re-additions of the boron until the boron ends up on the least sterically hindered carbon.

Diols from reactions with alkenes

The reaction of either cold, dilute, aqueous potassium permanganate ($KMnO_4$) or osmium tetroxide (OsO_4), to an alkene will form, through syn addition, a *cis*-glycol. Using osmium tetroxide requires a second step using hydrogen peroxide, H_2O_2, to regenerate the expensive reagent.

Figure 3-12 gives an example of this process. The reaction of the alkene with a peroxyacid can lead to the *trans*-glycol (anti addition).

Figure 3-12: Production of 1,2-propanediol from propene.

$$CH_3-CH=CH_2 \xrightarrow[\text{Cold, dil}]{KMnO_4(aq)} CH_3-\underset{|}{\overset{OH}{C}}H-\overset{OH}{C}H_2$$

Preparation of alcohols by the reduction of carbonyls

The reduction of carboxylic acids or esters requires very powerful reducing agents such as lithium aluminum hydride ($LiAlH_4$) or sodium (Na) metal. Aldehydes and ketones are easier to reduce, so they can use sodium borohydride ($NaBH_4$). Examples of these reductions are shown in Figure 3-13.

Grignard reagents and production of alcohols

A Grignard reagent is an organometallic compound containing magnesium. Grignard reagents are very useful in organic synthesis. (In most cases, an organolithium compound can be substituted for a Grignard reagent.)

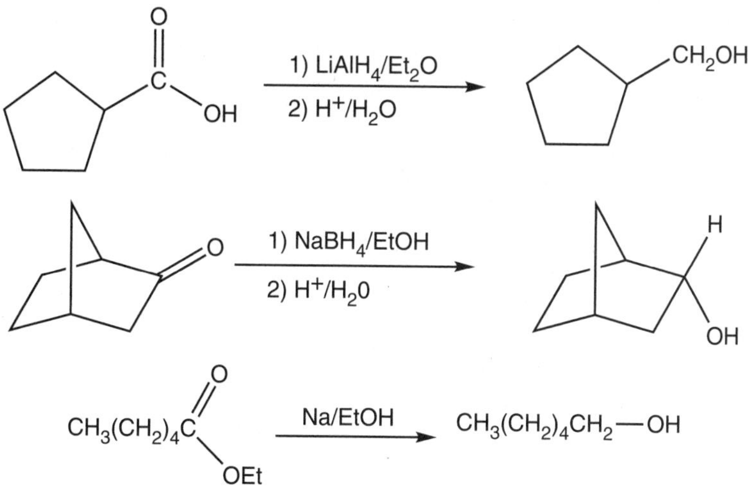

Figure 3-13: Examples of the preparation of alcohols by the reduction of carbonyls.

The synthesis often begins with the preparation of the Grignard reagent by reacting an alkyl halide with magnesium metal in dry ether. Then the Grignard reagent is used to attack a carbonyl group to form a magnesium salt. The mechanism is shown in Figure 3-14. The addition of water causes hydrolysis of the salt to produce an alcohol, some examples of which are shown in Figure 3-15.

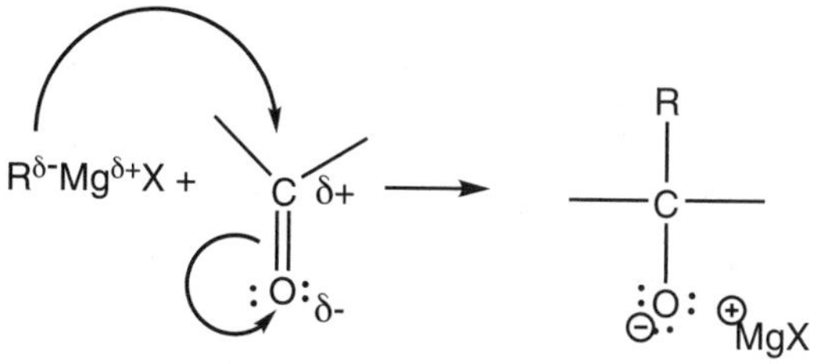

Figure 3-14: Mechanism for the Grignard reaction.

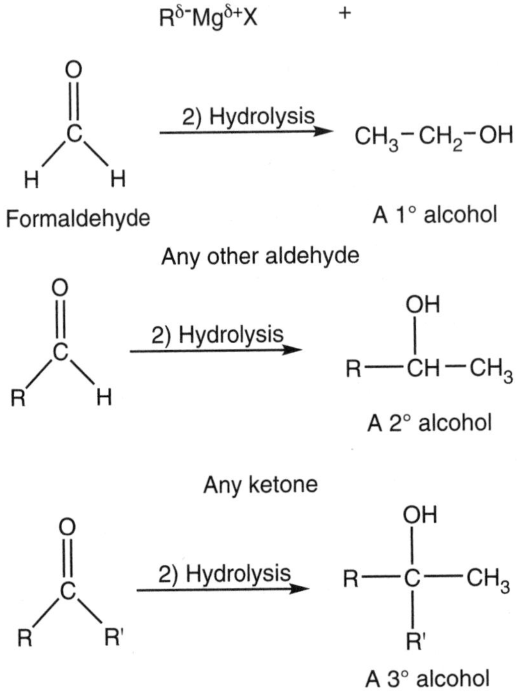

$R^{\delta-}Mg^{\delta+}X$ +

Formaldehyde (H–C(=O)–H) —2) Hydrolysis→ CH_3-CH_2-OH

A 1° alcohol

Any other aldehyde

R–C(=O)–H —2) Hydrolysis→ R—CH(OH)—CH_3

A 2° alcohol

Any ketone

R–C(=O)–R' —2) Hydrolysis→ R—C(OH)(R')—CH_3

A 3° alcohol

Figure 3-15: Synthesis of alcohols by the Grignard reaction.

What will they do besides burn? Reactions of alcohols

Alcohols represent a very versatile class of compounds for organic synthesis, so a variety of synthetic pathways utilize alcohols. Some examples appear in Figure 3-16. The conversion of alcohols to ethers appears later in this chapter.

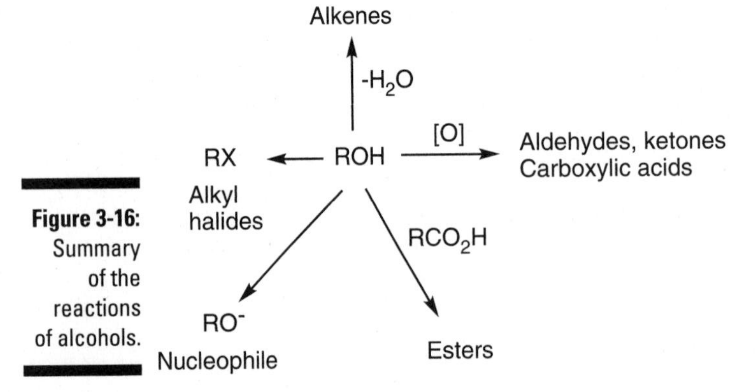

Figure 3-16: Summary of the reactions of alcohols.

Dehydration of alcohols ($-H_2O$)

Alcohols can be dehydrated by a number of procedures. One method is to use a dehydrating agent such as phosphorus oxychloride ($POCl_3$), and another method is acid-catalyzed dehydration using sulfuric acid (H_2SO_4) or phosphoric acid (H_3PO_4). An example of acid-catalyzed dehydration is in Figure 3-17, and the mechanism is shown in Figure 3-18.

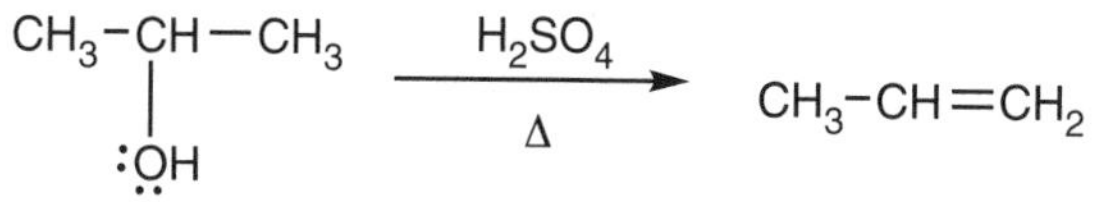

Figure 3-17: Acid-catalyzed dehydration of 2-propanol to propene.

$$CH_3{-}CH(OH){-}CH_3 \xrightarrow[\Delta]{H_2SO_4} CH_3{-}CH{=}CH_2$$

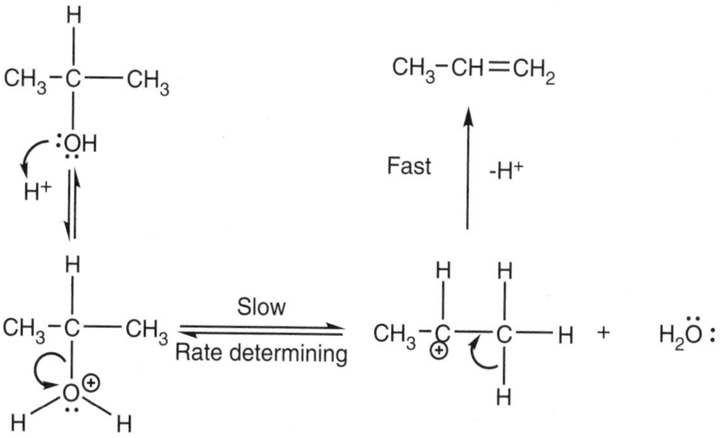

Figure 3-18: Mechanism of the acid-catalyzed dehydration of 2-propanol to propene.

$$CH_3{-}CH(\ddot{O}H){-}CH_3 \underset{}{\overset{H^+}{\rightleftharpoons}} CH_3{-}CH(\overset{\oplus}{O}H_2){-}CH_3 \underset{\text{Rate determining}}{\overset{\text{Slow}}{\rightleftharpoons}} CH_3{-}\overset{\oplus}{C}H{-}CH_3 + H_2\ddot{O}: \xrightarrow[\text{Fast}]{-H^+} CH_3{-}CH{=}CH_2$$

Tertiary alcohols dehydrate most readily, primary alcohols least readily, and, unsurprisingly, secondary alcohols are intermediate. This relates to the relative stability of the intermediate carbocation. The temperature and concentration of the acid depends upon the type of alcohol. A primary alcohol, such as ethanol, requires concentrated acid and a very high temperature (180 degrees Celsius), while a tertiary alcohol, such as t-butyl alcohol, requires 20 percent sulfuric acid at 85 to 90 degrees Celsius. The process follows an E1 mechanism and produces the thermodynamically more stable product.

Oxidation of alcohols

Both primary and secondary alcohols can be oxidized, but tertiary alcohols won't undergo simple oxidation. Oxidation of a primary alcohol gives an aldehyde; however, preventing further oxidation of the aldehyde to a carboxylic acid is difficult. Secondary alcohols oxidize to a ketone without the problem of additional oxidation occurring.

The symbol [O] often appears in organic reactions to indicate oxidation, and the symbol [H] indicates reduction.

Most organic compounds, including alcohols, undergo oxidation in the form of a combustion reaction. The general reaction for the oxidation of a primary alcohol is in Figure 3-19. Oxidizing agents that aren't strong enough to oxidize the aldehyde formed in the first step include PCC (pyridinium chlorochromate), Sarett's reagent (CrO_3•2py), and acidic, aqueous potassium dichromate ($K_2Cr_2O_7$). The reaction with potassium dichromate requires the distillation of the aldehyde from the mixture before further oxidation can occur. Two examples are in Figure 3-20.

$$R\text{-}CH_2OH \longrightarrow RCHO \longrightarrow RCOOH$$

Figure 3-19: Oxidation of a primary alcohol to an aldehyde and then to an carboxylic acid.

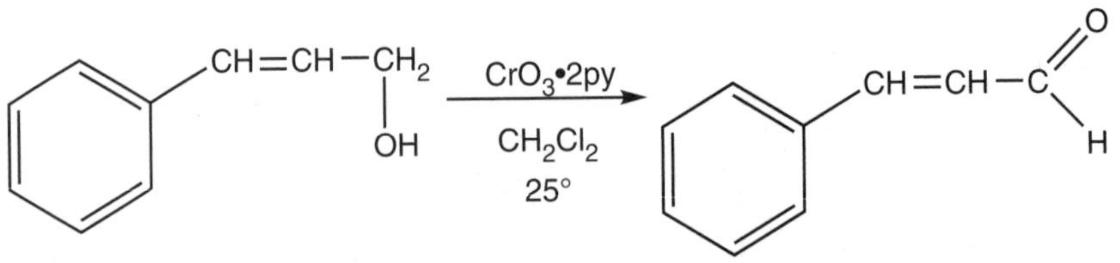

$$C_6H_5\text{-}CH{=}CH{-}CH_2OH \xrightarrow[CH_2Cl_2,\ 25^\circ]{CrO_3\text{•}2py} C_6H_5\text{-}CH{=}CH{-}CHO$$

Figure 3-20: Example of the oxidation of a primary alcohol to an aldehyde.

Stronger oxidizing agents, such as hot basic potassium permanganate or Jones reagent (acidic CrO_3), oxidize a primary alcohol to a carboxylic acid as illustrated in Figure 3-21. The characteristic dark purple color of the permanganate ion changes to brown, indicating the formation of MnO_2.

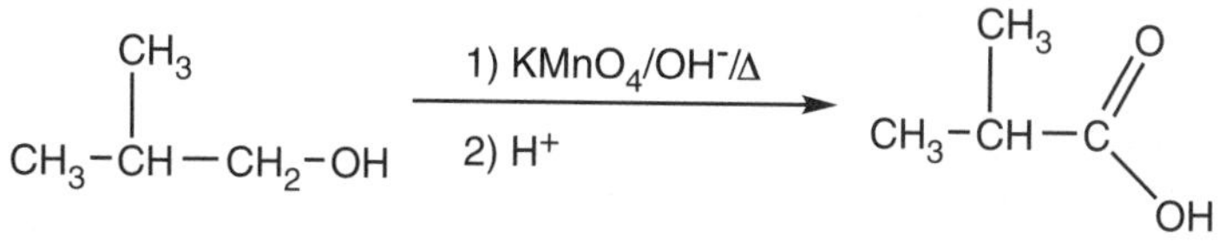

$$CH_3-CH(CH_3)-CH_2-OH \xrightarrow[2)\ H^+]{1)\ KMnO_4/OH^-/\Delta} CH_3-CH(CH_3)-C(=O)OH$$

Figure 3-21: Oxidation of a primary alcohol to a carboxylic acid.

The oxidation of a secondary alcohol to a ketone can also occur through the use of a variety of oxidizing agents. Figure 3-22 illustrates the oxidation of a secondary alcohol using Jones reagent.

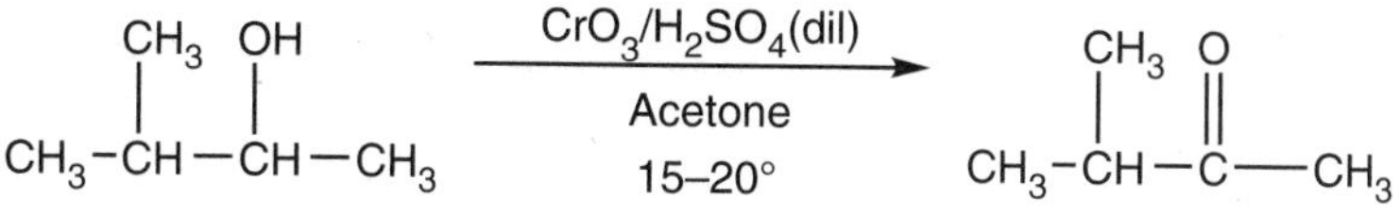

$$CH_3-CH(CH_3)-CH(OH)-CH_3 \xrightarrow[\text{Acetone, } 15\text{–}20^\circ]{CrO_3/H_2SO_4(dil)} CH_3-CH(CH_3)-C(=O)-CH_3$$

Figure 3-22: Oxidation of a secondary alcohol to a ketone.

Conversion of alcohols to esters

In the presence of an acid catalyst, alcohols react with carboxylic acids to form esters. The general reaction is in Figure 3-23, and you can find a discussion of the reaction in Chapter 12.

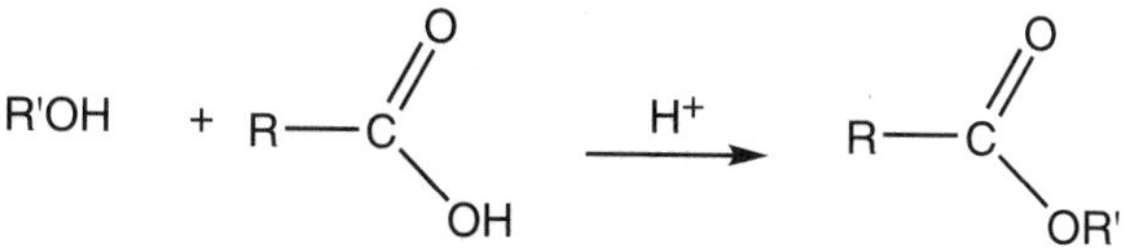

$$R'OH + R-C(=O)OH \xrightarrow{H^+} R-C(=O)OR'$$

Figure 3-23: Acid catalyzed reaction of an alcohol with a carboxylic acid to form an ester.

Reaction of alcohols as acids

In general, alcohols are neutral compounds; however, in the presence of extremely strong bases or an active metal, it's possible to force the loss of a hydrogen ion to leave an alkoxide (RO^-) ion. The general reaction is shown in Figure 3-24. The amide ion, NH_2^-, is a strong enough base to force the loss, as is the hydride ion, H^-; however, the hydroxide ion, OH^-, isn't strong enough. (Poor little guy.) Active metals include sodium and potassium. Specific examples are in Figure 3-25. The alkoxide ions are useful in organic synthesis because they're good nucleophiles.

Figure 3-24: Formation of an alkoxide ion from an alcohol.

$$R-\overset{\delta-}{\ddot{O}}:-H^{\delta+} \xrightarrow[\text{active metal}]{\text{Strong base or}} R-\ddot{O}:^{\ominus}$$

Alkoxide ion

Figure 3-25: Examples of the formation of an alkoxide ion from an alcohol.

$$K + HO-C(CH_3)_2-CH_3 \longrightarrow K^{\oplus}\,{:}\ddot{O}{:}^{\ominus}-C(CH_3)_2-CH_3 + 1/2\ H_2$$

$$NaH + EtOH \longrightarrow NaOEt + H_2$$

Conversion of alcohols to alkyl halides

Sometimes the alcohol group can be replaced with a halogen. A number of reagents cause this conversion, including the hydrohalic acids (HCl, HBr, and HI), sodium bromide with sulfuric acid ($NaBr/H_2SO_4$), hydrochloric acid with zinc chloride ($HCl/ZnCl_2$), phosphorus tribromide (PBr_3) in base, or refluxing with thionyl chloride ($SOCl_2$). The acidic reactions utilizing the general mechanism appear in Figure 3-26. Methanol and most primary alcohols follow an S_N2 mechanism, while most other alcohols react by S_N1. The S_N1 reactions may involve rearrangement such as a hydride shift.

Don't forget that S_N2 mechanisms involve an inversion of configuration.

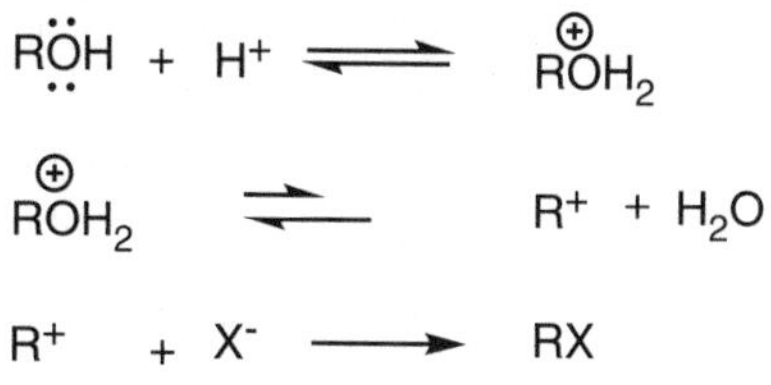

$$R\ddot{O}H + H^+ \rightleftharpoons R\overset{\oplus}{O}H_2$$

$$R\overset{\oplus}{O}H_2 \rightleftharpoons R^+ + H_2O$$

$$R^+ + X^- \longrightarrow RX$$

Figure 3-26: General steps for the conversion of an alcohol to an alkyl halide.

Figure 3-27 illustrates two examples using $HCl/ZnCl_2$. In the second example, a rearrangement takes place.

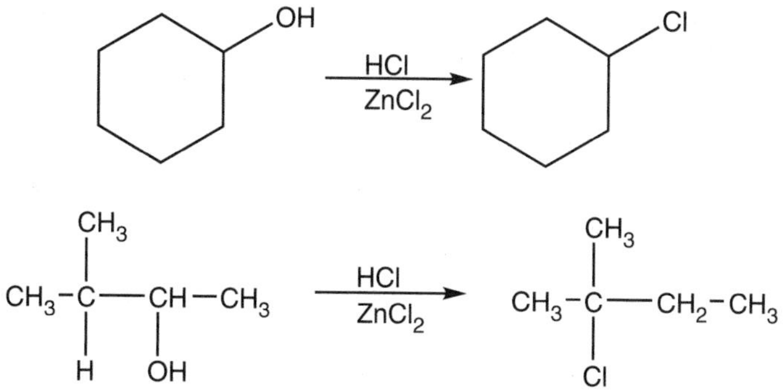

Figure 3-27: Examples of the formation of an alkyl halide from an alcohol.

A common analytic test used to classify alcohols is the *Lucas test.* This test is for alcohols with six or fewer carbon atoms (and therefore soluble in reagent mixture). They react with $HCl/ZnCl_2$ to form an alkyl chloride, which is insoluble in the reaction mixture (causing the solution to turn cloudy). The time required for the reaction to occur indicates the type of alcohol. Table 3-1 summarizes the time required for each type of alcohol.

Table 3-1 Lucas Test for Classifying Alcohols

Type of Alcohol	*Time Required to Form an Alkyl Chloride*
Tertiary (3°)	Immediately
Secondary (2°)	Five minutes
Primary (1°)	24 hours while heating

Introducing Ether (Not the Ether Bunny)

Ethers are compounds closely related to the alcohols. You probably think of anesthetics when you think of ethers, but these compounds are a lot more versatile than that. In this section, we look at the structures, properties, and reactions of ethers.

Structure and nomenclature of ethers

The general formula of an ether is R-O-R', where R and R' may be any alkyl or aryl group. You can name an ether simply by identifying the R groups and adding the word *ether* or by naming one R group and the oxygen atom as an alkyoxy group attached to the other R group. Examples of both methods are in Figure 3-28. A cyclic ether containing two carbon atoms and an oxygen atom is an epoxide.

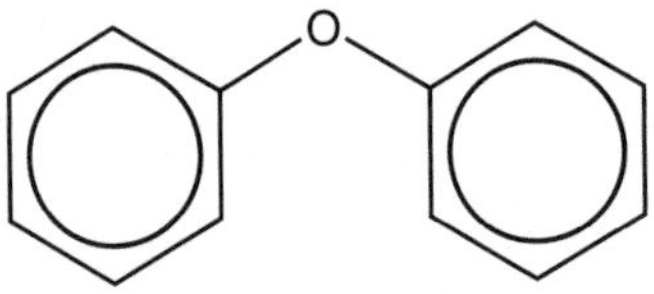

O

Diphenyl ether or phenyl ether

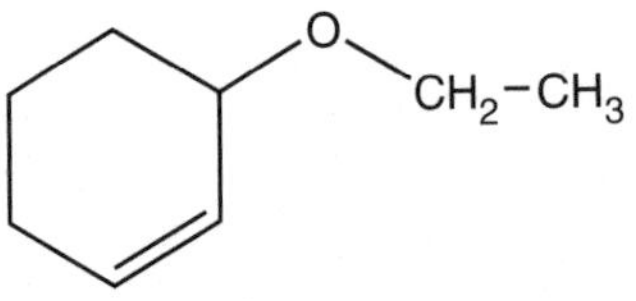

O

CH_2-CH_3

3-ethoxycyclohexene

Figure 3-28: Examples of naming ethers.

Sleepy time: Physical properties of ethers

In this section we examine some of the physical properties of ethers.

Melting and boiling points

The ethers are only slightly polar; therefore their melting and boiling points are only slightly above those of the corresponding alkanes. The corresponding alcohols have significantly higher melting and boiling points.

Solubility

Ethers are much less soluble in water than the alcohols are. The solubility is normally less than 8 grams (g) per 100 milliliters (mL). The addition of various inorganic salts to the solution further decreases the solubility of the ether. This process is called *salting out.* Ethers are soluble in concentrated sulfuric acid.

Ethers are good for extracting organic compounds from aqueous solutions because the density of ethers is less than water, causing the ether layer to float on top of the water layer.

Synthesis of ethers

Symmetric ethers (R = R') can be prepared by the acid-catalyzed dehydration of primary alcohols. However, this reaction competes with the acid-catalyzed dehydration of the alcohol to form an alkene. Lower temperatures favor ether formation over alkene formation. Secondary and tertiary alcohols favor alkene formation. The general reaction is shown in Figure 3-29.

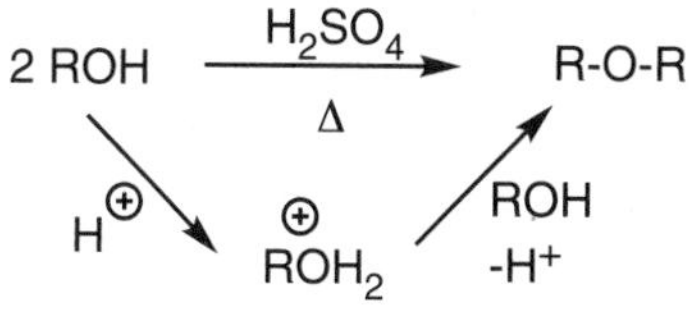

$$2\,ROH \xrightarrow[\Delta]{H_2SO_4} R\text{-}O\text{-}R$$

$$2\,ROH \xrightarrow{H^{\oplus}} ROH_2^{\oplus} \xrightarrow[-H^+]{ROH} R\text{-}O\text{-}R$$

Figure 3-29: Preparation of symmetric ethers from an alcohol.

The *Williamson ether synthesis* is a general method for producing both symmetric and asymmetric (R ≠ R') ethers. This is an S_N2 process following the general procedure in Figure 3-30. The process involves the reaction of an alkoxide ion with an alkyl halide.

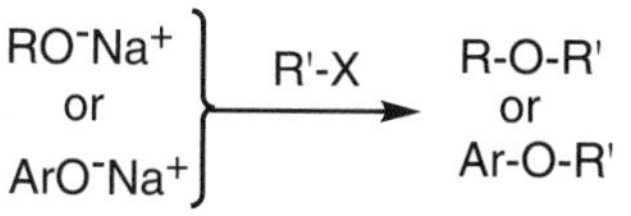

$$\left.\begin{matrix} RO^-Na^+ \\ \text{or} \\ ArO^-Na^+ \end{matrix}\right\} \xrightarrow{R'\text{-}X} \begin{matrix} R\text{-}O\text{-}R' \\ \text{or} \\ Ar\text{-}O\text{-}R' \end{matrix}$$

Figure 3-30: Williamson synthesis of an ether from an alkoxide ion and an alkyl halide.

The synthesis of cyclic ethers (especially epoxides) provides important reagents for organic synthesis. Industrially, ethylene oxide is the most important ether because it is used in the synthesis of many other organic compounds. This compound forms by the catalytic oxidation of ethylene as seen in Figure 3-31.

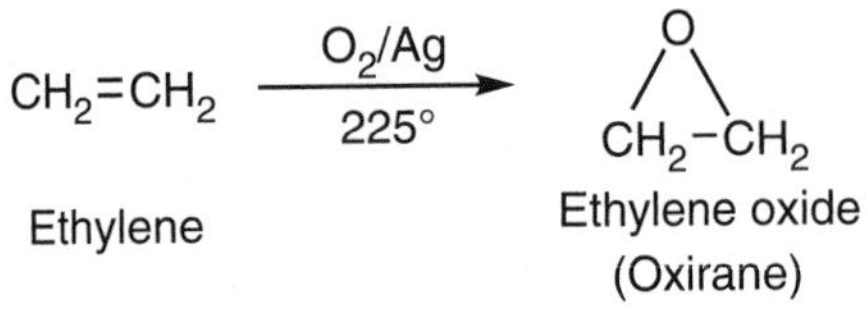

Figure 3-31: Production of ethylene oxide by the catalytic oxidation of ethylene.

$$CH_2{=}CH_2 \xrightarrow[225°]{O_2/Ag} CH_2{-}CH_2 \text{ (with O bridging)}$$

Ethylene → Ethylene oxide (Oxirane)

An epoxide can also be formed by peroxidation of an alkene. This reaction typically employs MCPBA (*meta*-chloroperoxybenzoic acid — now you see why we abbreviated it); however, CF_3COO_2H also works. Figure 3-32 illustrates one example and Figure 3-33 presents a generic mechanism.

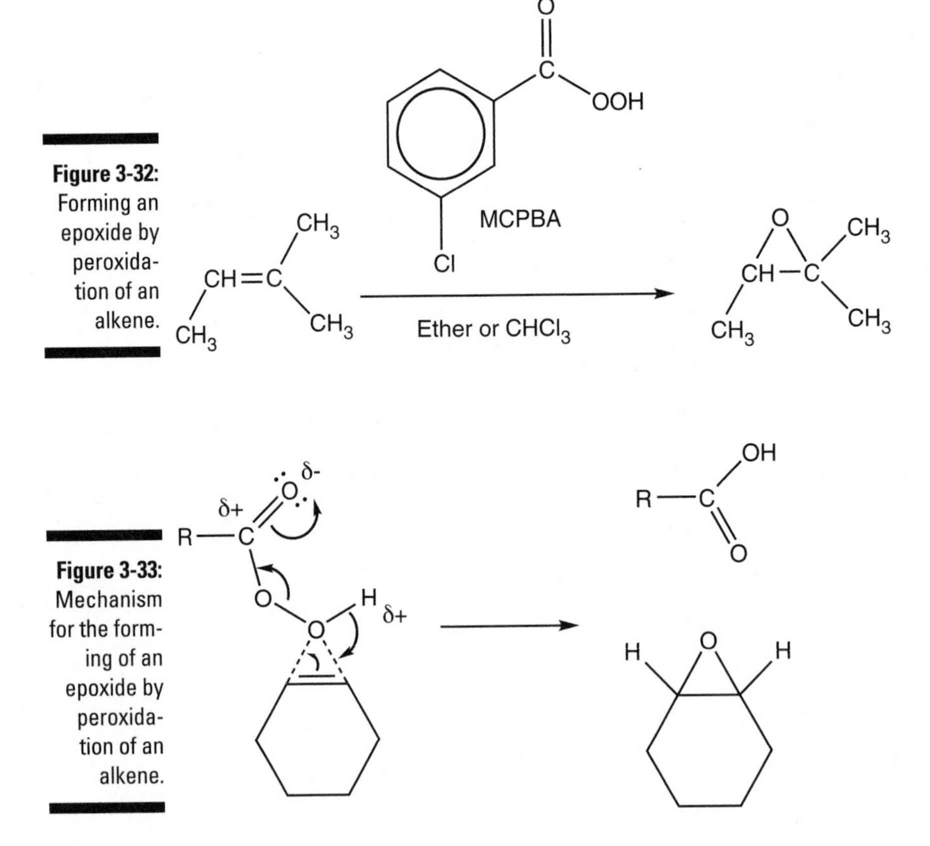

Figure 3-32: Forming an epoxide by peroxidation of an alkene.

Figure 3-33: Mechanism for the forming of an epoxide by peroxidation of an alkene.

Reactions of ethers

Ethers are relatively unreactive compounds; however, they will react under certain circumstances. In this section we describe situations where reactions occur.

In addition to reacting under certain conditions, ethers will slowly air oxidize to produce explosive peroxides.

Acidic cleavage of ethers

Hydrohalic acids can cause cleavage of the ether. Hydrofluoric acid, HF, doesn't work as well as the other acids in the group (HCl, HBr, or HI). Secondary and tertiary ethers undergo S_N1 reaction, while methyl and primary ethers undergo S_N2 reaction. A generic example is in Figure 3-34, and the mechanism is in Figure 3-35.

Figure 3-34: Cleavage of an ether by a hydrohalic acid.

$$ROR \xrightarrow{HI} ROH + RI \xrightarrow{HI} RI + H_2O$$

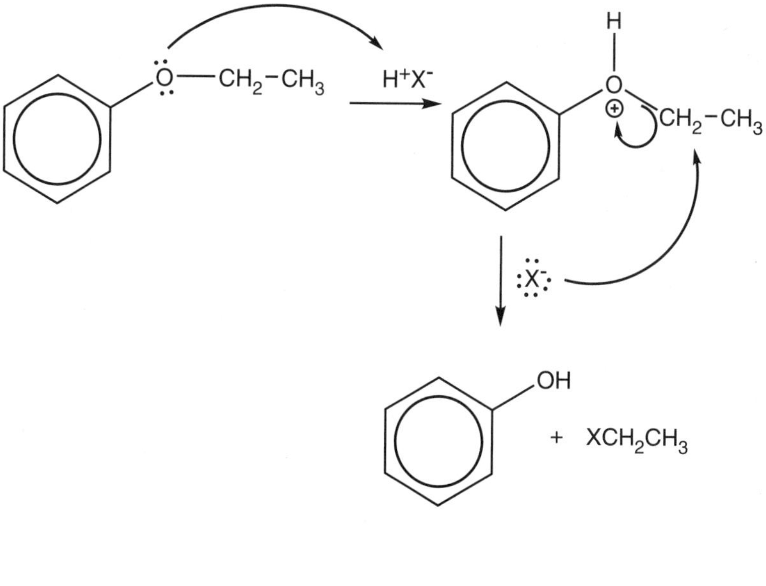

Figure 3-35: Mechanism for the cleavage of an ether by a hydrohalic acid.

Ph–O–CH_2–CH_3 $\xrightarrow{H^+X^-}$ Ph–O^+(H)–CH_2–CH_3 $\xrightarrow{:X^-}$ Ph–OH + XCH_2CH_3

Sulfuric acid and ethers

Cold concentrated sulfuric acid will react with ethers to give a soluble product. An example of this process is in Figure 3-36.

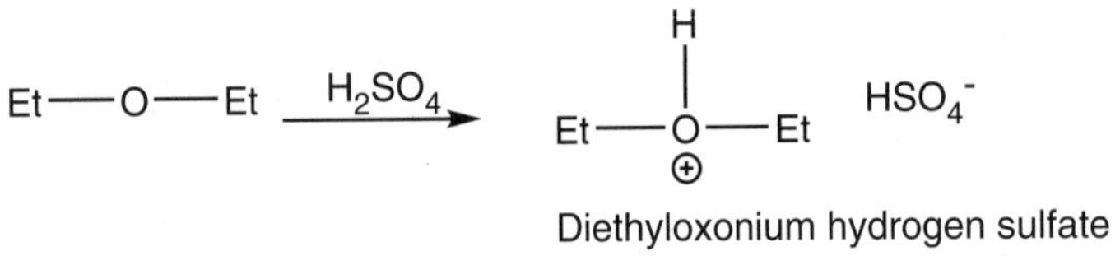
Et—O—Et
H_2SO_4
H
Et—O—Et
⊕
HSO_4^-
Diethyloxonium hydrogen sulfate

Figure 3-36: Reaction of ethers with cold concentrated sulfuric acid.

Reactions of epoxides

Epoxides are more reactive than other ethers. This is due to ring strain inherent in any three-atom ring system. The most useful reactions are acidic cleavage and nucleophilic cleavage.

Acidic cleavage is an S_N2 mechanism with a pseudo-carbocation ion. The reaction produces the *trans* product (anti addition). Figure 3-37 shows a typical reaction, and Figure 3-38 illustrates the mechanism.

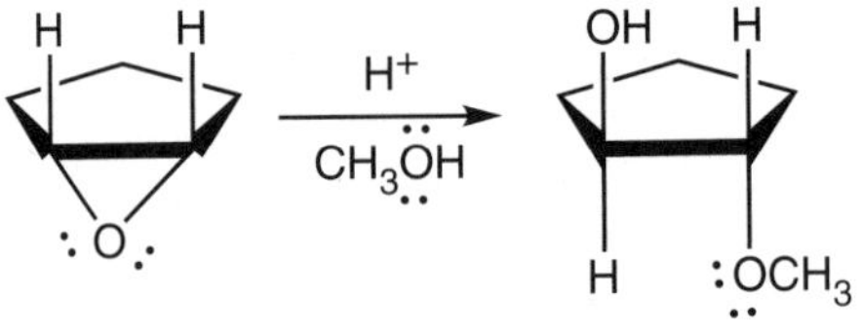
H
H
O
H+
CH_3OH
OH
H
H
OCH_3

Figure 3-37: Acid cleavage of an epoxide.

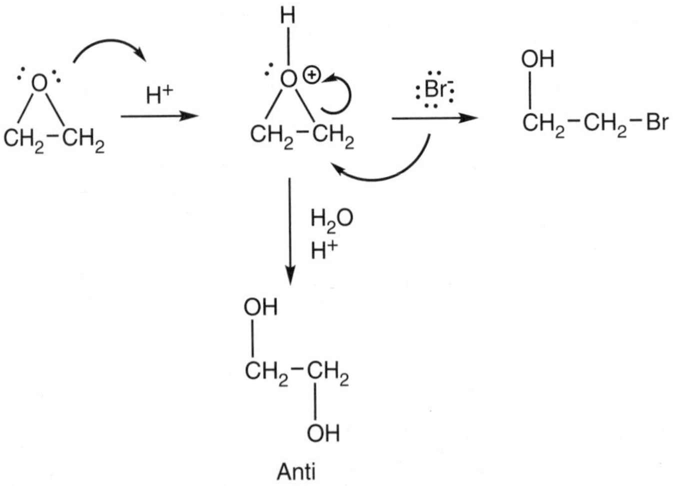
O
CH_2-CH_2
H+
H
O
CH_2-CH_2
:Br:
OH
CH_2-CH_2-Br
H_2O
H+
OH
CH_2-CH_2
OH
Anti

Figure 3-38: Mechanism for the acid cleavage of an epoxide.

A generic nucleophilic cleavage appears in Figure 3-39, and Figure 3-40 shows two reactions. The three-membered ring in the intermediate in Figure 3-40 is an example of a pseudo-carbocation.

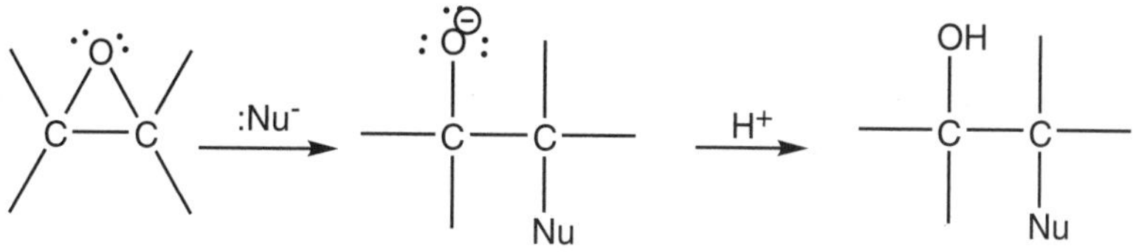

Figure 3-39: Nucleophilic cleavage of an epoxide.

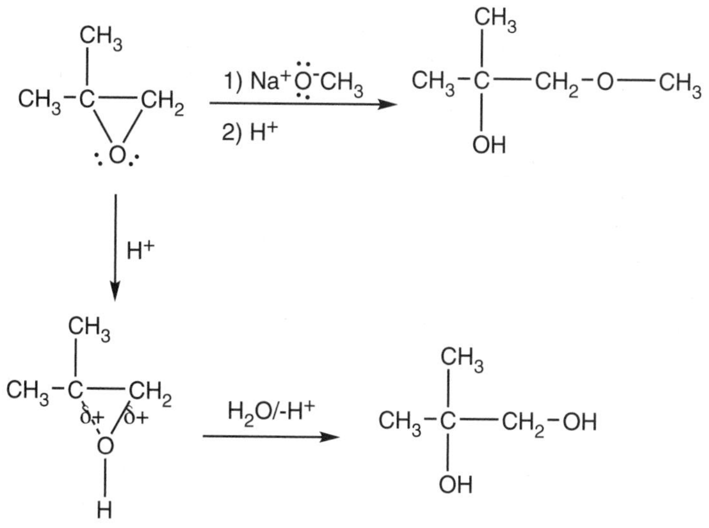

Figure 3-40: Examples of nucleophilic cleavage of an epoxide.

Summarizing the Spectra of Alcohols and Ethers

Tables 3-2 and 3-3 summarize the infrared and proton-NMR (nuclear magnetic resonance) spectroscopic properties of alcohols and ethers. In the proton NMR, the oxygen atom is deshielding. Phenols and alcohols rapidly exchange protons so their NMR spectra are solvent dependant. The alcohol and ether groups don't have any characteristics absorptions in UV-vis spectra.

Table 3-2 Infrared Data

Functional Group	*Absorption Location*
C-H stretch	~3000
O-H stretch	~3600–3200 cm^{-1}
C-O stretch	~1150 (3°)–1050 (1°) cm^{-1} (Strong for ethers)
Ar-C-O stretch (C-O stretch)	~1275–1200 cm^{-1}

Table 3-3 Proton NMR Data

Functional Group	*Chemical Shift (δ) (In parts per million)*
-CHO	3.3–4
ArOH	6–8
ROH	1–5

Chapter 4

Conjugated Unsaturated Systems

In This Chapter

- Understanding unsaturated systems
- Tackling delocalization and resonance
- Deciphering reactions of conjugated unsaturated systems

The presence of a double or triple bond generally makes a molecule more reactive. However, if the system is conjugated, some stability is restored. In this chapter you look at conjugated unsaturated systems, especially their reactions. And to help prepare you for an exam, we also go through some example problems. Have fun!

When You Don't Have Enough: Unsaturated Systems

An unsaturated system has one or more double or triple bonds. You probably learned quite a bit about systems containing one double or one triple carbon-carbon bond in Organic Chemistry I. Now you're looking at systems containing more than one multiple bond. Molecules containing more than one multiple carbon-carbon bond can be classified into one of two categories — those with isolated multiple bonds and those with conjugated multiple bonds. Isolated multiple bonds are simply an extension of the behavior of systems containing only one double or triple bond. However, systems with conjugated multiple bonds behave differently. In conjugated systems, resonance plays a significant role.

Conjugated systems

A conjugated system is a species with alternating double and single bonds and/or a p-orbital next to a π-bond. The p-orbital may contain zero, one, or two electrons. Some examples of conjugated systems are shown in Figure 4-1.

These systems have a high degree of stability and undergo some unique and unusual reactions.

$CH_2{=}CH{-}CH{=}CH_2$

H

p-orbital with 2 electrons

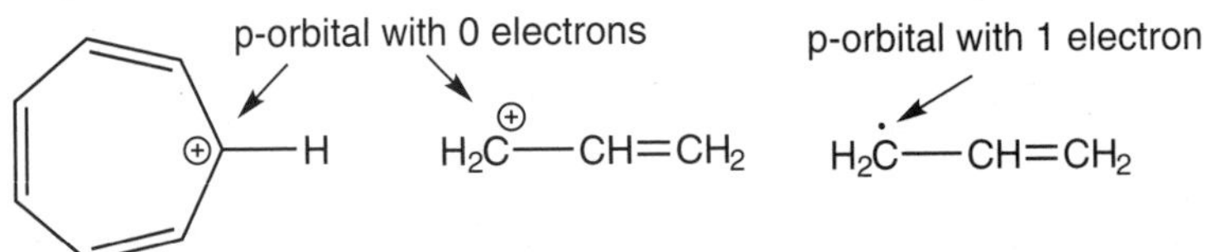

Figure 4-1: Some examples of conjugated systems.

The allylic radical

Although propene has only one double bond, it can become part of a conjugated system in the form of a radical or an ionic species. The radical formed from propene is the allylic radical, and it can be formed through the reaction of propene with another free radical such as a chlorine atom. Resonance stabilizes the radical as illustrated in Figure 4-2, and the resonance hybrid is shown in Figure 4-3. In Figure 4-2, the resonating electrons (in p-orbitals) are shown, while the nonresonating electrons (in σ-bonds) are part of the lines indicating the bonds.

Figure 4-2: Resonance of the allylic radical.

Figure 4-3 shows that a partial radical character is on the first and third carbon atoms. Both of the carbon-carbon bonds are equal and intermediate in length between a single and a double bond. The allylic radical is more stable than a tertiary carbon radical.

The stability of radicals is in the series allylic > 3° > 2° > 1° > methyl.

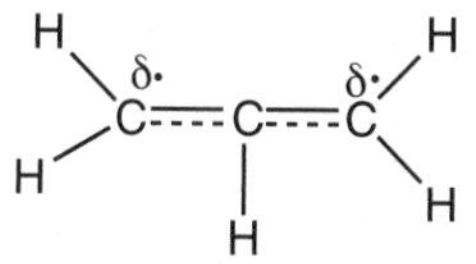

Figure 4-3: Allylic radical resonance hybrid.

Butadiene

A pair of carbon-carbon double bonds separated by a carbon-carbon single bond is the basis for a conjugated system. The simplest conjugated system is 1,3-butadiene, so we use it here to illustrate the behavior of all conjugated systems. The structure of 1,3-butadiene is shown in Figure 4-4. In this compound, the single bond is shorter than the single bond in ethane (CH_3-CH_3). The sp^2 hybridized carbon atoms are more electronegative than the sp^3 hybridized carbon atoms. They pull the electrons closer to the nuclei, making the atoms smaller and causing the carbon atoms to move closer together. Molecular orbital theory predicts some double-bond character between the middle carbon atoms.

Figure 4-4: Structure of 1,3-butadiene.

CH_2=CH-CH=CH_2

At room temperature, the *cis* and *trans* forms of 1,3-butadiene are in equilibrium. The equilibrium favors the *trans* form, with the distribution 5 percent *cis* and 95 percent *trans*. Figure 4-5 shows the equilibrium and the structures of the two forms.

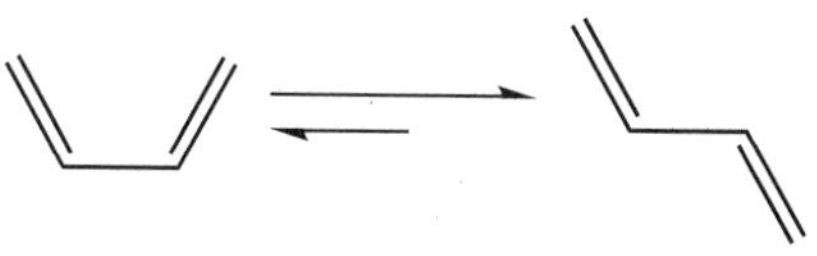

Figure 4-5: The equilibrium between *cis* and *trans* 1,3-butadiene.

Delocalization and Resonance

Everyone receives an introduction to the basic concepts of resonance in Organic Chemistry I. Organic Chemistry II requires an extension of the basic rules of resonance to other systems. In addition to constructing reasonable resonance structures, you also need to understand which structures are more stable.

Resonance rules

Resonance, as you saw in Organic I, occurs in many systems, and you need to be able to recognize when it's going to affect the outcome of a reaction. In general, resonance makes a species more stable by delocalizing the electrons. Delocalization, among other things, reduces electron-electron repulsion.

Following is an expanded set of rules for drawing resonance structures. This list is an expansion of the rules necessary to understand resonance for Organic Chemistry I. You may want to bookmark this list because these rules apply throughout the remainder of this book.

1. Resonance structures are simply a way of understanding stability; they don't really exist.
2. When writing the resonance structures, you are only permitted to move lone-pair electrons or π-electrons (think of the π-electrons as a swinging gate).

   Never, ever move any atoms.
3. All the structures must have reasonable Lewis structures. This includes the same overall charge (same number of electrons). For example, carbon should never have five bonds.

   Second period elements can never exceed an octet of electrons. Also keep the following rules in mind:
   - A negative charge on a more electronegative atom is more stable than a negative charge on a less electronegative atom.
   - Structures that place unlike charges close together or like charges apart are more stable. Structures doing the opposite are less important (less stable). However, such structures are not necessarily impossible.
   - Structures with all atoms having complete valence shells are very important. (In most cases, this is an octet.)
4. No change in the number of unpaired electrons should occur.

5. All atoms that are part of the delocalized π-system must be coplanar or nearly coplanar.

   An sp^3 hybridized carbon atom will not be involved in resonance.

6. The presence of resonance leads to stabilization, which means that the species is more stable than any of the contributing structures.
7. Equivalent resonance structures are equally important to the overall structure.
8. The more stable the resonance structure, the more important it is (the more it contributes) to the hybrid. In general, the more stable resonance structure is the one with more bonds.

Stability of conjugated unsaturated systems

You can estimate the stability of a conjugated versus nonconjugated system by comparing the energy changes. For example, the enthalpy change for the hydrogenation of 1-butene is –127 kJ/mol (kilojoules per mole). If conjugation doesn't lead to an increase in stability, the energy change for the hydrogenation of 1,3-butadiene would be about twice this value (–254 kJ/mol). However, the observed value for the hydrogenation of 1,3-butadiene is –239 kJ/mol. The difference between the predicted (2 × 1-butene) and the observed value indicates that resonance stabilizes 1,3-butadiene by about 15 kJ/mol.

Reactions of Conjugated Unsaturated Systems

Many students make the mistake of reacting conjugated systems as extensions of the reactions they learned for simple alkenes and alkynes. While in some cases this is acceptable, the unique nature of these systems makes their chemistry different. You investigate some of these features in the next few pages.

Put in the second string: Substitution reactions

The reaction of chlorine with propene illustrates one difference caused by conjugation. The products of the reaction depend upon the reaction conditions, as illustrated in Figure 4-6.

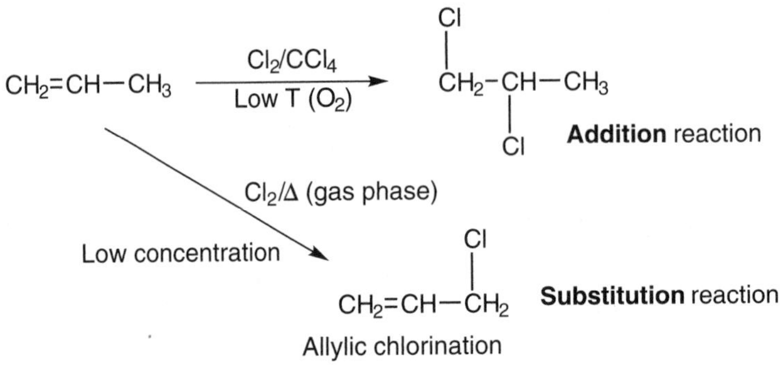

$CH_2{=}CH{-}CH_3 \xrightarrow[\text{Low T } (O_2)]{Cl_2/CCl_4} CH_2Cl{-}CHCl{-}CH_3$ **Addition** reaction

$CH_2{=}CH{-}CH_3 \xrightarrow[\text{Low concentration}]{Cl_2/\Delta \text{ (gas phase)}} CH_2{=}CH{-}CH_2Cl$ **Substitution** reaction

Allylic chlorination

Figure 4-6: Two reactions of propene with chlorine.

At low temperature, propene behaves like another alkene and undergoes a simple addition of a halogen across the double bond to form 1,2-dichloropropane. These conditions minimize the possibility of forming chlorine atoms (chlorine free radicals), and the presence of oxygen traps the few that do form. However, when the conditions promote the formation of chlorine atoms, a substitution occurs to produce 3-chloropropene.

The mechanism

Allylic halogenation is a substitution reaction involving a free-radical mechanism. The general mechanism is in Figure 4-7. The final X·cycles back to the beginning (shown with the large curved arrow).

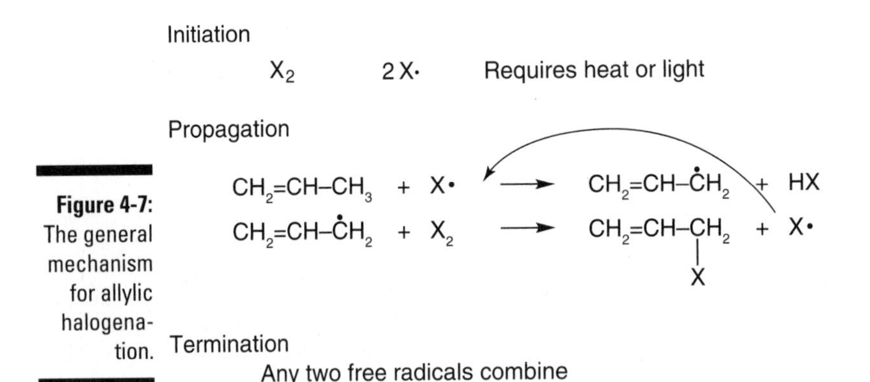

Figure 4-7: The general mechanism for allylic halogenation.

Understanding the reaction

The question is, why doesn't the X in Figure 4-7 add to the double bond to form the following radical shown in Figure 4-8? The answer is that this would give a secondary free radical, which is less stable than the allylic free radical because it doesn't have resonance stabilization.

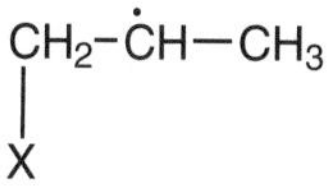

$CH_2-\dot{C}H-CH_3$ with X on CH_2

Figure 4-8: A secondary free radical.

Allylic bromination

Another example of allylic halogenation is shown in Figure 4-9.

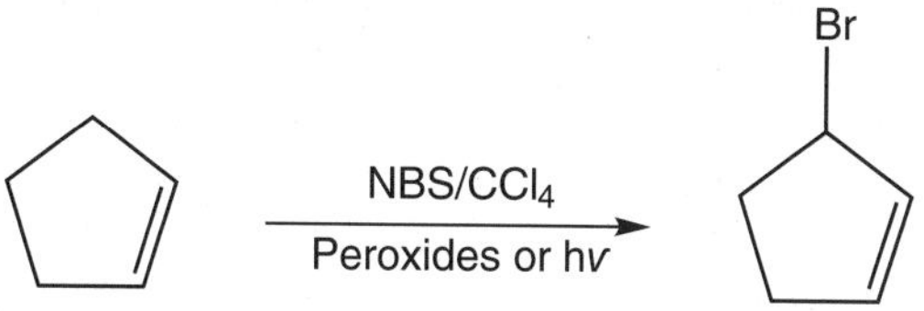

Cyclopentene —(NBS/CCl_4; Peroxides or hν)→ 3-bromocyclopentene (Br)

Figure 4-9: Allylic bromination.

The reactant NBS (N-bromosuccinimide) shown in Figure 4-10 is a good source of low concentrations of bromine atoms (free radicals).

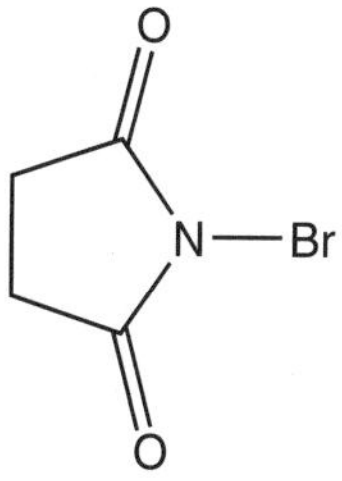

N—Br, O, O

Figure 4-10: The structure of NBS.

Electrophilic addition

The addition of a hydrogen halide, such as HBr, is an important addition reaction for alkenes often seen in Organic Chemistry I. However, conjugated dienes may behave differently. An example is the reaction of HBr with 1,3-butadiene as illustrated in Figure 4-11.

In this case, no stereochemistry is implied. The distribution of the products depends on the reaction conditions shown in Table 4-1. The information in the last column of the table indicates the process is reversible and an equilibrium results upon heating. The equilibrium leads to the production of the more stable product.

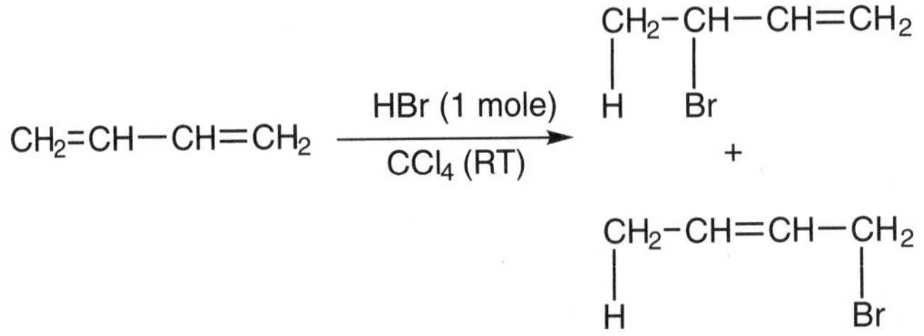

$$CH_2{=}CH{-}CH{=}CH_2 \xrightarrow[CCl_4\ (RT)]{HBr\ (1\ mole)} CH_2(H){-}CH(Br){-}CH{=}CH_2 + CH_2(H){-}CH{=}CH{-}CH_2(Br)$$

Figure 4-11: The reaction of HBr with 1,3-butadiene.

Table 4-1 Distribution of Products in the Reaction of HBr with 1,3-Butadiene

	40°C	*–80°*	*–80°/40°**
1,2-addition	15%	80%	15%
1,4-addition	85%	20%	85%

**Distribution resulting after heating the –80° reaction products to 40° in the presence of HBr*

The mechanism

Examining the mechanism can help you understand the different results. The reaction begins with the protonation of one of the carbon-carbon double bonds (see Figure 4-12) by the hydrogen ion from the HBr. A primary or a secondary carbocation can be formed by this reaction. As seen in Organic Chemistry I, a secondary carbocation is more stable than a primary carbocation. Also, this secondary carbocation is even more stable because it's allylic and resonance stabilized.

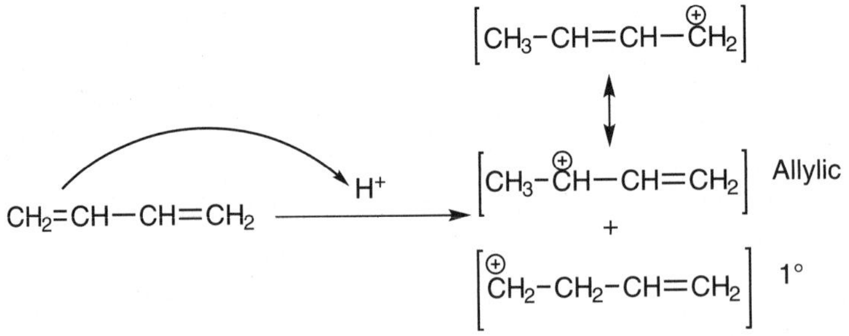

$$CH_2{=}CH{-}CH{=}CH_2 \xrightarrow{H^+} [CH_3{-}\overset{\oplus}{C}H{-}CH{=}CH_2 \leftrightarrow CH_3{-}CH{=}CH{-}\overset{\oplus}{C}H_2]\ \text{Allylic} + [\overset{\oplus}{C}H_2{-}CH_2{-}CH{=}CH_2]\ 1°$$

Figure 4-12: The protonation of 1,3-butadiene.

In the second step of the mechanism (shown in Figure 4-13), the bromide ion from the HBr attacks the allylic carbocation at one or the other of the partially positive carbon atoms. Attack on the second carbon gives 1,2-addition, while attack on the fourth carbon gives 1,4-addition.

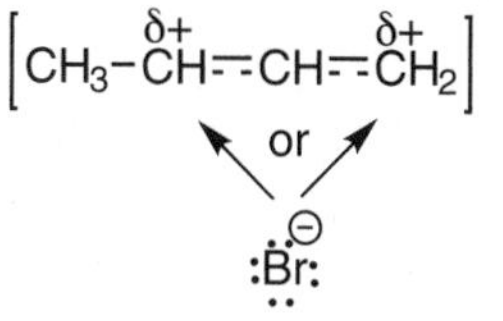

δ+ δ+
[CH_3–CH=CH=CH_2]
or
:Br:⊖

Figure 4-13: Bromide attack on the allylic carbocation.

Understanding the reaction

The bromide has an equal probability of attacking either carbon atom two or carbon atom four, so why is the product mixture not 50 percent of each? At low temperatures, the bromine doesn't move very far after giving up its H, so it's near carbon two (1,2-addition). At high temperatures, the bromine donates an H^+ and can move, so it's able to form the more stable product (a disubstituted C=C).

Figure 4-14 uses a reaction diagram to illustrate this situation. The two-step process requires a diagram with two hills. The first step is the same for both products, so the second step is the one that makes the difference. At low temperatures, fewer molecules have sufficient kinetic energy to get over the higher barrier. Therefore, the 1,2-addition product (lower barrier) is likely to form.

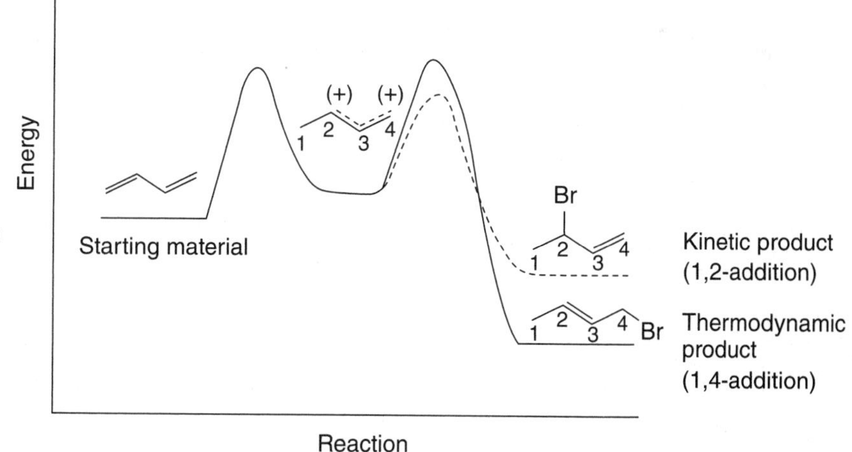

Figure 4-14: The reaction diagram for 1,2-addition and 1,4-addition.

The 1,2-addition is rate controlled, which leads to the formation of the kinetic product (fastest formed). This formation is especially noticeable at low temperatures because they always favor the kinetic product, which is the reaction product with the lower activation energy barrier. In addition, few molecules have sufficient energy to surmount the barrier in the reverse direction to allow the establishment of an equilibrium.

The 1,4-addition is thermodynamically controlled. This reaction forms the thermodynamic product. At higher temperatures, more molecules have sufficient energy to cross the second barrier in the reverse direction and establish an equilibrium. The equilibrium allows the less stable 1,2-addition product to convert to the more stable 1,4-addition product.

Low temperatures favor the kinetic product and high temperatures favor the thermodynamic product.

More than a tree: Diels-Alder reactions

In the Diels-Alder reaction, a diene, such as 1,3-butadiene, reacts with a dienophile, such as ethylene, to form a product with a six-membered ring. This is an important reaction, not only to students trying to pass Organic Chemistry, but also in organic synthesis.

Any of a number of dienes react as long as a conjugated system is present. Substituents attached to the conjugated system alter the reactivity of the diene. The dienophile is typically a substituted alkene; however, as you see later in this chapter, other species may react. The substituents around the double bond also alter the reactivity of the dienophile. Figure 4-15 illustrates the general reaction. In this figure, the arrows are for apparent, not actual, electron movement as a means of keeping track of the process. The final double bond (between carbons two and three) is between the positions of the double bonds in the original diene.

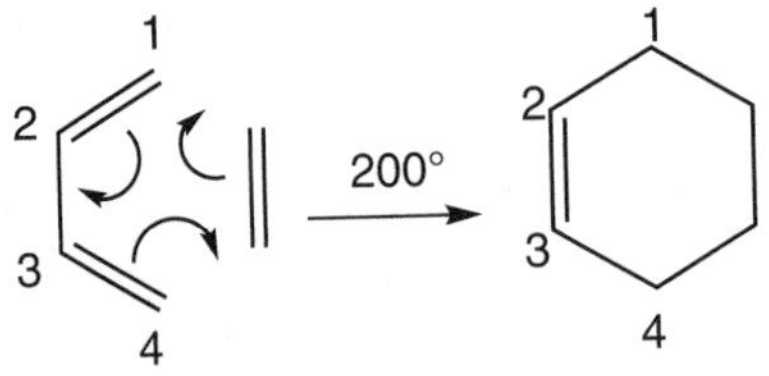

Figure 4-15: A general Diels-Alder reaction.

The basic reaction remains the same when substituents are present, as illustrated in Figure 4-16. In this example, the aldehyde is an electron-withdrawing group (the electronegative oxygen pulls electron density away from the double bond). The polarity arrow illustrates this electron shift. This shift of electron density speeds up the reaction (a lower temperature is necessary).

Conditions

As long as a diene and a dienophile are present, a Diels-Alder reaction occurs. However, the yield of the reaction can be improved by adjusting the reactivity of the reactants.

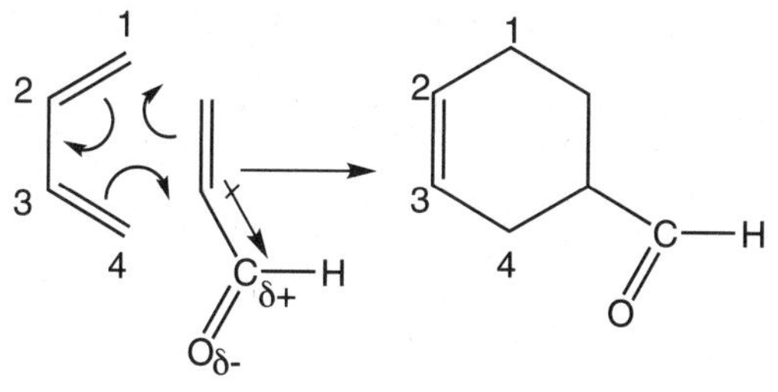

Figure 4-16: Diels-Alder reaction of a substituted dienophile.

The presence of good electron-donating groups (EDG) on the diene results in the diene reacting faster. Examples of electron-donating groups are alcohols, ethers, and amines. The dienophile has the opposite requirement; that is, an electron-withdrawing group (EWG) facilitates the reaction. Examples of electron-withdrawing groups are carbonyl groups, cyano groups, and nitro groups.

The EDG and EWG groups must be directly attached to the diene or dienophile. If another carbon atom is between the group and the diene or dienophile, the group doesn't count.

In order for a reaction to occur, the diene must be in the *cis* configuration and not the *trans* configuration (refer to Figure 4-5). Normally, both forms are in equilibrium. However, the diene can be locked into the *cis* conformation and facilitate the reaction. One way to lock the conformation is to use a ring system. The compound 1,3-cyclopentadiene contains a diene locked in the *cis* conformation. Figure 4-17 illustrates the reaction of 1,3-cyclopentadiene with ethylene.

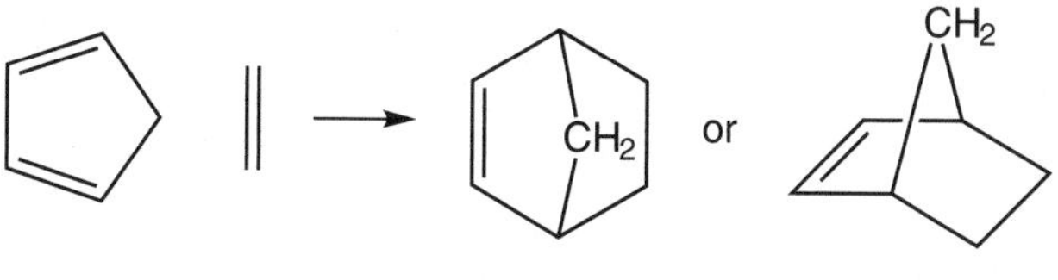

Figure 4-17: The reaction of 1,3-cyclopentadiene with ethylene.

Stereochemistry

The Diels-Alder reaction is a concerted syn addition (meaning the addition is on one side) with the stereochemistry of the dienophile preserved in the stereochemistry of the product. If the dienophile is *cis* then the product is also *cis*, and if the dienophile is *trans*, the product is also *trans*. See Figure 4-18 for the attack by a *cis* dienophile. The reaction in Figure 4-18 is the reaction of isoprene with maleic anhydride *(cis)*.

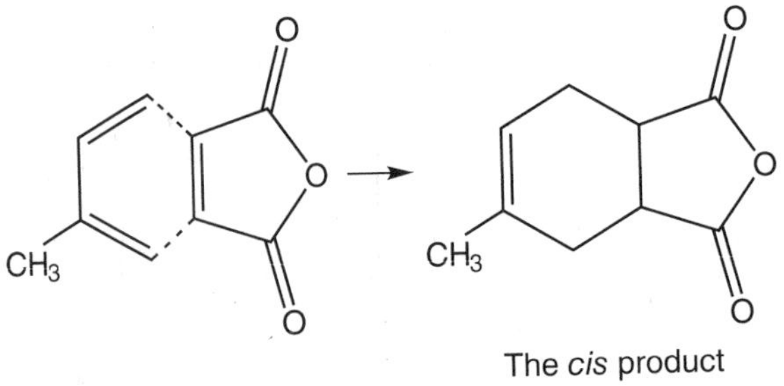

Figure 4-18: The reaction of a *cis* dienophile.

Figure 4-19 illustrates the attack by a *trans* dienophile.

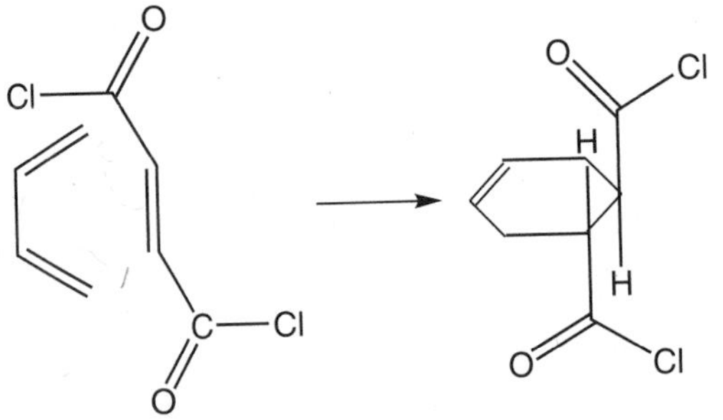

Figure 4-19: The reaction of a *trans* dienophile.

In some situations, such as in Figure 4-20, two different *cis* products may form. The two products are the *endo*- and the *exo*-product. The *endo*-product is the major or only product of the reaction. The process leading to the *endo*-product is Alder *endo*-addition. The *endo*-form is more stable than the *exo*-form.

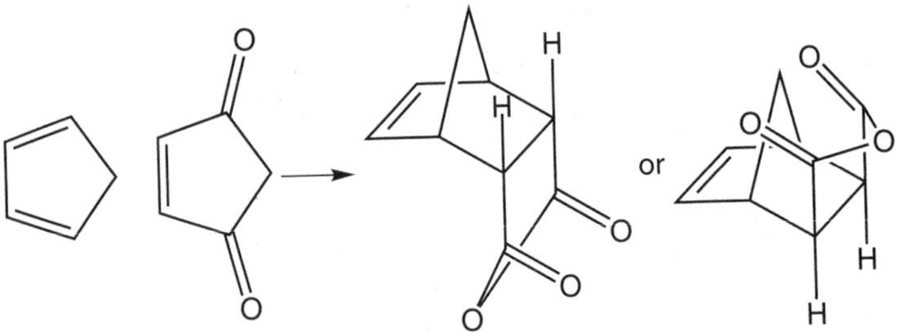

Figure 4-20: Two possible *cis* products.

Figure 4-21 illustrates the difference between the two forms of the product.

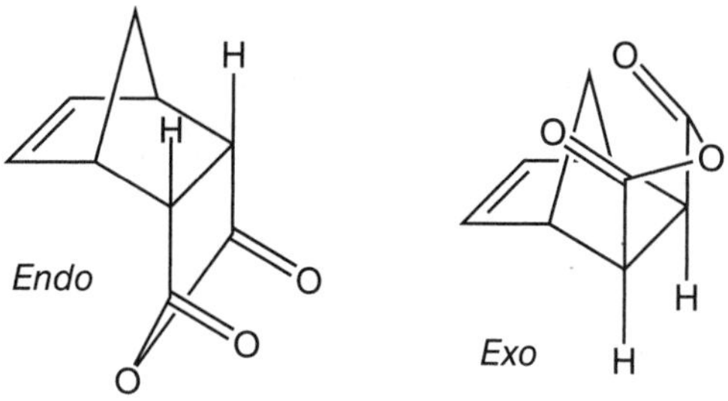

Figure 4-21: The *endo* and *exo* forms.

Passing an Exam with Diels-Adler Questions

Questions concerning Diels-Alder reactions commonly appear on organic chemistry exams. In order to gain insight into the reaction itself (and to increase your score on your organic exams), in this section we look at some typical questions that you might face.

Two general types of Diels-Alder questions commonly appear on organic chemistry exams. These may be simple stand-alone questions or part of a larger question. One type of question involves identifying the product, and the other type of question involves identifying the reactants.

Indentifying the product

What is the Diels-Alder product from the reaction in Figure 4-22?

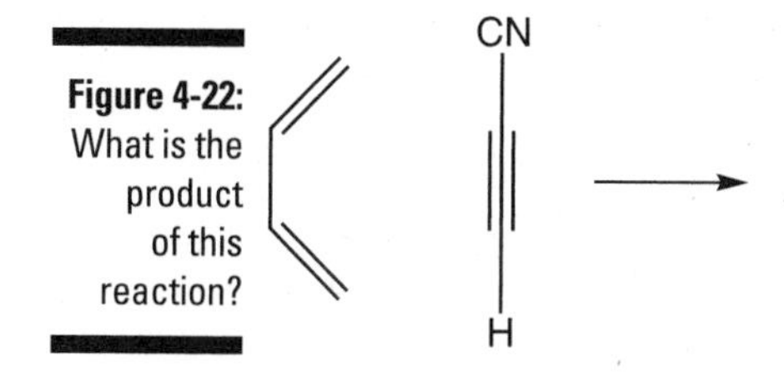

Figure 4-22: What is the product of this reaction?

The product of the reaction in Figure 4-22 is shown in Figure 4-23.

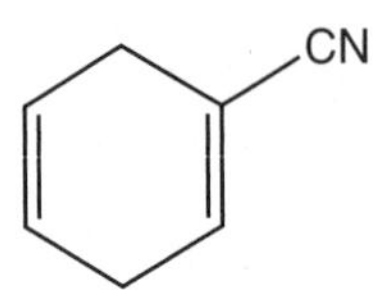

Figure 4-23: The product of the reaction in Figure 4-22.

If the product has two C=C, then the double bond next to the electron-withdrawing group was originally part of the dienophile.

Identifying the reactants

What reactants are necessary to form the product in Figure 4-24?

Figure 4-24: The product of a Diels-Alder reaction.

The reactants necessary to form the product in Figure 4-24 are in Figure 4-25.

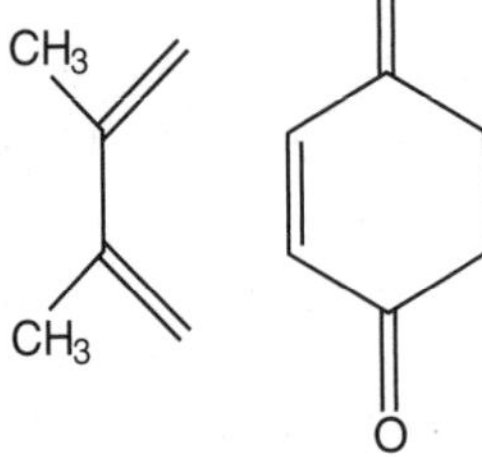

Figure 4-25: A diene and a dienophile.

Chapter 5

"Seeing" Molecules: Spectroscopy Revisited

In This Chapter

- Investigating infrared spectroscopy
- Understanding ultraviolet and visible spectroscopy
- Mastering mass spectroscopy
- Looking at NMR spectroscopy

A variety of instrumental methods supply information about the structure of molecules. In general, the absorption of energy (and many times its subsequent release of energy) leads to evidence about the presence of some feature of the molecule. In infrared (IR) spectroscopy, the absorption of energy in the infrared region of the spectrum gives information about the types of bonds present. Ultraviolet-visible (UV-vis) spectroscopy provides info about the molecular orbital arrangement in a molecule. Mass spectroscopy uses energy to *ionize,* or to break up a molecule into ions. The masses of these ions give information concerning the size and structure of the original molecule. Finally, nuclear magnetic resonance (NMR) spectroscopy utilizes the absorption of energy in the radio wave portion of the spectrum to give information concerning the environments occupied by certain nuclei, especially ^{1}H and ^{13}C. In this chapter you find out how these different types of spectroscopy can be used to learn about organic molecules.

You may want to get yourself a copy of *Organic Chemistry I For Dummies* by Arthur Winter (Wiley). It has some detailed sections on the theory behind instrumentation and exactly how it works.

In principle, you can determine the complete structure of an organic molecule from just the IR, NMR, or mass spectrum of a compound. However, the process of determination can be very tedious. Organic chemists use all the data available when attempting to determine the structure of a compound.

For example, they gather some information from each of the spectra they have available and combine this data to produce a structure consistent with all the available data. When looking at a compound such as $C_{10}H_{20}O$, they may look at the IR to determine whether the oxygen atom is part of a carbonyl group, an alcohol group, or some other group. Then they may look at the NMR spectrum to get some idea about the carbon backbone.

Later in this book you can see the specific functional groups and explore other features of the spectra that are characteristic of the group.

The methods in this chapter apply to the common types of organic compounds. Unusual compounds have their own characteristic behavior. The behavior of these unusual compounds normally isn't relevant until advanced courses in organic chemistry.

Chemical Fingerprints: Infrared Spectroscopy

Light with energy in the infrared region of the electromagnetic spectrum has enough energy to cause a covalent bond to vibrate. (The vibrations are due to the stretching and bending of the bonds.) If the vibration causes a change in the dipole moment of the molecule, energy will be absorbed. Most organic compounds have several absorptions in the 4,000–600 cm^{-1} region of the spectrum. All of the absorptions give information about the structure of the molecule, but some absorptions are more useful than others are. The region below 1,500 cm^{-1} is the *fingerprint region* of the IR spectrum. The fingerprint region gets its name because this region tends to be unique for every compound, no matter how similar it may be to other molecules.

Don't get too distracted by the mess of closely-spaced peaks in the fingerprint region. Instead, look primarily in the important places (between 1,500 and 2,800 cm^{-1} and above 3,000 cm^{-1}).

The following sections offer a summary of some of the more important IR absorptions that occur in typical organic compounds.

Double bonds

The carbon-oxygen double bond and, to a lesser degree, the carbon-carbon double bond are found in many organic compounds. The carbonyl (carbon-oxygen double bond) has a very sharp and intense band around 1,700 cm^{-1}.

In many cases, this band is the most prominent feature of the IR spectrum of compounds containing a carbonyl group.

The peak due to a carbon-carbon double bond is characteristic of alkenes. It's normally around 1,650 cm^{-1}.

Triple bonds

Though not as common as double bonds, both carbon-carbon triple bonds (alkynes) and carbon-nitrogen triple bonds (nitriles) are important. Both occur in the 2,600–2,100 cm^{-1} region of the spectrum. They are usually very sharp. The carbon-nitrogen triple bond tends to give a more intense peak than the alkyne peak.

O-H and N-H stretches

Compounds containing a hydroxyl group (OH) have a strong very broad absorption in the 3,600–2,500 cm^{-1} region of the spectrum. The most common examples are the alcohols and the carboxylic acids. The combination of a broad 3,600–2,500 cm^{-1} band with a 1,700 cm^{-1} peak often indicates a carboxylic acid (or amide — keep reading).

The N-H stretch in amines occurs in the same general region (3,500–3,100 cm^{-1}) as the hydroxyl group; however, the intensity tends to be less. In addition, primary amines usually have two bands. The observation of a 3,500–3,100 cm^{-1} band with a 1,700 cm^{-1} peak is often indicative of an amide — or carboxylic acid. (If you've been paying attention, you saw that coming! If not, see the preceding paragraph.)

C-H stretches

Nearly every organic compound has one or more carbon-hydrogen bonds. For this reason, the C-H stretch isn't as useful as you might think. Some guidelines are helpful:

- The C-H stretch for hydrogen bound to an sp^3 hybridized carbon is in the 3,000–2,850 cm^{-1} region.
- For an sp^2 hybridized carbon, the C-H stretch is in the 3,150–3,000 cm^{-1} region.
- For an *sp* hybridized carbon, the C-H stretch is near 3,300 cm^{-1}.

See Figure 5-1 for samples of the IR spectra of various functional groups.

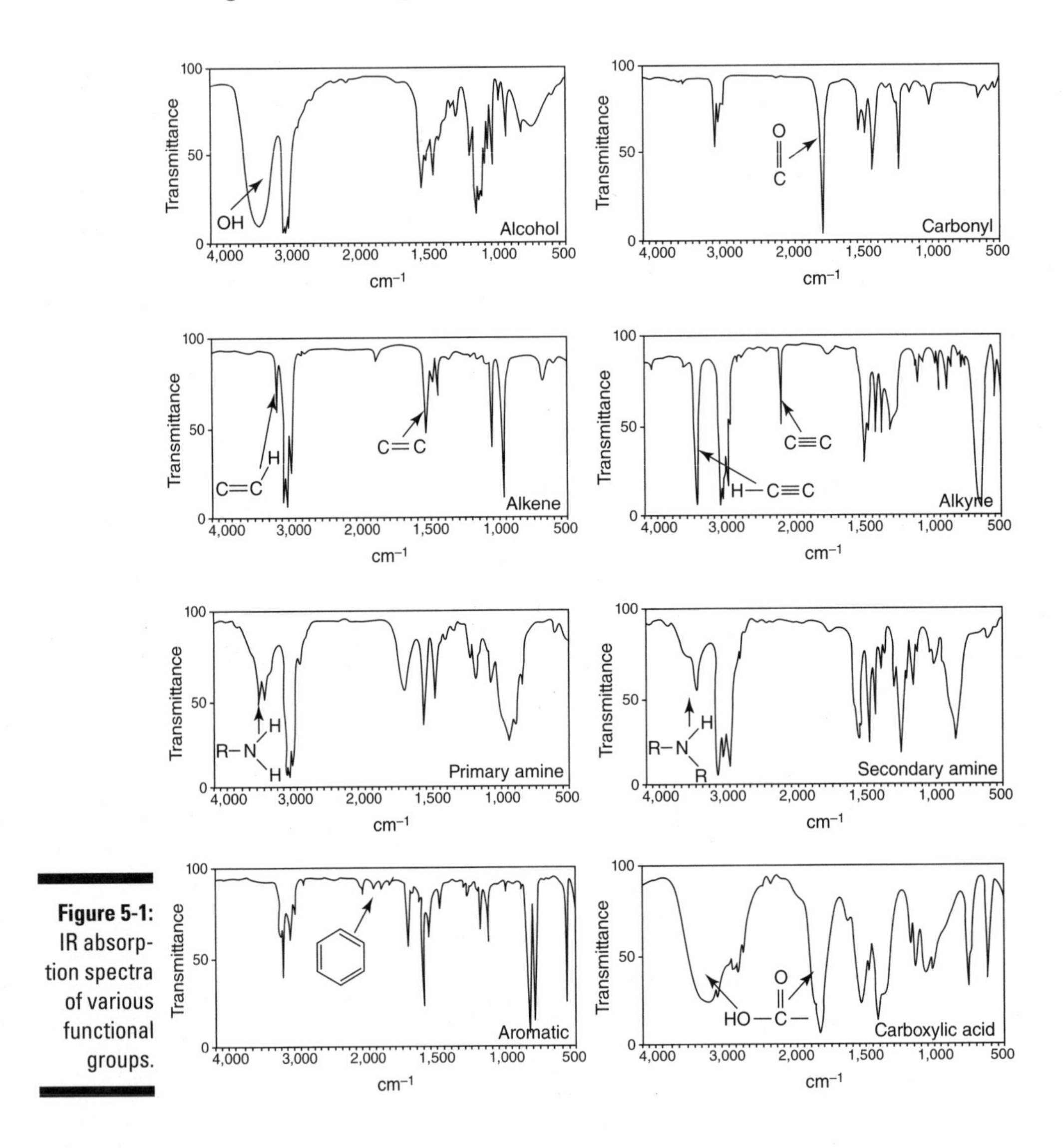

Figure 5-1: IR absorption spectra of various functional groups.

Suntans and Beyond: Ultraviolet and Visible Spectroscopy

Ultraviolet and visible spectroscopy (UV-vis) is an analytical technique useful in the investigation of some organic molecules. Absorption of energy in this region of the electromagnetic spectrum can excite an electron from the ground state to an excited state — usually from the HOMO (highest occupied molecular orbital) to the LUMO (lowest unoccupied molecular orbital). This technique is particularly useful for compounds containing multiple bonds.

The useful spectral range is usually found between 700 and 200 nm. (The region below 200 nm is the vacuum ultraviolet, which requires special instrumentation and, for this reason, is less important.) The spectrum is the result of measuring the absorption of light versus wavelength. The position of the absorption maximum (wavelength = peak position), λ_{max}, is important, as is related to the amount of radiation absorbed (molar absorptivity, ε).

UV-vis spectra tend to be much simpler than IR spectra. Most UV-vis spectra contain only a few peaks; in many cases, only one or two. Organic compounds with no double bonds or only one double bond typically show absorption only in the vacuum ultraviolet portion of the spectrum. The main use of this method is for molecules that have conjugated double bonds. The more double bonds in a conjugated system, the longer the wavelength is where the molecule absorbs light. The presence of eight or more conjugated double bonds can shift the absorption maximum into the visible portion of the spectrum, which occurs in many organic dyes. Absorptions in the UV-vis region of the spectrum is evidence for a conjugated π-electron system.

Compounds containing carbon-oxygen double bonds also exhibit absorptions in the UV-vis region. In general, these absorptions occur at longer wavelengths than absorptions that are due to carbon-carbon double bonds. Conjugation shifts the absorption maximum to still longer wavelengths.

In most cases, information from the UV-vis spectrum of a molecule is useful to corroborate data and information from other sources. This method is also useful in the quantitative determination of the concentration of an absorbing molecule. Figure 5-2 shows a typical UV-vis spectrum.

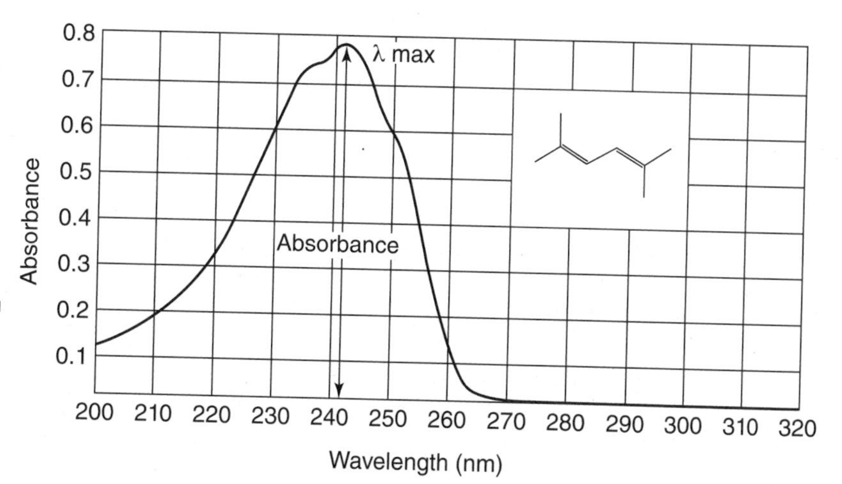

Figure 5-2: A typical UV-vis spectrum.

Not Weight Watchers, Mass Watchers: Mass Spectroscopy

In mass spectroscopy an organic molecule is vaporized and is injected into a mass spectrometer where it undergoes bombardment by electrons. This assault causes ionization of some of the molecules, producing a molecular ion (M^+). Other molecules fragment to produce pieces that may or may not be ions. The fragments with a positive charge are important. Finally, some of the molecules or fragments may rearrange and/or undergo further fragmentation to produce other species, some of which are positive ions. The mass spectrometer then takes the positive ions and sorts them according to their masses. The relative intensity (number of ions) is plotted versus the mass-to-charge ratio (m/e). The result is a mass spectrum.

In the following sections, we investigate the properties of the molecular ion (M^+) and the various other ions that result.

The molecular ion

The molecular ion represents the molecular mass of the compound. Specifically, the molecular mass is the sum of the masses of the most abundant isotope of each element present. You can derive an approximate formula for a compound based on its molecular mass, especially if only some or all of the following elements are present: C, H, N, O, F, or I.

If the compound contains or may contain nitrogen, the *nitrogen rule* is applicable. According to this rule, any molecule with an odd number of nitrogen atoms has an odd mass. For example, in the compound $N(CH_3)_3$ the mass is 59 g/mole.

In some cases the $(M + 1)^+$ ion is important. The primary source of this ion is the presence of carbon-13. This isotope is about 1.1 percent of all the carbon atoms present. The relative intensities of the $(M + 1)^+$ to the M^+ ions indicates the number of carbon atoms present in the molecule. To determine the number of carbon atoms, you have to calculate the relative intensities of the $(M + 1)^+$ peak, which you do by multiplying the intensity of the $(M + 1)^+$ peak by (100/(intensity of the M^+ peak)). Dividing the relative intensity of the $(M + 1)^+$ peak by 1.1 gives the number of carbon atoms present in the formula.

The presence of a chlorine or bromine atom results in an intense $(M + 2)^+$ peak due to the presence of either two chlorine isotopes (chlorine-35 and chlorine-37) or two bromine isotopes (bromine-79 and bromine-81). In the case of chlorine, the $(M + 2)^+$ peak is about one-third the intensity of the M^+

peak, and in the case of bromine, the M^+ and $(M + 2)^+$ peaks are of approximately equal intensity. If more than one chlorine or bromine atom is present, the pattern is more complicated, but it includes a group of peaks separated by two mass units.

Sulfur has a less obvious $(M + 2)^+$ peak because the abundance of sulfur-34 is only 4.4 percent of sulfur-32. (A little sulfur-33 is also present, which contributes to the $[M + 1]^+$ peak.)

Fragmentation

In addition to the molecular ion, the molecule generates a number of fragments. In general, the fragments result by breaking the weakest bonds. Different types of compounds often have characteristic fragmentation patterns.

The alkyl portion of organic compounds gives a number of $C_xH_y^+$ fragments. If it's possible to form a particularly stable carbocation, such as a tertiary carbocation like $(CH_3)_3C^+$, an especially intense peak results.

The loss of 15 mass units from a molecular ion generally indicates the loss of a methyl (CH_3) group. The loss of 29 mass units often indicates the loss of an ethyl (CH_2CH_3) group.

Cleavage (the breaking of a bond) next to a heteroatom (any atom other than carbon or hydrogen) is also relatively common. This is particularly important if fragmentation involves the loss of a very stable molecule such as H_2O or CO_2 since this would indicate the presences of very particular functional groups and thus would give clues to the structure of the molecule.

No Glowing Here: NMR Spectroscopy

NMR spectroscopy allows the organic chemist to "see" the environment surrounding the nuclei in a molecule. NMR isn't applicable to all nuclei; however, most elements have one or more isotopes for which NMR is applicable. NMR spectroscopy may be used on nuclei that behave as small magnets. Organic chemists usually rely on ^1H (proton) and ^{13}C as the most important isotopes because most organic compounds contain hydrogen and all organic compounds contain carbon.

Nuclei that behave like small magnets (have a magnetic moment) are subject to the influence of other magnets. In an NMR spectrometer, the sample resides in a large external magnetic field. This external field forces the nuclei to align themselves either with the field or against the field. Nuclei with one alignment can absorb energy and switch to the other alignment and vice versa in a process called *spin flipping*. When spin flipping occurs, the nucleus is in resonance. The energy required to induce this transition is in the radio frequency region of the electromagnetic spectrum.

In addition to the external magnetic field, other magnetic fields influence the nuclei. The electrons in the molecule also have their own magnetic field. The field due to the electrons tends to oppose the external magnetic field, which results in electron shielding. The amount of shielding depends on the number of electrons. The more electron-shielding taking place, the lower the energy requirement for resonance (resulting in a downfield shift). The position of the absorption is referred to as the *chemical shift*. For proton NMR, the chemical shift is normally in the 0–15 ppm (parts per million) region relative to the standard TMS (tetramethyl silane, $Si(CH_3)_4$). For ^{13}C NMR, the chemical shift is normally in the 0–200 ppm region.

Chemically equivalent nuclei absorb at the same energy level. Consider, for example, the structure of ethanol (see Figure 5-3). Three distinct types of hydrogen atoms appear in this structure. In the proton NMR spectrum of ethanol (discussed in the following section, "Proton"), the three hydrogen atoms of the CH_3 group are chemically equivalent, as are the two hydrogen atoms of the CH_2 group, and both are different from the hydrogen atom attached to the oxygen. Therefore, the proton NMR spectrum of ethanol begins with three signals. (Later in this chapter, you see that there's more than that to the NMR spectrum of ethanol.)

```
     H   H
     |   |
 H — C — C — O — H
     |   |
     H   H
```

Figure 5-3: The structure of ethanol.

Neighboring magnetic nuclei also contribute to the magnetic field surrounding a nucleus, which gives rise to coupling. (A bit suggestive, I know, but that's what it's called.) Coupling results in the splitting of some of the absorptions.

Proton

Proton NMR spectra follow the generalizations expressed in the previous sections, but this section discusses some additional factors. You find that some of these complications don't occur in ^{13}C NMR spectra.

It's important to know the chemical shift of each chemically equivalent set of hydrogen atoms. A proton adjacent to an atom that has a high electronegativity has a lower electron density than a proton adjacent to an atom with a low electronegativity. Therefore, a proton adjacent to an oxygen atom, for example, comes into resonance at a higher frequency than a proton adjacent to, for example, a carbon atom (lower electronegativity).

Often, drawing all the hydrogens on the molecule can help you see the different chemical environments (different interaction due to nearby hydrogen atoms) each hydrogen experiences.

Integration

One important additional aspect of proton NMR spectra is the ability to integrate the spectra. Integration provides a means to determine how many hydrogen atoms are in each equivalent set, and it involves determining the area under each of the peaks in the spectrum. The area is proportional to the number of hydrogen atoms contributing to that particular peak. For example, in the proton NMR of ethanol, the integration of the peak due to the alcoholic hydrogen would represent one hydrogen atom. The integration of the peak due to the CH_2 group would be twice this value (twice the area), because twice as many hydrogen atoms contribute to the intensity. Finally, the integration of the CH_3 absorption would be three times the OH absorption.

The total of all the integrations must equal the total number of hydrogen atoms in the compound.

Coupling

Coupling is an old BBC comedy that's sure to make you laugh, but never mind that. In organic chemistry, coupling is another feature of proton NMR spectra. Coupling is the result of the interaction of the magnetic field of some hydrogen nuclei with other hydrogen nuclei. In general, hydrogen atoms attached to the same atom (chemically equivalent) don't couple. Hydrogen atoms on adjacent carbon atoms or separated by a double bond do couple.

Coupling follows the $n + 1$ rule. According to this rule, a peak splits into $n + 1$ peaks due to neighboring hydrogen atoms, where n is the number of equivalent hydrogen atoms. The amount of splitting is expressed by the coupling constant (J).

In the proton NMR of ethanol, the hydrogen atom attached to the oxygen doesn't couple and appears as a single peak (a singlet). The peak due to the hydrogen atoms in the CH_3 group couples with the two hydrogen atoms of the adjacent CH_2 group. The result is the splitting of the CH_3 peak into three peaks (a triplet), which corresponds to $n = 2$ (because CH_2 has two hydrogen atoms). At the same time, the peak due to the CH_2 group splits into four peaks (a quartet), which corresponds to $n = 3$. (A doublet occurs when $n = 1$.)

The different types (doublet, triplet, and so on) exhibit a characteristic ratio of intensities. Doublets are equally intense. Triplets have a more intense central peak flanked by two equal peaks of lesser intensity. A quartet has two equally intense central peaks with two smaller outer peaks that are equal to each other in intensity. See Figure 5-4 for the NMR spectrum of ethanol.

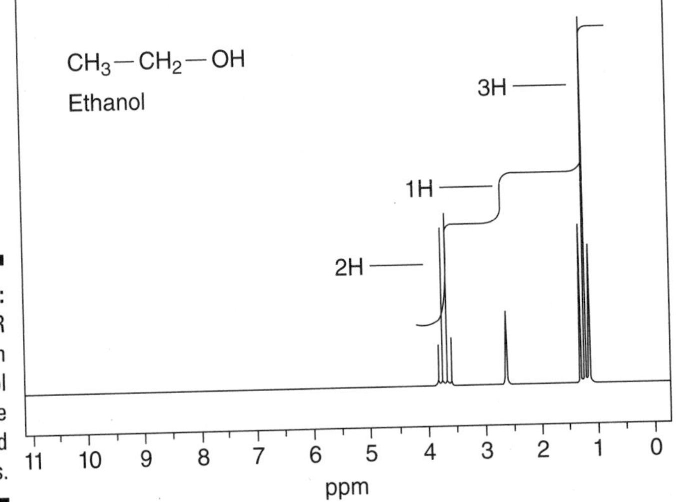

Figure 5-4: NMR spectrum of ethanol showing the splitting and intensities.

Ethanol has a simple NMR spectrum because the hydrogen atoms attached to the carbon atoms can only couple with the hydrogen atoms on one adjacent carbon atom. However, in most organic compounds the hydrogen atoms on one carbon couple with hydrogen atoms on two or three adjacent carbon atoms, resulting in a multiplet of indistinguishable peaks appearing in the spectrum. For example, in the compound $CH_3CH_2CHClBr$, the CH_2 hydrogen atoms have four hydrogen atoms on the adjacent carbon atoms. In this case, the CH_3 hydrogen atoms split the CH_2 hydrogen atoms into a quartet and the remaining hydrogen atom (on the CHClBr) splits each of the members of the quartet into a doublet to give a total of eight peaks.

Carbon-13

In general, ^{13}C NMR spectra are simpler than proton NMR spectra. Instead of integration, the height of each absorption peak is approximately proportional to the number of carbon atoms contributing to the peak. It's only approximately proportional because the more hydrogen attached to the carbon atom, the greater the peak height.

Normally, no coupling is present in ^{13}C NMR spectra because the abundance of carbon-13 is so low that it's unlikely to have two carbon-13 atoms close enough together to couple. This lack of coupling greatly simplifies the spectra. You can see this simplification in the carbon-13 spectra of butyric acid in Figure 5-5.

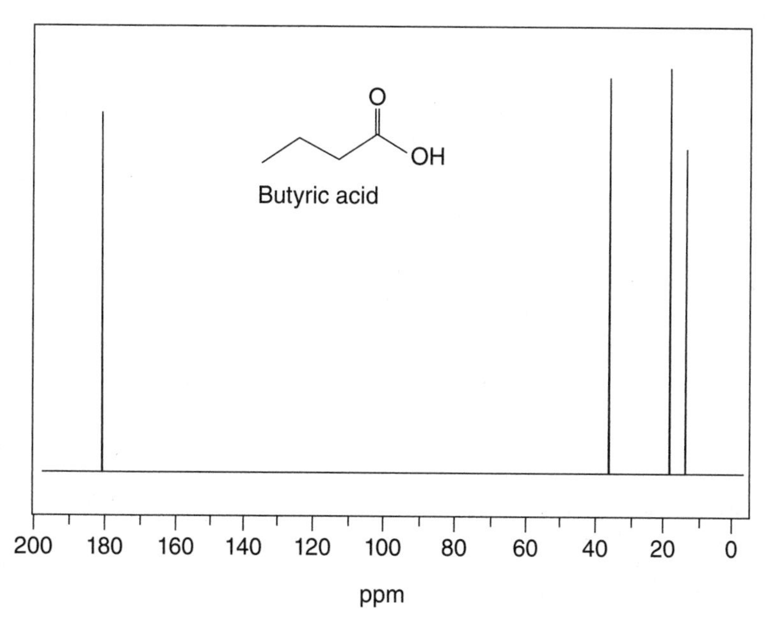

Figure 5-5: Carbon-13 spectrum of butyric acid.

Part II
Discovering Aromatic (And Not So Aromatic) Compounds

The 5th Wave
By Rich Tennant
PROF. BLOWFISH HAD THE REPUTATION OF BEING SOMEWHAT UNAPPROACHABLE.
©RICHTENNANT

In this part . . .

In Part II we spend a lot of time and pages on aromatic systems, starting with benzene. You examine benzene's structure, its resonance stabilization, and its stability. Next you study benzene derivatives and heterocyclic aromatic compounds, and then we address the spectroscopy of these aromatic compounds. And in Chapters 7 and 8 we introduce you to aromatic substitution by both electrophiles and nucleophiles, and you get to see a lot of reactions and a lot of examples. In this part you also start working with many more named reactions.

Chapter 6

Introducing Aromatics

In This Chapter

- Exploring benzene
- Coping with the aromatic family
- Getting acquainted with heterocyclic aromatic compounds
- Shedding light on the spectroscopy of aromatic compounds

Organic chemistry has two main divisions. One division deals with aliphatic (fatty) compounds, the first compounds you encountered in Organic Chemistry I. Methane is a typical example of this type of compound. The second division includes the aromatic (fragrant) compounds, of which benzene is a typical example.

Compounds in the two groups differ in a number of ways. The two differ chemically in that the aliphatic undergo free-radical substitution reactions and the aromatic undergo ionic substitution reactions. In this chapter you examine the basics of both aromatic and heterocyclic aromatic compounds, concentrating on benzene and related compounds.

Benzene: Where It All Starts

Benzene is the fundamental aromatic compound. An understanding of the behavior of many other aromatic compounds is much easier if you first gain an understanding of benzene. For this reason, you may find it useful to examine a few key characteristics of benzene, which we discuss in the following sections.

Figuring out benzene's structure

Benzene was first isolated in 1825 from coal tar. Later, chemists determined that it had the molecular formula C_6H_6. Further investigation of its chemical behavior showed that benzene was unlike other hydrocarbons in both structure and reactivity.

Chemists proposed many structures for benzene. However, the facts didn't support any of the possibilities until Kekulé proposed a ring structure in 1865. Some of the proposed structures, including Kekulé's, are in Figure 6-1.

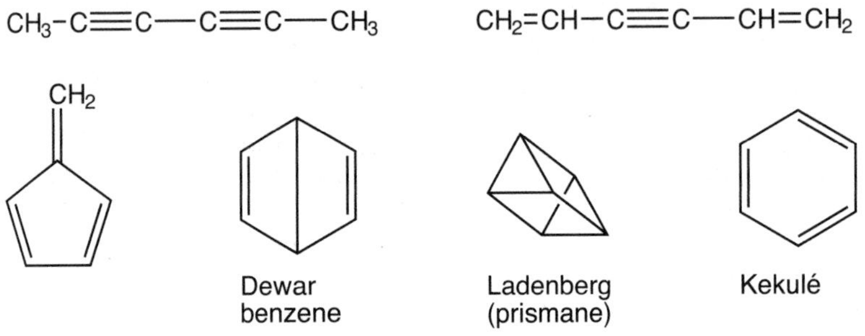

$CH_3-C\equiv C-C\equiv C-CH_3$ $CH_2=CH-C\equiv C-CH=CH_2$

CH_2

Dewar benzene

Ladenberg (prismane)

Kekulé

Figure 6-1: Some proposed structures of benzene.

The reaction of benzene with bromine in the presence of an iron catalyst eliminated most of the proposed structures. This reaction produced only one monobromo product and three distinct dibromo products (ortho, meta, and para).

Kekulé reasoned that ortho-dibromobenzene (1,2-dibromobenzene) existed in two forms that were in rapid equilibrium and couldn't be isolated. These two forms are in Figure 6-2.

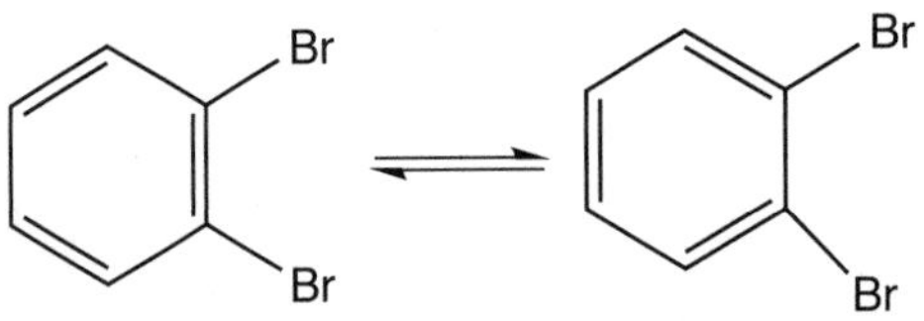

Br Br Br Br

Figure 6-2: Kekulé's proposed structures for 1,2-dibromobenzene.

The ring structure with the rapidly moving double bonds explained many of the facts known about benzene at the time. However, as more information became available and as chemistry advanced, it became obvious that more was going on in this system than just rapidly interconverting structures. Chemists determined that only one benzene structure existed, not an equilibrium between two related structures.

Understanding benzene's resonance

Recall from Organic I that the concept of resonance was developed to describe the electron structure of a molecule having delocalized bonding by writing all the possible Lewis structures of that molecule. The term *delocalized bonding* refers to a situation in which one or more bonding pairs of electrons are spread out over a number of atoms. The development of the concept of resonance came after Kekulé had proposed the equilibrium structures for benzene. The presence of resonance explains why the carbon-carbon bonds in benzene are of equal length and strength. The original Kekulé structures are, in reality, not equilibrium structures but contributing structures to the resonance hybrid. As contributing structures, they have no independent existence. The only form present is the hybrid. The two resonance forms are shown in Figure 6-3, and the resonance hybrid is shown in Figure 6-4. Notice the use of the double-headed arrow in Figure 6-3. (To review the resonance arrow and others, see Chapter 2.)

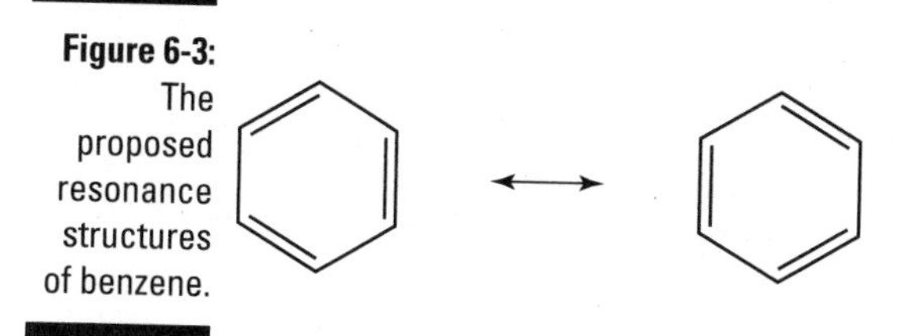

Figure 6-3: The proposed resonance structures of benzene.

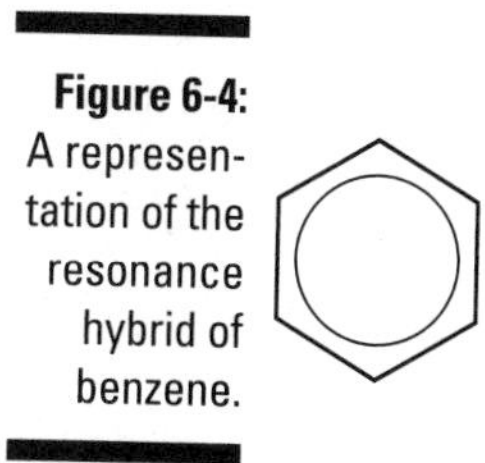

Figure 6-4: A representation of the resonance hybrid of benzene.

The presence of the double bonds in the resonance structures typically implies that benzene should react like an alkene in terms of addition reactions and similar reactions. However, as Figure 6-5 shows, benzene does not react like alkenes.

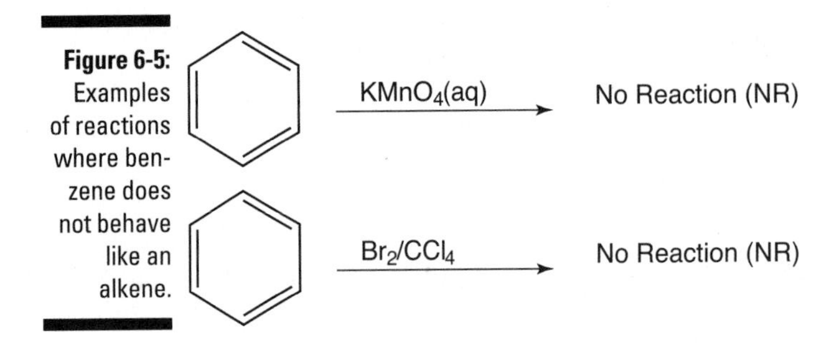

Figure 6-5: Examples of reactions where benzene does not behave like an alkene.

The pi-electrons are delocalized over the entire ring structure, not localized between two carbons. This contributes to the observed stability of benzene.

The stability of benzene

One way to investigate the stability of benzene is to compare the amount of heat produced by the reactions of benzene to similar compounds that are not aromatic. For example, a simple comparison of the heat of hydrogenation for a series of related compounds allows us to see the difference. Figure 6-6 shows the hydrogenation of cyclohexane, 1,3-cyclohexadiene, and benzene, which make a suitable set because all three yield cyclohexane.

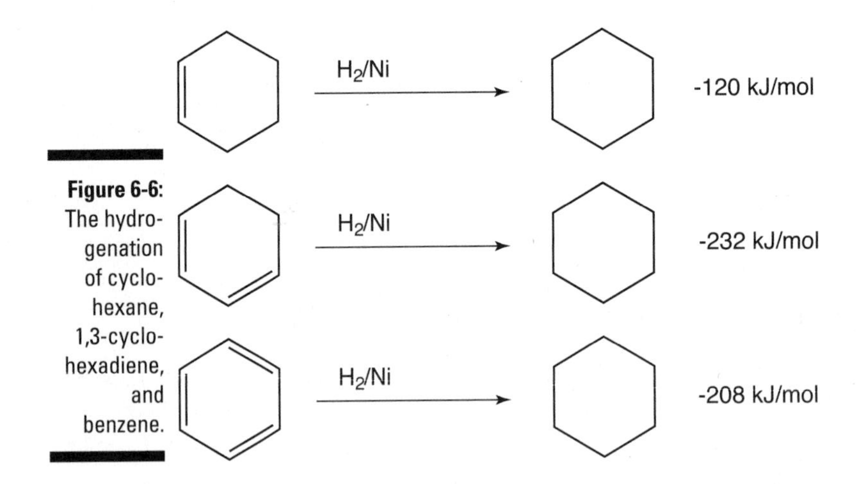

Figure 6-6: The hydrogenation of cyclohexane, 1,3-cyclohexadiene, and benzene.

The heat of hydrogenation of cyclohexene is about –120 kJ/mol (kilojoules per mole). If the reaction of one double bond releases this amount of energy, then the reaction of two double bonds (1,3-cyclohexadiene) should release about twice this amount of energy. The classical, three double-bond benzene should release three times as much energy. The energy change for 1,3-cyclohexadiene is –232 kJ/mol, which is close to the predicted value (–240 kJ/mol). However, the value for benzene, –208 kJ/mol, is far from the predicted value of –360 kJ/mol. The small value for benzene indicates that it is significantly more stable than a triene. This difference is a direct measure of the resonance energy.

Physical properties of benzene

Benzene is a typical nonpolar compound that, like other nonpolar compounds, has a low solubility in water. It has a characteristic odor which most people find unpleasant. It was widely used in academic labs as a solvent, but that use has been largely discontinued since it was found that benzene may be carcinogenic.

Table 6-1 compares the melting and boiling points of benzene to those of cyclohexane, which indicates some differences found in aromatic compounds. The values for benzene are higher than the corresponding values for cyclohexane. In most other situations, when dealing with compounds with similar structures, the lower molecular-weight compound has the lower melting and boiling point; however, when comparing benzene to cyclohexane, you see that the reverse is true. This is because of the delocalization of the electron pairs in benzene.

Table 6-1 Melting and Boiling Points of Benzene and Cyclohexane

	Benzene (78 g/mol)	*Cyclohexane (84 g/mol)*
Melting Point	6°C	–95°C
Boiling Point	80°C	69°C

Organic math — Hückel's Rule

Just what makes a substance, such as benzene, aromatic? We know that benzene is more stable than expected. The increased stability of benzene is due to resonance having a stabilizing influence. Following are the requirements for a species to be aromatic:

- The compound must contain a ring system.
- The system must be planar (or nearly planar).

- Each atom in the ring must have an unhybridized p-orbital perpendicular to the plane of the ring.

  This means that an sp^3 atom cannot be part of an aromatic system.
- The ring system must have a Hückel number of π-electrons.

This last requirement is an important characteristic of all aromatic systems. It's known as Hückel's rule, or the $4n + 2$ rule. To apply this rule, begin by assigning $4n + 2$ = number of π-electrons in a cyclic system. Next, solve for n, and if n is an integer (a whole number), the system is aromatic. In the case of benzene, $4n + 2 = 6$, so $n = 1$. One is an integer (a Hückel number), so the last requirement to be classified as an aromatic is satisfied. Figure 6-7 contains several aromatic species with $n = 1$.

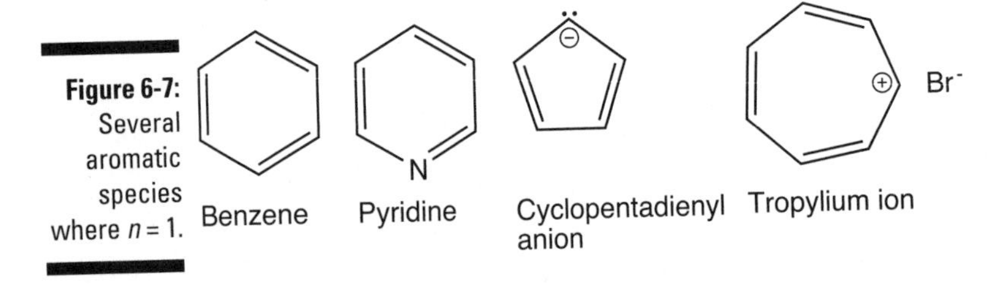

Figure 6-7: Several aromatic species where $n = 1$.

Naphthalene and anthracene (see Figure 6-8) are aromatic systems with $n = 2$ and 3, respectively.

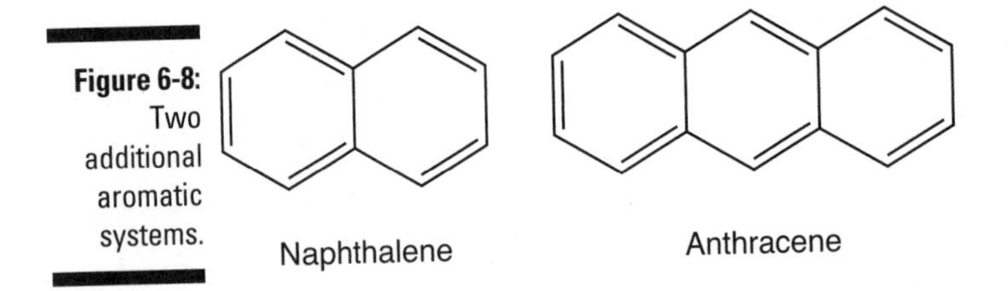

Figure 6-8: Two additional aromatic systems.

Common examples of systems often mistaken as being aromatic (because of their alternating double and single bonds) are cyclobutadiene and cyclooctatetraene (shown in Figure 6-9). In the case of cyclobutadiene, $4n + 2 = 4$, giving $n = 0.5$, while for cyclooctatetraene, $4n + 2 = 8$, so that $n = 1.5$. In these two compounds, n is not an integer, so these systems are anti-aromatic (non-aromatic). Anti-aromatic systems (non-Hückel systems) are less stable than aromatic or "normal" systems.

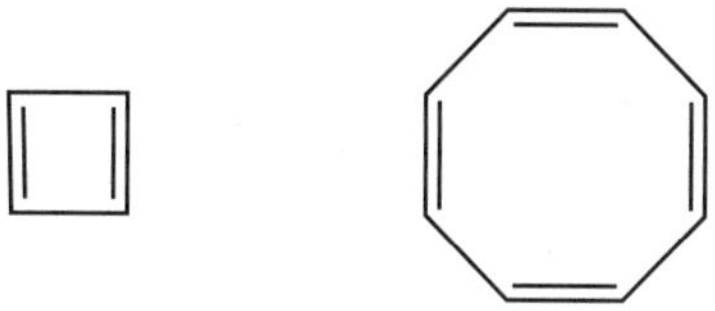

Cyclobutadiene Cyclooctatetraene

Figure 6-9: Two anti-aromatic systems.

Other aromatics

A number of other molecules in addition to those shown in Figures 6-7 and 6-8 are aromatic. The first five possible values of n are 0, 1, 2, 3, and 4. These numbers correspond to $4n + 2$ values of 2, 6, 10, 14, and 18, respectively. Pyridine (refer to Figure 6-7) illustrates the fact that aromatic compounds are not necessarily hydrocarbons. However, the replacement of the nitrogen in pyridine with oxygen places an sp^3 hybridized atom in the ring, so the system is no longer aromatic.

Two species in Figure 6-7 are not molecules but ions. Aromaticity is not restricted to molecules. The cyclopentadienyl ion and tropylium ion (cycloheptatrienyl ion) are aromatic species where $n = 1$.

Smelly Relatives: The Aromatic Family

Benzene is an important and common aromatic compound. However, many other aromatic compounds are based on benzene: the substituted benzenes. In this section we take a look at the properties and reactions of some of these compounds.

Nomenclature of the aromatic family

The nomenclature (naming) of aromatic compounds begins with numbering the ring positions. This numbering is similar to the numbering of cycloalkane systems. The key group is at the number one position (see Figure 6-10). Older usage uses ortho *(o-)* in place of 1,2-numbers, meta *(m-)* in place of 1,3-numbers, and para *(p-)* in place of 1,4-numbers.

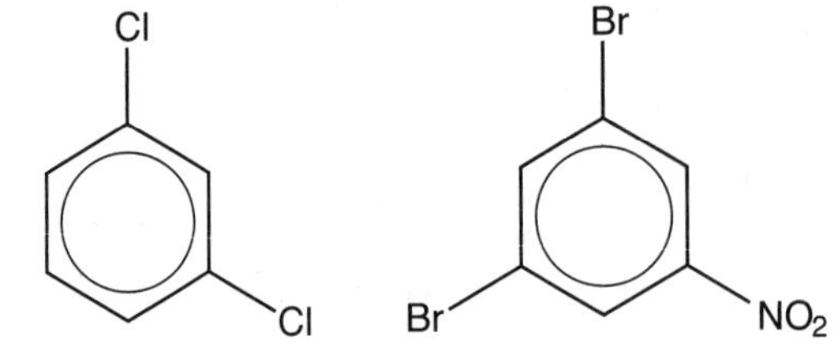

Cl

Cl

Br

Br

NO_2

m-dichlorobenzene

3,5-dibromonitrobenzene

Figure 6-10: Two examples demonstrating different ways of naming aromatic compounds.

Derivatives of benzene

The nomenclature adopted by the IUPAC (International Union of Pure and Applied Chemistry) for some additional aromatic systems is shown in Figure 6-11. The symbolism for xylene indicates that two methyl groups are present. The methyl groups may be at the one and two positions (ortho-xylene), the one and three positions (meta-xylene), or the one and four positions (para-xylene). Alternate names are *o*-xylene, *m*-xylene, and *p*-xylene. In the other cases, one group is attached at the number one position. All numbering begins at this position.

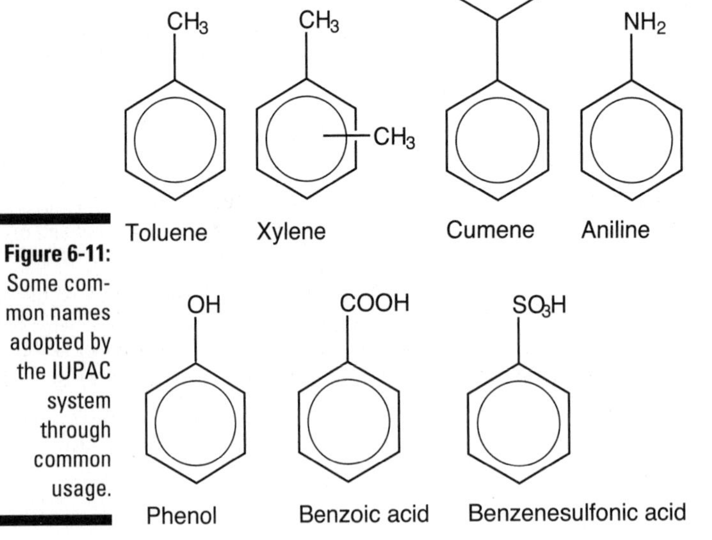

Figure 6-11: Some common names adopted by the IUPAC system through common usage.

Branches of aromatic groups

Aromatic groups may be branches on other systems. The general name for an aromatic branch is *aryl.* Two examples of aromatic branches are in Figure 6-12. Be careful to notice that the third example, the benzyl group, is not truly aromatic. The benzyl group consists of an aromatic ring and an anti-aromatic -CH_2– group. For a branch to be truly aromatic, the connection must be directly to the ring.

Additional ways of indicating the presence of a phenyl group include Ph, C_6H_5, and ϕ.

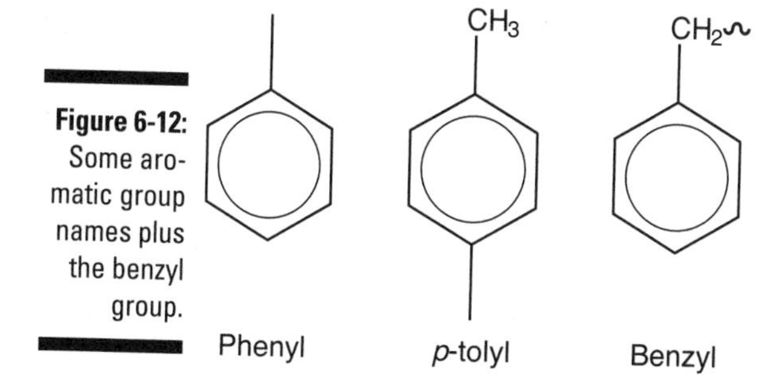

Figure 6-12: Some aromatic group names plus the benzyl group.

Black Sheep of the Family: Heterocyclic Aromatic Compounds

Up to this point, this chapter has discussed aromatic systems composed exclusively of rings of carbon atoms. But aromatic systems can contain *heteroatoms,* which in this case means any atoms in the ring other than carbon.

Heteroatoms (such as O, S, N) that are double-bonded to other atoms in a ring can't donate lone-pair electrons to the pi system because their *p*-electrons are already involved in the double bond. Single-bonded heteroatoms can donate a single lone-pair to the pi system but not two, because one lone-pair must be in an unhybridized *p* orbital orthogonal (at 90 degrees) to the sp^2 ring plane.

Aromatic nitrogen compounds

In addition to pyridine (refer to Figure 6-7), other aromatic nitrogen systems exist and show up on organic chemistry tests. Two examples of aromatic nitrogen systems are shown in Figure 6-13.

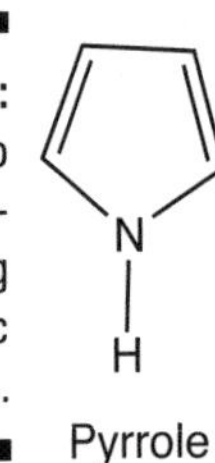

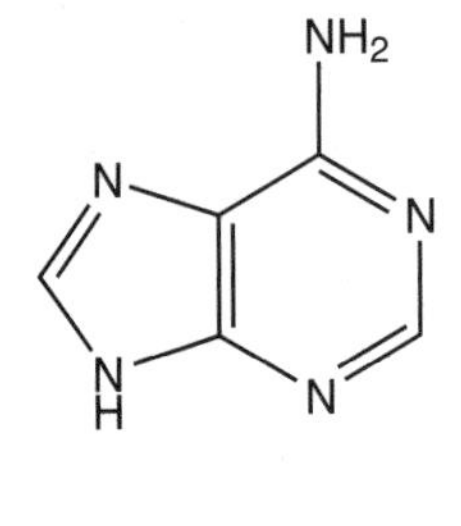

Figure 6-13: Two nitrogen-containing aromatic systems.

Aromatic oxygen and sulfur compounds

Two additional examples of aromatic compounds containing heteroatoms are shown in Figure 6-14. In both compounds, the heteroatom has two lone pairs. However, only one of the pairs is in a p-orbital perpendicular to the plane of the ring. The other electron pair is in the plane of the ring.

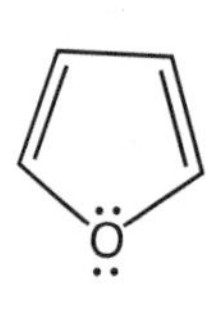

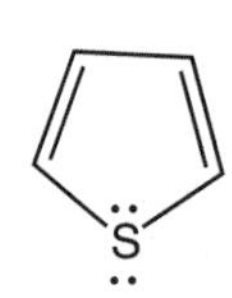

Figure 6-14: A sulfur- and an oxygen-containing aromatic compound.

Spectroscopy of Aromatic Compounds

Spectroscopy provides many clues to the identity of a compound. Aromatic compounds, because of the delocalization of the electrons, have unique features in their spectra. In fact, spectral evidence can indicate what atoms or functional groups are attached to the aromatic ring or whether the ring itself contains an atom other than carbon.

IR

The C-H bends are characteristic of the substitution around the aromatic ring. Aromatic compounds have a characteristic C-H peak near 3,030 cm^{-1}. In addition, the infrared spectra can contain the following features:

- Up to four ring stretches exist in the 1,450–1,600 cm^{-1} region and two stronger peaks are in the 1,500–1,600 cm^{-1} region.
- In the infrared spectra of monosubstituted benzenes, usually two very strong peaks appear: one between 690 and 710 cm^{-1} and one between 730 and 770 cm^{-1}.
- Ortho-disubstituted benzenes show a strong peak between 680 and 725 cm^{-1} and a very strong peak between 750 and 810 cm^{-1}.
- Meta-disubstituted benzenes show a strong peak between 735 and 770 cm^{-1}.
- Para-disubstituted benzenes show a very strong peak between 800 and 860 cm^{-1}.

UV-vis

The conjugated π-electrons of a benzene ring have characteristic absorptions in the ultraviolet-visible spectrum that include the following:

- A moderately intense band appears at 205 nm, and a weaker band shows in the 250–275 nm region.
- Conjugation outside the ring leads to additional absorptions.

NMR

The nuclear magnetic resonance spectrum of aromatic compounds commonly contain the following features:

- The proton NMR of aromatic species have characteristic peaks between $\delta = 6.0$ and $\delta = 9.5$ downfield. The downfield shift is due to the aromatic system. The aromatic ring has a ring current, which gives rise to an induced field.
- The ^{13}C NMR of aromatic species generally absorb in the δ 100–170 region.

Mass spec

The stability of the aromatic system leads to fragments containing aromatic species. The following tips help you interpret the mass spectral data of aromatic compounds:

- The mass spectra of monosubstituted benzenes often contain the $C_6H_5^+$ ion (m/z = 77).
- Other common species are the benzyl cation, $C_6H_5CH_2^+$ (m/z = 91), and the tropylium ion, $C_7H_7^+$ (m/z = 91).

Chapter 7

Aromatic Substitution Part I: Attack of the Electrophiles

In This Chapter

- Exploring electrophilic substitution
- Checking out what happens in benzene reactions
- Tackling Friedel-Crafts reactions two ways
- Figuring out how to modify the reactivity of an aromatic
- Understanding the limitations of electrophilic substitution

Aromatic systems are pretty stable; they resist reacting. Nevertheless, a number of reactions involving aromatic systems can be carried out. However, with the exception of combustion, the conditions required by the anti-aromatic systems for reactions that you studied in your first semester organic course are different than the conditions for aromatic systems.

In this chapter we focus on the substitution attack on an aromatic molecule by an electrophile. Throughout the chapter we show you how these electrophilic substitutions occur, first by using benzene as an example. Once you become familiar with electrophilic substitution on benzene, you're ready to see what happens when substituted aromatic molecules replace the benzene molecule.

Don't try to memorize the mechanisms in this chapter as separate unrelated entities. These mechanisms are closely related and fit together quite nicely. Look for the relationships; concentrate on understanding how and why the reactions occur as they do and avoid simple memorization.

Basics of Electrophilic Substitution Reactions

One type of reaction that can involve aromatic systems is an electrophilic substitution reaction. Like the substitution reactions you learned in your first semester of Organic Chemistry, this process involves the substitution of something for a hydrogen atom. In this reaction, a nucleophile (the aromatic system) attacks the electrophile. The stability of the aromatic system makes it a poor nucleophile, though, so a very strong electrophile is needed to force the reaction to occur. For example, the electrophile bromine (Br_2) is strong enough to attack the double bond in an alkene, but it's not strong enough to attack an aromatic system. However, the Br^+ ion works because it's a much stronger electrophile.

In general, E^+ is the symbol generally used for an electrophile. The electrophile attracts electrons from the π-system of the aromatic ring to form an intermediate. Loss of a hydrogen ion from the intermediate to a base completes the reaction. The general mechanism is shown in Figure 7-1. This mechanism is the key to all electrophilic substitution reactions. You need to grasp this basic mechanism and be able to recognize it in each of the mechanisms in this chapter and the next.

In the mechanism in Figure 7-1, the hydrogen ion (H^+) doesn't simply fall off the ring. For this loss to occur, a base must be present to pull it away. In this particular case a wide variety of bases would work.

Structure III in Figure 7-1 represents an arenium ion, more commonly called a sigma complex. Figure 7-2 shows the energy changes occurring during the reaction.

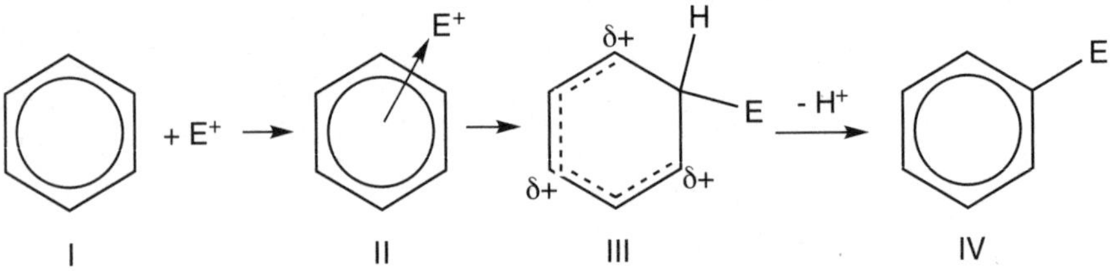

Figure 7-1: The general mechanism for an electrophilic substitution reaction.

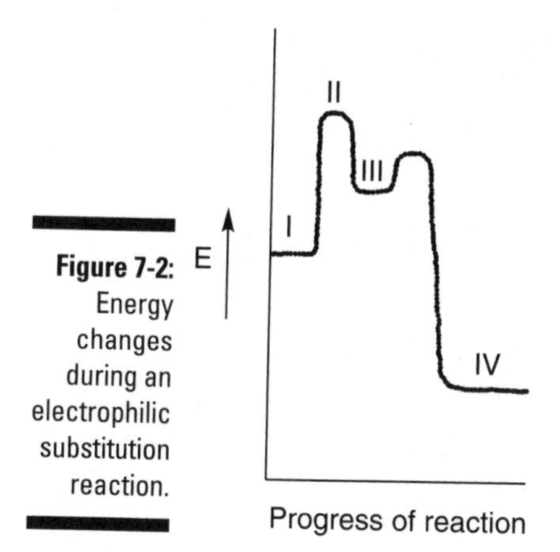

Figure 7-2: Energy changes during an electrophilic substitution reaction.

Reactions of Benzene

In this section we use benzene as a typical aromatic compound to study three basic reactions: halogenation, sulfonation, and nitration. In the case of halogenations, the electrophile is the X^+ ion (X = Cl or Br). In sulfonation and nitration, the electrophiles are SO_3 and NO_2^+, respectively. In each case, part of the mechanism involves the generation of the electrophile.

Don't try to memorize a nitration mechanism as a separate mechanism from halogenations or sulfonation. Let your understanding of one mechanism reinforce your understanding of other mechanisms.

Halogenation of benzene

In a halogenation reaction, a catalyst is necessary to generate the electrophile. The most common catalysts are the Lewis acids AlX_3 and FeX_3. For chlorination, X is Cl, and for bromination, X is Br. (Adding iodine (I) requires slightly different reaction conditions.) Figure 7-3 illustrates the general reaction for the chlorination of benzene. Figure 7-4 shows a partial mechanism for the reaction.

$+ AlCl_3 + Cl_2 \longrightarrow$ (Cl) $+ HCl$

Figure 7-3: The chlorination of benzene.

The reaction works equally well with $AlCl_3$, $AlBr_3$, $FeCl_3$, $FeBr_3$, and a number of other Lewis bases. Some catalysts can also be generated through reactions like $2\ FeBr_2(s) + Br_2(l) \rightarrow 2\ FeBr_3(s)$.

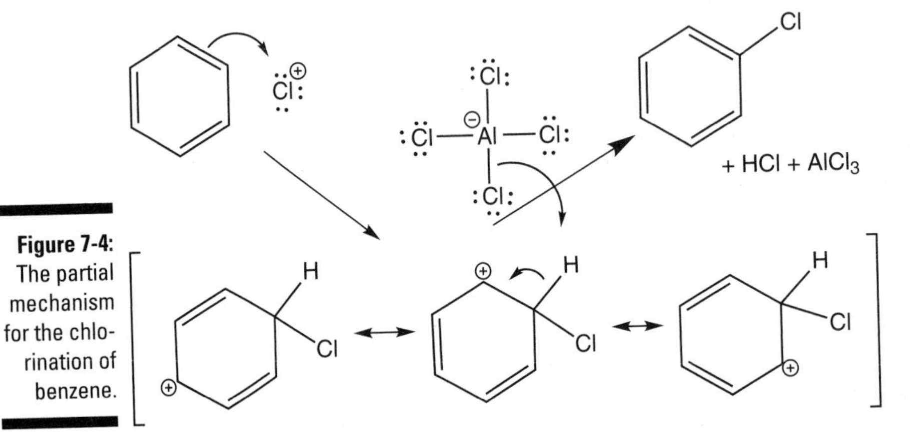

Figure 7-4: The partial mechanism for the chlorination of benzene.

The mechanism begins with the attack on the chlorine molecule by aluminum chloride. (This step would be the same if iron(III) chloride were the catalyst.) The Cl^+ ion attracts a pair of electrons from the benzene to form an intermediate species. The presence of resonance in this intermediate stabilizes it and helps the reaction along.

These resonance forms and similar forms are important to all the electrophilic substitution mechanisms in this chapter.

As noted in the earlier section "Basics of Electrophilic Substitution Reactions," the loss of the hydrogen ion (H^+) requires the presence of a strong base. The chloride ion (Cl^-) is a base, but it isn't strong enough to accomplish this task. However, as shown in the mechanism, the tetrachloroaluminate ion ($AlCl_4^-$) is a sufficiently strong base. This process also regenerates the catalyst so that it's available to continue the process.

Nitration of benzene

The generation of the electrophile for a nitration reaction begins with the reaction of nitric acid with sulfuric acid. Even though nitric acid is a strong acid, it's weaker than sulfuric acid and therefore more baselike, so nitric acid acts as a base and donates an OH^-. This process forms the nitronium ion (NO_2^+) and water (H^+ from the sulfuric acid and OH^- from the nitric acid), as shown in Figure 7-5. The remainder of the reaction is shown in Figure 7-6.

The nitrogen atom in all structures in Figure 7-5 has four bonds. This means the nitrogen has a positive formal charge (+1). Nitrogen can't have five bonds because no second-period element can ever exceed an octet. Nitrogen has a zero formal charge when it has three bonds (and a lone pair) and a negative formal charge (–1) when it has two bonds (and two lone pairs).

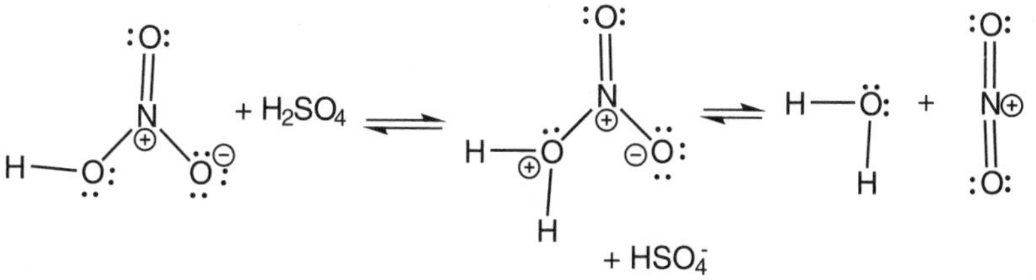

Figure 7-5: The formation of the nitronium ion.

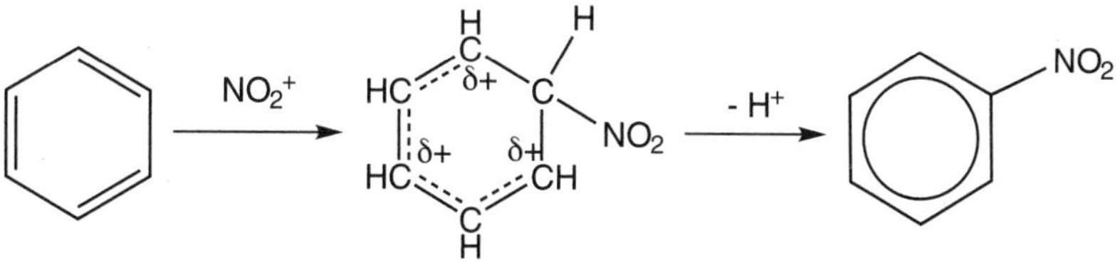

Figure 7-6: The nitration of benzene.

The only differences between this mechanism and the halogenation mechanism (see the preceding section for more information on halogenation) are the identity of the electrophile and the identity of the base used to remove the hydrogen ion. Unlike the base that causes the loss of hydrogen ion in the halogenation reaction, the base that removes the hydrogen ion in this mechanism is the hydrogen sulfate ion (HSO_4^-).

Nitroaromatic compounds are useful in synthesis because converting the nitro ($-NO_2$) group to an amino ($-NH_2$) group is relatively easy. For example, the reaction of nitrobenzene with acidic tin(II) chloride ($SnCl_2$) converts nitrobenzene to aniline, an important industrial chemical used in the production of medicines, plastics, and dyes, to name but a few.

Sulfonation of benzene

The generation of the electrophile for the sulfonation generally begins with adding *fuming sulfuric acid* (a mixture of sulfuric acid and sulfur trioxide) to benzene. The electrophile can then attack the benzene ring in a manner analogous to the attack by the nitronium ion. The general reaction is shown in Figure 7-7 and the mechanism is in Figure 7-8. Notice the similarity between the mechanism in Figures 7-4 and 7-8.

This reaction appears extensively in synthesis problems. Keep this reaction in mind when dealing with any synthesis problem involving an aromatic system.

Sulfonation is easily reversible. Simply diluting the fuming sulfuric acid leads to the removal of the -SO_3H. This is an important synthetic technique for protecting certain sites from reaction. Sulfonation can act as a placeholder while other reactions are performed, and then the easy removal of the sulfonic acid group makes the site available for reaction in a later step in a series of reaction steps.

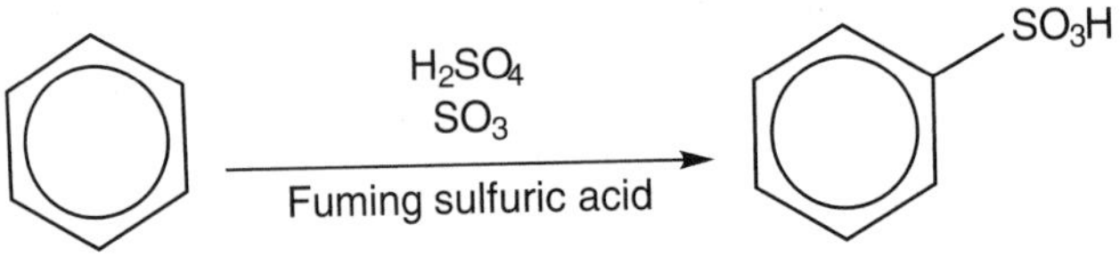

Figure 7-7: The sulfonation of benzene.

H_2SO_4
SO_3
Fuming sulfuric acid

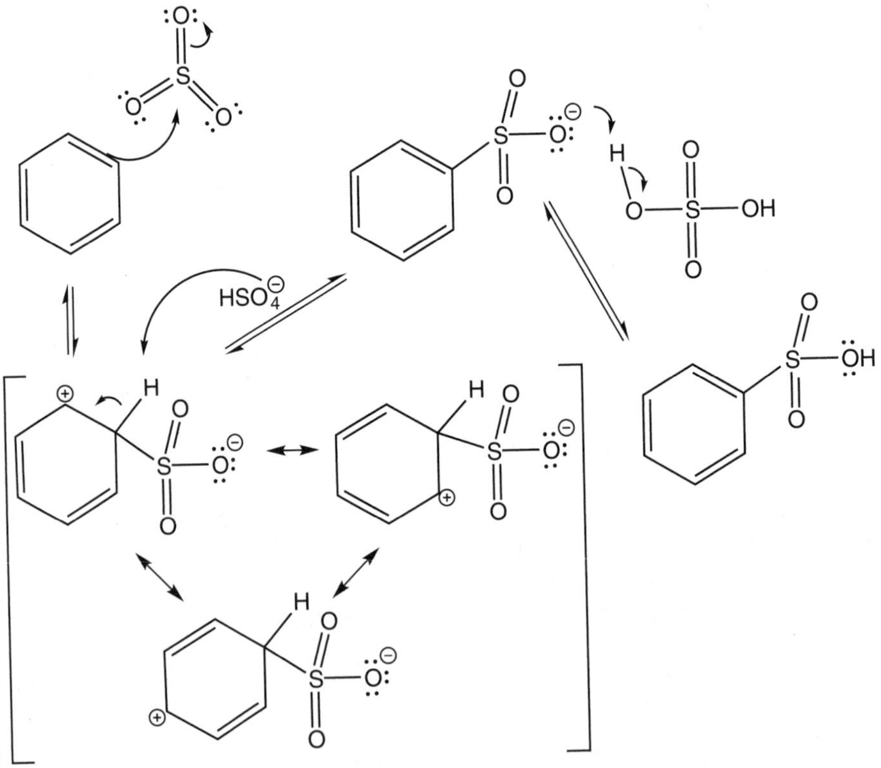

Figure 7-8: The mechanism for the sulfonation of benzene.

Friedel-Crafts Reactions

Friedel-Crafts reactions are electrophilic substitution reactions in which the electrophile is a carbocation or an acylium ion. The removal of a halide ion from an alkyl halide is the means of generating the carbocation. An acylium ion is created by removing a chloride ion from an acid chloride (R-CO-Cl). Both of these processes require a Lewis acid as a catalyst. The most commonly used Lewis acid is aluminum chloride.

Alkylation

Figure 7-9 illustrates a typical Friedel-Crafts alkylation. Once formed, the carbocation is a very strong electrophile. A complication that may occur is the rearrangement of the carbocation to a more stable carbocation, as seen in S_N1 mechanisms of alkyl halides. These rearrangements may involve a hydride or other shift.

Tertiary cations are more stable than either secondary, primary, or methyl cations. Methyl and primary cations are, in fact, the least stable.

Figure 7-9: A typical Friedel-Crafts alkylation.

Here is the general mechanism of a Friedel-Crafts alkylation:

1. $RCl + AlCl_3 \rightarrow R^+ + AlCl_4^-$
2. $R^+ + C_6H_6 \rightarrow [R\text{-}C_6H_6]^+$ Slow
3. $[R\text{-}C_6H_6]^+ + AlCl_4^- \rightarrow HCl + R\text{-}C_6H_5 + AlCl_3$

In simple mechanisms you encountered in Organic Chemistry I, methyl and primary carbocations were seriously frowned upon. However, $AlCl_3$ and $FeCl_3$ are such good Lewis acids that even these elusive primary carbocations can form. You should still avoid primary carbocations in S_N1 mechanisms.

In reality, a true carbocation doesn't form during this reaction. The reaction begins with the formation of a complex with the catalyst. The complex is of the form $AlCl_3$-Cl-R, where R has a partial positive charge (δ+), not a full positive charge (+1).

Acylation

A Friedel-Crafts acylation is a synthetic method that avoids the problem of rearrangement of the cation. Figure 7-10 illustrates the generation of the electrophile (the acylium ion) from an acid chloride. The presence of resonance stabilizes the acylium ion, and that reduces the possibility of rearrangement.

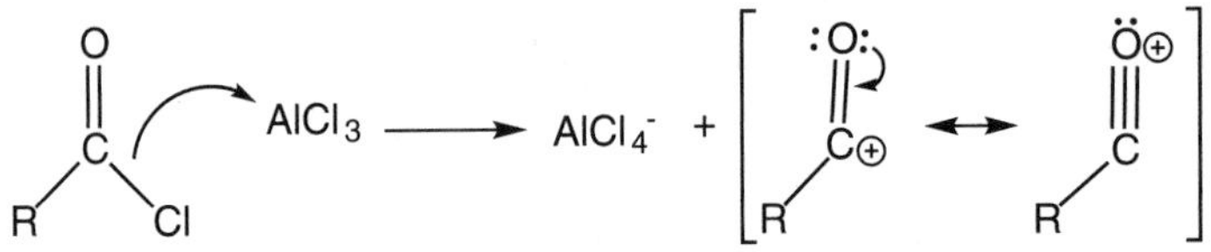

$AlCl_3$ → $AlCl_4^-$ +

Figure 7-10: The generation of an acylium ion.

Figure 7-11 shows a Friedel-Crafts acylation reaction. The reaction produces an aryl ketone, which is useful in synthesis because it makes it relatively easy to convert the ketone (RCOR) group to an alkyl (R) group. The procedure involves the catalytic hydrogenation of the aryl ketone, and it's particularly useful when the electrophile in a Friedel-Crafts alkylation is susceptible to rearrangement.

Once formed, the electrophile behaves like any other electrophile, so the mechanism of the attack is the same as that for the previous situation where a nucleophile attacked the electrophile (described in the earlier section "Basics of Electrophilic Substitution Reactions"). The attack leads to the formation of the resonance-stabilized sigma complex, followed by the loss of a hydrogen ion to a base.

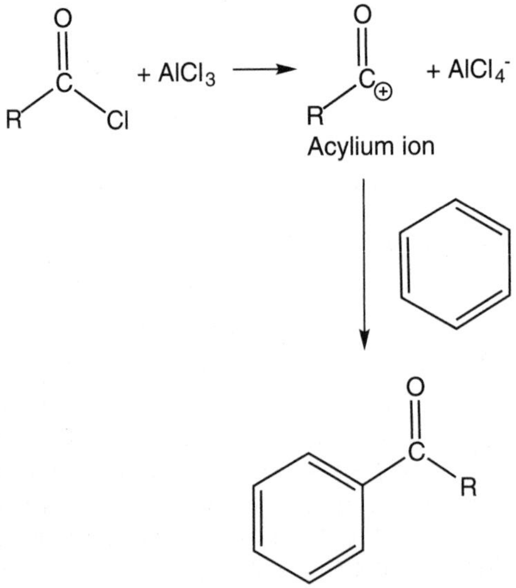

+ $AlCl_3$ → + $AlCl_4^-$

Acylium ion

Figure 7-11: A Friedel-Crafts acylation reaction.

The ketone can be reduced with a hot mixture of HCl and zinc amalgam (zinc metal dissolved in mercury).

Why Do an Alkylation?

Alkylated aromatics are useful in organic synthesis. If the alkyl group has a hydrogen atom (benzylic hydrogen) on the carbon adjacent to the ring, the alkyl group is susceptible to oxidation. A powerful oxidizing agent, such as acidic potassium permanganate ($KMnO_4$) or acidic dichromate ($Cr_2O_7^{2-}$), converts the alkyl group to a carboxylic acid group with the elimination of all carbon atoms except the one connected to the ring. Figure 7-12 illustrates this reaction. Any R group with a benzylic hydrogen gives the same product. If the ring is disubstituted, this can produce a number of isomers. Xylenes give phthalic acid, isophthalic acid, and terephthalic acid.

$C_6H_5CH_2CH_3 \xrightarrow[\text{or } Cr_2O_7^{2-}/H^+/\Delta]{KMnO_4(aq)/\Delta} C_6H_5COOH$

Figure 7-12: The oxidation of an alkylated benzene.

An alkylation using 1-chloropropane gives a mixture of products containing both the propyl and the isopropyl group, because the primary carbocation (propyl) rearranges to give the secondary carbocation (isopropyl). An acylation beginning with the acid chloride (CH_3CH_2COCl) followed by hydrogenation yields only the propyl product.

Changing Things: Modifying the Reactivity of an Aromatic

An electrophilic substitution reaction can take place on a substituted benzene. You can replace one of the hydrogen atoms on the substituted benzene with an electrophile. A monosubstituted benzene has two ortho-hydrogen atoms, two meta-hydrogen atoms, and one para-hydrogen atom (see Figure 7-13). Based on those ratios, substitution of a monosubstituted benzene should be 40 percent ortho, 40 percent meta, and 20 percent para. However, you never see this distribution, so something else must be happening. The group already present on the aromatic ring must be influencing further substitution on the aromatic ring.

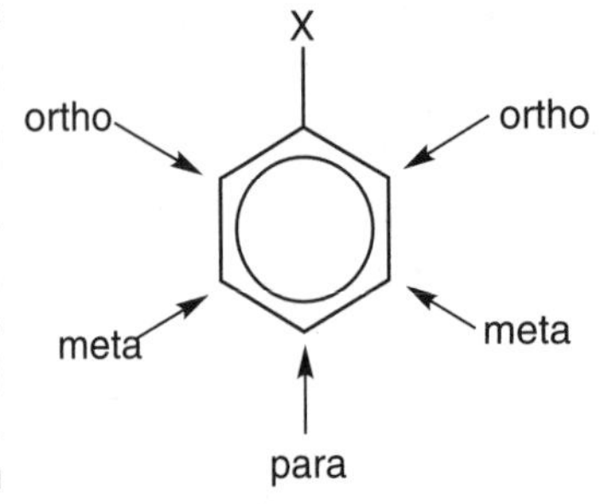

Figure 7-13: The ortho, meta, and para positions of a monosubstituted benzene.

Modifying the reactivity of the aromatic system doesn't invalidate the mechanisms from earlier in this chapter, it just means that you have to go further.

All substituents influence the resonance in any aromatic system, and they may do this in a number of ways:

- Some substituents withdraw electrons from and others donate electrons to the system.
- Some substituents may allow extension of the resonance beyond the ring.
- Substituents may increase the reactivity of the aromatic system (activating) or decrease the reactivity of the aromatic system (deactivating).
- Substituents may direct the electrophile to specific positions around the aromatic ring.
    - Some substituents are meta-directors, encouraging electrophilic attack upon the *meta* position. The presence of a meta-director means the major (or only) product of the reaction will be a meta-disubstituted benzene.
    - Other substituents are ortho-para-directors, which facilitate attack at these positions.

In the next section, you see why ortho- and para-directors are a single group.

Lights, camera, action: Directing

A group (G) attached to a ring makes its presence known in one of the two following ways:

- An inductive effect due to the "pushing" or "pulling" electrons through σ-bonds. Groups that increase the electron density of the ring are activating. This helps direct an attacking group to certain positions on the ring.

- A resonance effect due to the group donating or withdrawing pairs of electrons by creating new π-bonds. Groups that withdraw electron density from the ring are deactivating, which also helps direct an attacking group to certain positions on the ring.

Figure 7-14 illustrates activation and deactivation.

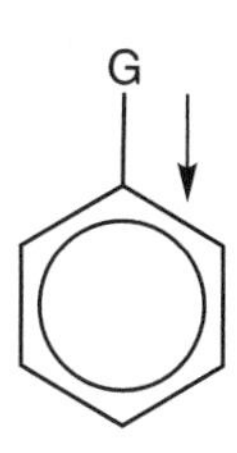

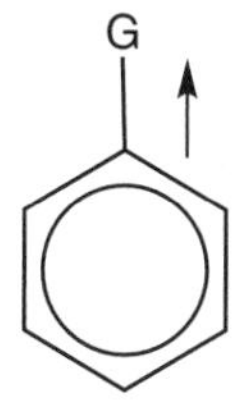

Figure 7-14: How a group (G) influences the reactivity of an aromatic system.

In order to understand any chemical process, you need to remember that stability is the key. In the case of aromatic systems, resonance is important to stability. To find out more about stability and resonance, we begin by examining the resonance in each of the different sigma complexes that may form, listed below and shown in Figure 7-15. The entering group can attack in one of three relative positions:

- Ortho, with the two substituents on adjacent carbons
- Meta, with the two substituents separated by one carbon
- Para, with the two substituents separated by two carbons

Sigma complexes always have a positive charge.

Figure 7-15 shows that each attack yields three equivalent resonance structures. This implies that the simple electrophilic attack is not the key to what product will form. It must be the identity of the original substituent (G).

All attacks will give three stable resonance structures.

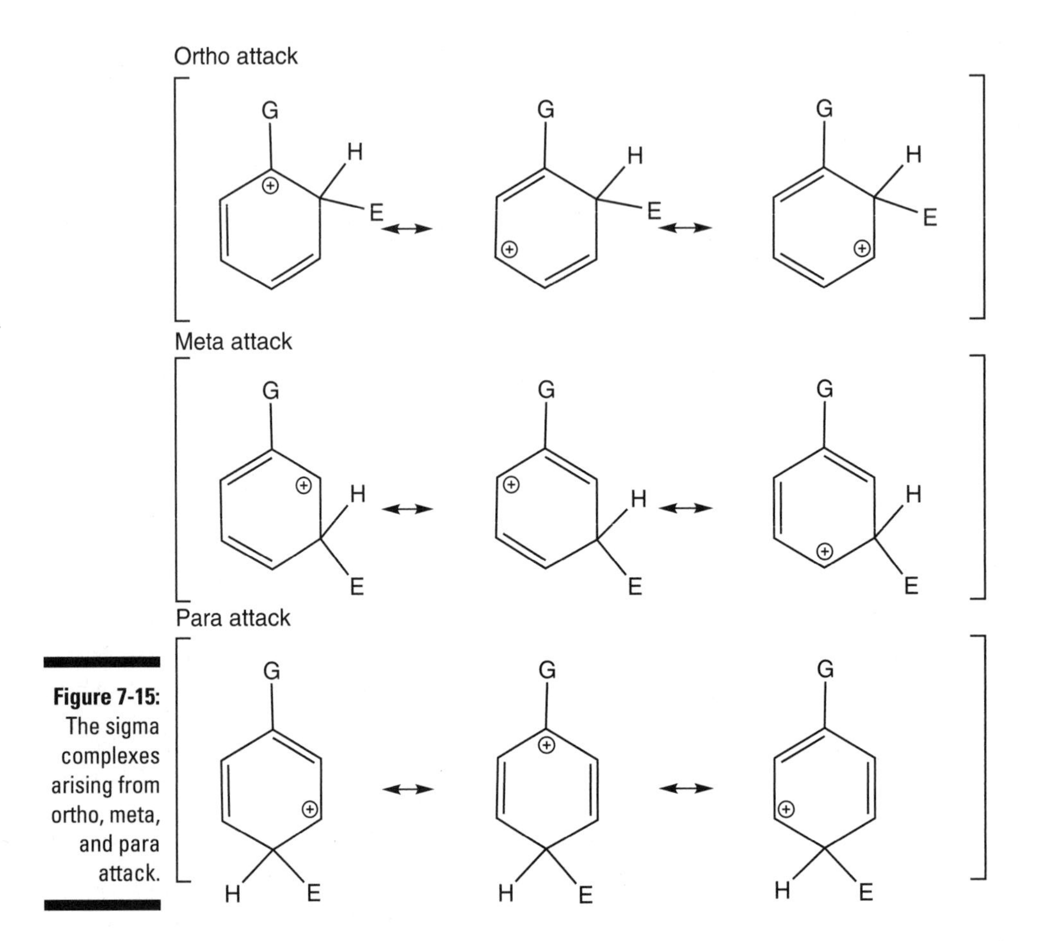

Figure 7-15: The sigma complexes arising from ortho, meta, and para attack.

Making a difference with the substituent

How can the substituent influence the resonance shown in Figure 7-15? The answer is that if the substituent can create another resonance structure, the sigma complex is further stabilized. This additional stabilization leads to a preference for a certain attack.

If G has a free electron pair next to the ring, a positive charge next to it is stabilized by a resonance. Two resonance structures in Figure 7-15 place a positive charge next to G. In these two cases, an additional resonance form (Figure 7-16) is possible.

Figure 7-16: Additional resonance forms arising when G has a lone pair of electrons.

This makes a fourth reasonable resonance structure. The more contributing structures present, the greater the stabilization of the sigma complex. The more stable the hybrid, the greater the chances of it being formed. Therefore, the presence of a substituent capable of donating a pair of electrons favors not only ortho attack, but also para attack. Such a substituent is an ortho-para-director (or in abbreviated form, an *o-p*-director). (Refer to Figure 7-15, paying special attention to the position of the charge.)

If G were an alkyl group, it could stabilize an adjacent positive charge by an inductive effect. You saw this aspect of alkyl groups in Organic Chemistry I. This inductive affect also stabilizes resonance structures with a positive charge on the carbon atom next to G.

An electron-donating substituent stabilizes an additional resonance form (as shown in Figure 7-16). Thus, any electron-donating substituent is an *o-p*-director.

What happens if the substituent is electron withdrawing? Figure 7-17 shows what happens in this case. When G attempts to pull electron density from an electron-poor atom (+), the result is a destabilization of the structure for both ortho and para sigma complexes, greatly reducing the probability that they will form.

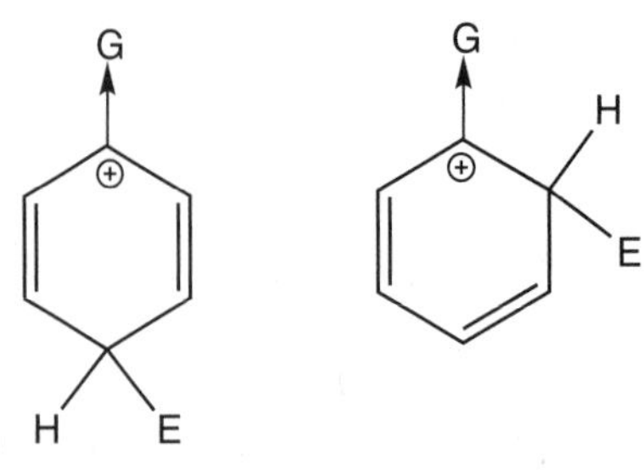

Figure 7-17: The destabilizing influence of an electron-withdrawing substituent.

Making predictions

Many times, you can look at the hybrid of the starting material in order to predict where the electrophile will attack. For example, start by looking at Figure 7-18, which shows the resonance structures for the phenolate ion. The resonance hybrid of the phenolate ion is shown in Figure 7-19.

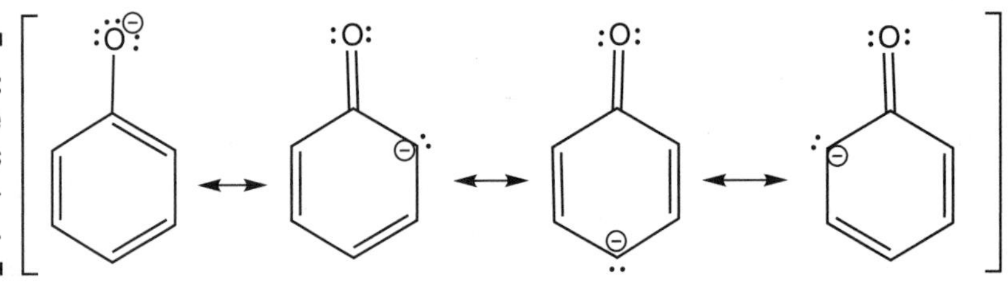

Figure 7-18: Resonance structures of the phenolate ion.

When drawing resonance structures the overall charge doesn't change from structure to structure.

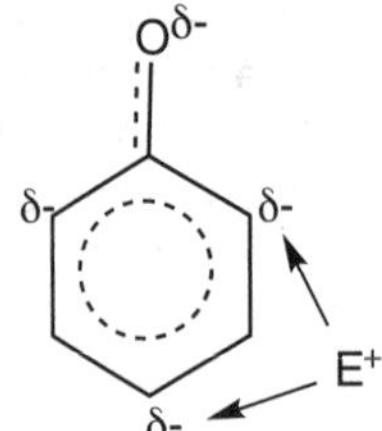

Figure 7-19: The resonance hybrid of the phenolate ion.

Figure 7-19 illustrates where the electrophile will attack: the electron-rich region (δ–) on the ring. (The two ortho positions are equivalent.) Otherwise, it wouldn't live up to its name.

If you look at the resonance structures arising when G is a nitro group, you see the structures given in Figure 7-20. The corresponding resonance hybrid is shown in Figure 7-21.

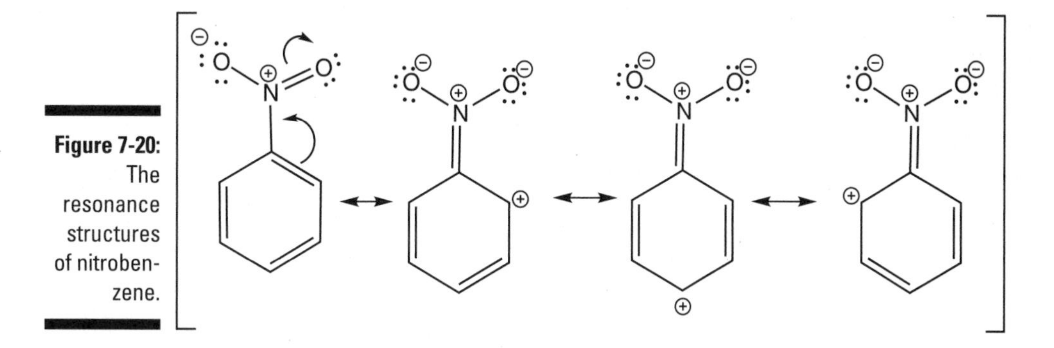

Figure 7-20: The resonance structures of nitrobenzene.

Figure 7-21: The resonance hybrid of nitrobenzene.

The presence of the partial positive charges (δ+) at the ortho and para positions makes an attack at these positions by an electrophile unlikely. This leaves the highest relative electron density at the meta position, which forces the electrophile to attack meta. The lack of a δ– at the meta-position indicates that electrophilic attack is not ideal (the ring is said to be deactivated), but it will still occur because that's the only option available.

Any substituent whose atom attached to the benzene contains a lone pair of electrons is ortho-para directing (but not necessarily a ring activator). Substituents without a lone pair on the atom attached to the ring are likely meta directors (with the exception of alkyl groups and aromatic rings, which turn out to be ortho-para directors).

Turning it on, turning it off: Activating and deactivating

Activation and deactivation deal with how the reactivity of the substituted benzene compares to benzene. If the aromatic system is more reactive than benzene, the substituent is said to cause activation. However, if the aromatic system is less reactive than benzene, the substituent causes deactivation.

Electrophiles seek regions of high electron density (– or δ–). Any substituent that gives rise to regions of high electron density facilitates electrophilic attack. This is activation. Referring back to Figure 7-19, you see that electron-donating groups concentrate a δ– at the ortho and para positions. For this reason, *o-p*-directors are activators. The degree of activation reflects how well the *o-p*-directors are capable of donating electron density to the aromatic resonance.

Electrophiles prefer not to attack if no regions of high electron density are available. In such cases, the ring is deactivated. A sufficiently strong electrophile still can, with difficulty, attack the ring, but never in an electron-deficient region (δ+). As shown in Figure 7-21, electron-withdrawing groups leave the ortho and para positions with slightly positive charges and the meta position with no charge.

Activators are *o-p*-directors (and vice versa), while deactivators are *m*-directors (and vice versa). The exception, in both cases, is halogens.

Halogens are unusual in that they are *o-p*-directors and deactivators. This means you should take care any time a halogen is present. Halogens are exceptions because they are electronegative (electron withdrawing) but also have electron pairs next to the ring (electron donating).

Table 7-1 summarizes what we know about directors and activators and deactivators. When using this table, remember two things:

- *O-p*-directors always beat *m*-directors.
- Strong activators always beat weak activators.

Table 7-1 Classification of Various Aromatic Substituents

Ortho-Para-Directors	
Very strong activators	$-NH_2$, -NHR, $-NR_2$, -OH, $-O^-$
Moderate activators	-OR, -NH-CO-R, -O-CO-R
Weak activators	-R, $-C_6H_5$
Mild deactivators	-F, -Cl, -Br, -I
Meta-Directors	
Very strong deactivators	$-N^+R_3$, $-NO_2$, $-CCl_3$, $-CF_3$
Moderate to mild deactivators	-CN, $-SO_3H$, -CO-R, -COOH, -COOR, $-CONH_2$, $-N^+H_3$

Pi electron-withdrawing groups direct substitution to the meta position, while electron-donating groups direct substitution to the ortho and para position.

In your first semester of organic chemistry you studied regiochemistry and retrosynthesis. The type of director (*o-p* or *m*) is an important aspect of this regiochemistry that you need to consider in any synthesis or retrosynthetic analysis problem.

Steric hindrance

Recall from your first semester of organic chemistry that steric hindrance is the blocking of a reactive site by part of a molecule.

The presence of a directing group indicates what the major product(s) of an electrophilic substitution reaction are. In most cases, the reaction also gives a small amount of product(s) not predicted by the directing group.

Prediction of the major product works well for *m*-directors, but what about *o-p*-directors? Will the ortho or the para product be the major product? Two positions are ortho and one is para, so the ortho product should be the major product (67 percent), and the para product should be the minor product (33 percent). Indeed, in the absence of steric factors, this prediction comes true.

However, groups larger than an ethyl group tend to interfere with attack at the ortho position. For example, in the case of the halogenation of propylbenzene, the para product is the major product.

Many students get into trouble when writing synthetic procedures because they overlook steric effects. For example, if the task is to synthesize the ortho product starting with propylbenzene, you can't rely on the *o-p*-directing ability of the propyl group. One solution is to place an easily removable substituent in the para position, leaving only the ortho position available for further attack. (If the para substituent is a meta-director, then both the meta-director and the propyl group favor attack at the same position.)

Limitations of Electrophilic Substitution Reactions

In nitration, halogenation, sulfonation, and acylation, the reactions are easy to control with temperature. The processes deactivate the ring toward further substitution, so the reactions inhibit further reaction.

However, a Friedel-Crafts alkylation can get out of hand: The process can continue until it replaces all the hydrogen atoms. For example, the alkylation of benzene can lead to the product pictured in Figure 7-22. To minimize the possibility of multiple alkylations, use a large excess of the aromatic compound.

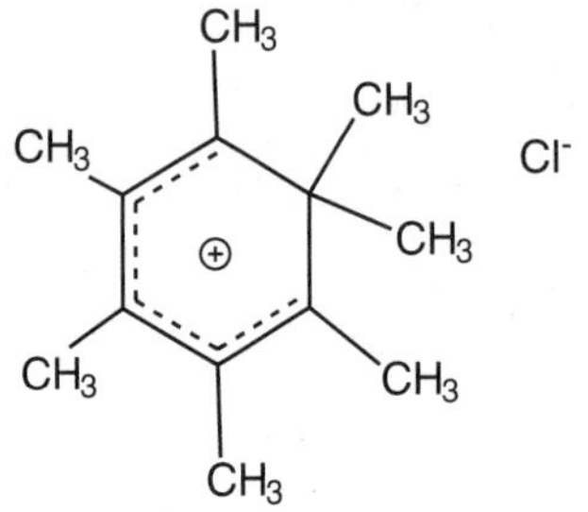

Figure 7-22: The result of uncontrolled alkylation of benzene.

An amine group limits Friedel-Crafts reactions because it reacts with the catalyst so the reaction can't proceed. Friedel-Crafts alkylation or acylation doesn't take place with groups more deactivating than halogen.

Chapter 8

Aromatic Substitution Part II: Attack of the Nucleophiles and Other Reactions

In This Chapter

- Going over the basics and mechanisms of nucleophilic substitution reactions
- Mastering mechanisms of elimination/addition reactions
- Determining synthesis strategies for aromatic systems

In this chapter we start by filling you in on nucleophiles before moving on to elimination/addition reactions. If your professor doesn't cover elimination/addition reactions, do a happy dance and feel free to skip that part of this chapter. If you do get to study these reactions, we're here to help you out. We end the chapter by exploring some important aromatic reactions that don't fit in other categories.

Coming Back to Nucleophilic Substitution Reactions

The basic concepts of nucleophilic substitution reactions appeared in the first semester of organic chemistry. These reactions follow S_N1 or S_N2 mechanisms. (In aromatic nucleophilic substitution mechanism, we use the designation S_NAr.) In S_N1 and S_N2 mechanisms, a nucleophile attacks the organic species and substitutes for a leaving group. In aromatic systems, the same concepts remain applicable, but with some differences that result from the inherent stability of aromatic systems.

While aromatic systems often undergo electrophilic substitution reactions, they can also undergo nucleophilic substitution. In electrophilic substitution the aromatic system behaves as a nucleophile, whereas, in nucleophilic substitution, the aromatic system behaves as an electrophile. To make an aromatic system change from a nucleophile into an electrophile, a strong electron-withdrawing group must be present. In addition, the aromatic system must have a leaving group similar to the leaving groups seen in Organic Chemistry I.

Nucleophilic substitution reactions on aromatic systems must have a strong electron-withdrawing group *and* a good leaving group, and the leaving group must be at the ortho or para position relative to the electron-withdrawing group.

An example of a strong electron-withdrawing group is the nitro group (-NO_2). Refer back to Figure 7-21 in the previous chapter to see the resonance hybrid of nitrobenzene. This hybrid has a partial positive charge (δ+) at the ortho and para positions. These partially positive regions are suitable for nucleophilic attack, and the meta position is unsuitable. So while the nitro group deactivates the ring towards electrophilic attack, it activates the ring towards nucleophilic attack. Of course, no matter how much activation occurs, if no leaving group is present, no reaction can take place.

H^- is *never* the leaving group.

Mastering the Mechanisms of Nucleophilic Substitution Reactions

A traditional S_N1 or S_N2 mechanism doesn't work for aromatic systems, so a new mechanism is necessary. This new mechanism is the S_NAr or addition-elimination mechanism. Like an S_N2 mechanism, an S_NAr mechanism begins with the attack of the nucleophile. After the initial attack the mechanism is quite different.

We use the reaction of 3-chloronitrobenzene with the hydroxide ion to illustrate in Figure 8-1 the mechanism for a nucleophilic substitution reaction. The different resonance structures represent a Meisenheimer complex. The loss of the leaving group reestablishes the stable aromatic system.

Unlike the positively charged sigma complex, the Meisenheimer complex has a negative charge.

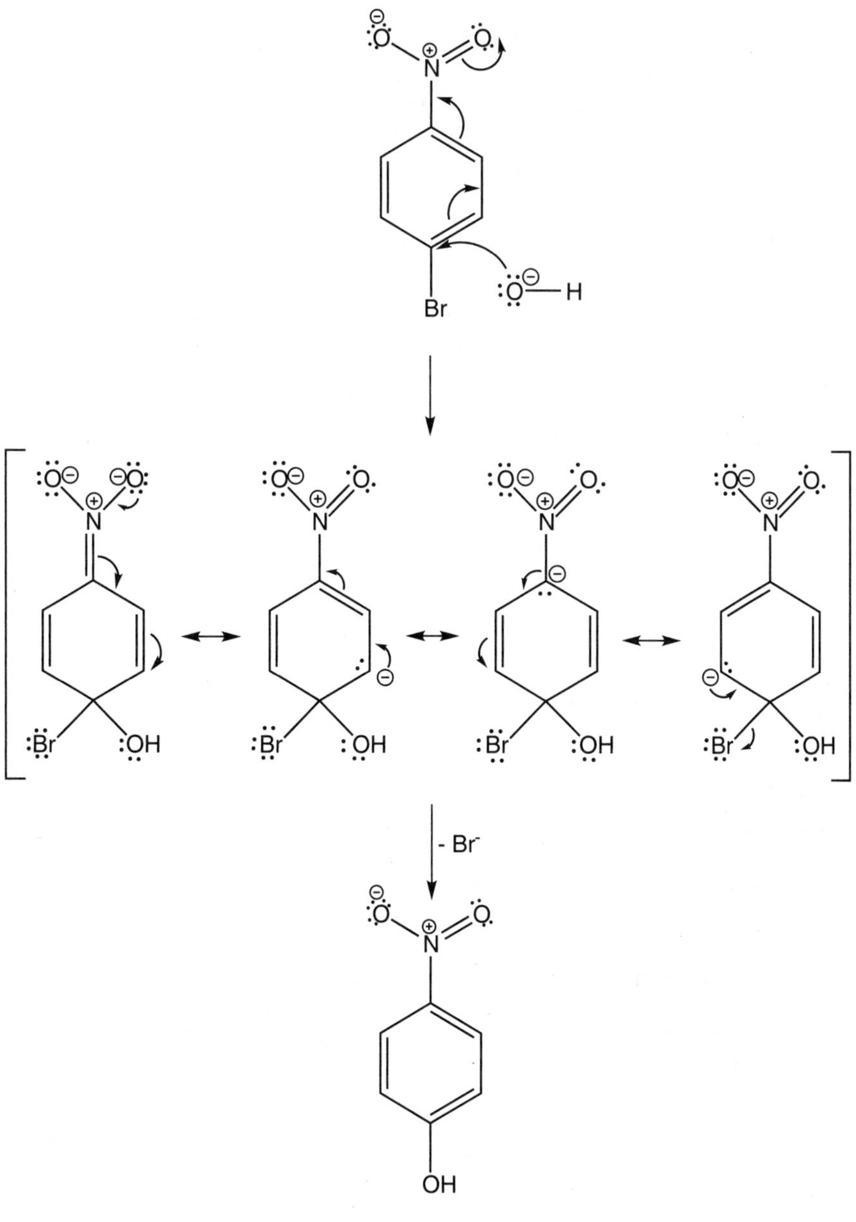

Figure 8-1: The mechanism for a nucleophilic substitution reaction.

Nucleophilic substitution occurs only when the benzene ring is activated by a strong electron-withdrawing group.

Losing and Gaining: Mechanisms of Elimination/Addition Reactions

An elimination/addition reaction is another distinct type of reaction mechanism that occurs in aromatic systems. In these mechanisms, the elimination involves the loss of an HX molecule. While this may seem like a dehydrohalogenation as seen in Organic Chemistry I, it really is a different reaction. The HX loss leads to the formation of a benzyne intermediate (see Figure 8-2). The mechanism ends with addition to the bond formed by the loss of HX.

An S_NAr mechanism is an addition/elimination, not an elimination/addition reaction.

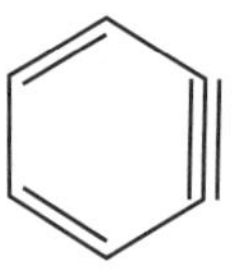

Figure 8-2: The structure of benzyne.

Benzyne

The benzyne molecule (refer to Figure 8-2) is a highly unstable and therefore highly reactive intermediate that forms at high temperatures and high pressures. The formation of the carbon-carbon triple bond requires two *sp* hybridized carbon atoms, the presence of which makes the structure unstable. The bond angle about the *sp* hybridized carbon atoms is 180 degrees, which is significantly larger than the 120 degrees of the sp^2 present in the ring of benzene.

A variety of experiments have shown that benzyne is a real molecule. However, no one yet has been able to isolate this unstable substance.

The elimination/addition mechanism

The Dow Process utilizes an elimination/addition reaction to convert chlorobenzene to phenol. The proposed mechanism for this reaction is shown in Figure 8-3. The high-temperature reaction begins with chlorobenzene and aqueous sodium hydroxide. Note that this mechanism starts with the hydroxide attacking as a base, beginning dehydrohalogenation to form benzyne. The second hydroxide ion attacks as a nucleophile to form a carbanion intermediate, which behaves as a base in the last step to yield the final product.

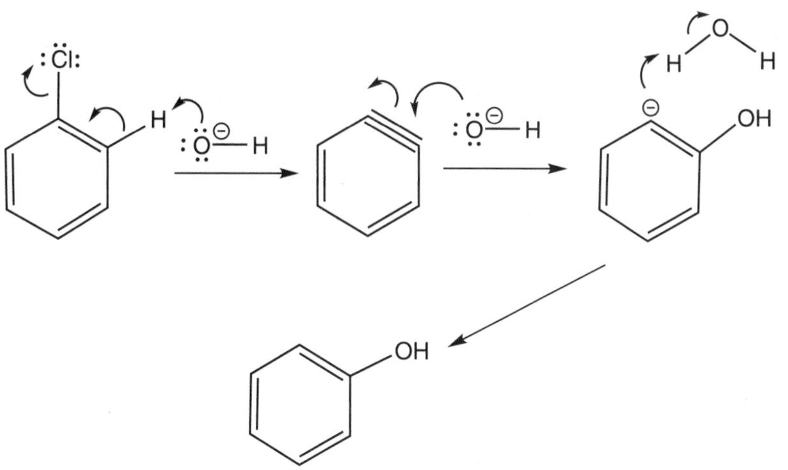

Figure 8-3: An elimination/addition reaction mechanism.

Experiments with the chlorine attached to a radioactively labeled carbon-14 atom produce phenol with 50 percent of the OH attached to the carbon-14 atom and 50 percent attached to the adjacent carbon atom. This distribution indicates that the second attack by the hydroxide ion has equal probability of attacking either side of the triple bond, which is evidence of the existence of the triple bond, and, therefore, of the benzyne molecule.

The NaOH in the reaction can be replaced with $NaNH_2$, because the amide ion, NH_2^-, is similar to the hydroxide ion in that both are strong bases and good nucleophiles (the amide ion is a much stronger base). The use of sodium amide yields aniline as the product.

Synthetic Strategies for Making Aromatic Compounds

Now you have three basic mechanisms for aromatic rings — electrophilic aromatic substitution, S_NAr, and elimination/addition. How do you choose among these? The first consideration is what types of other reagents are present. If the reagents include an electrophile, then the reaction will be electrophilic aromatic substitution. The presence of a nucleophile may lead to either S_NAr or elimination/addition. If the system meets the three requirements for S_NAr, then the reaction will follow that mechanism. If not, it will be an elimination/addition.

Nucleophilic substitution reactions on aromatic systems must have a strong electron-withdrawing group *and* a good leaving group, and the leaving group must be at the ortho or para position relative to the electron-withdrawing group.

Adding the substituents in the correct order is crucial in mastering aromatic synthesis problems. Examine the two reaction sequences given in Figure 8-4. Both sequences involve the same reagents; however, the order is reversed. This shows that the reaction sequence is important. You need to plan ahead in any multistep reaction sequence.

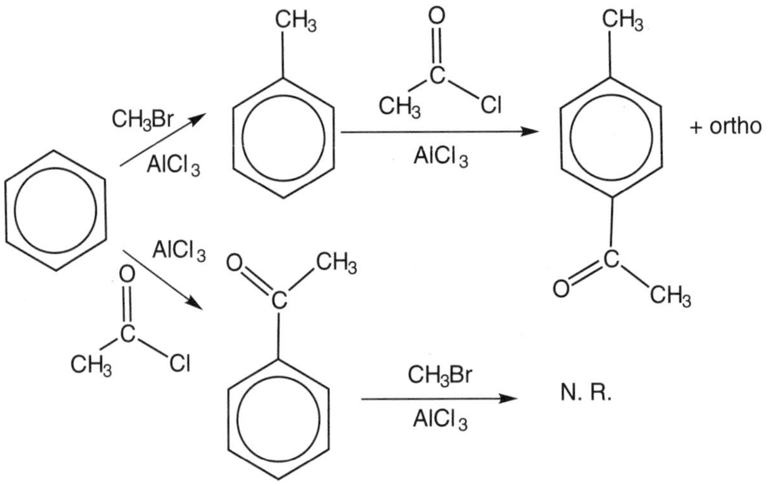

Figure 8-4: Two possible synthetic routes.

Another thing to consider when designing a reaction is the conditions. For example, are you promoting the formation of the kinetic or thermodynamic product? Figure 8-5 illustrates this concern. Figure 8-6 illustrates the arrangement of one of the kinetic products (*o*-xylene) to the thermodynamic product (*m*-xylene).

Oxidation of an alkylated benzene (making an aryl carboxylic acid) is a method of converting an ortho-para activator into a meta director. The reduction of a nitro group to make an aryl amine is a way of changing a meta director into an ortho-para activator.

For example, on an exam, you may be asked to prepare 1-bromo-3-nitrobenzene from benzene in two steps. In this case, you must know not only what reactants to use but also the order to use them.

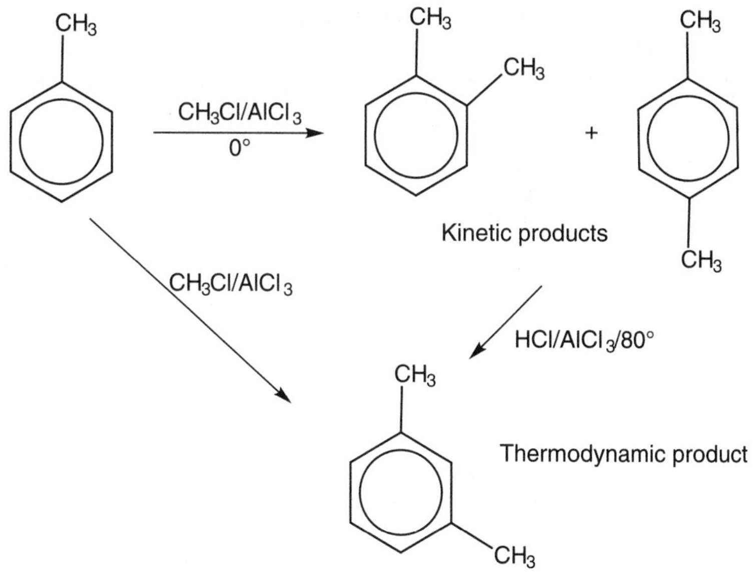

Figure 8-5: The formation of the kinetic and thermodynamic product.

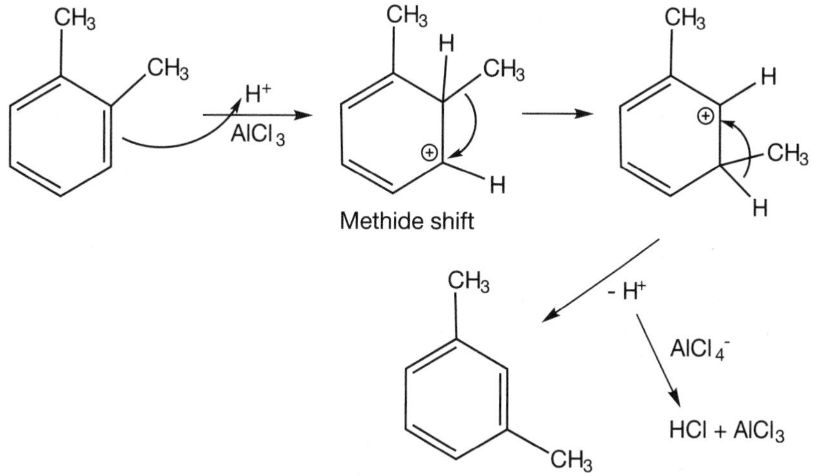

Figure 8-6: The mechanism for the conversion of *o*-xylene to *m*-xylene.

Briefly Exploring Other Reactions

Many other reactions involving aromatic systems are possible. Many of them are extensions of the reactions you learned in your first semester of organic chemistry. For example, you may have the hydrogenation reaction of a side chain (shown in Figure 8-7) or the dehydrohalogenation of a halogenated side chain (Figure 8-8). The reaction in Figure 8-8 gives only one product because this product extends the conjugated system. Figure 8-9 shows how Markovnikov and anti-Markovnikov additions take place.

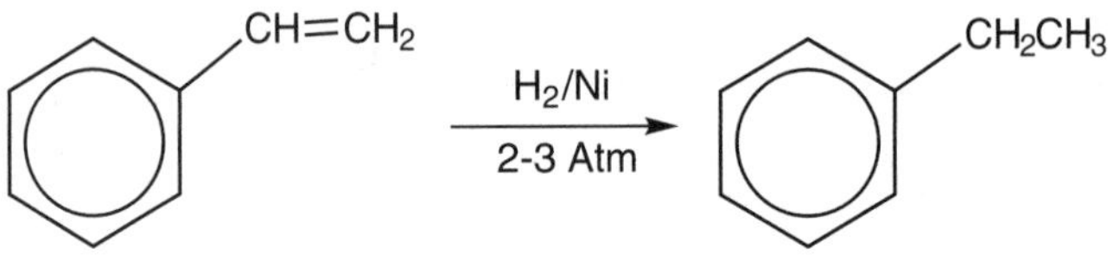

$CH{=}CH_2$ → (H_2/Ni, 2-3 Atm) → CH_2CH_3

Figure 8-7: The hydrogenation of the side chain.

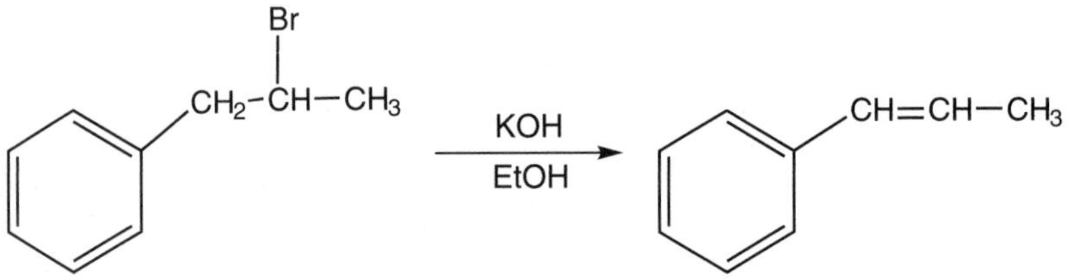

$CH_2{-}CH(Br){-}CH_3$ → (KOH, EtOH) → $CH{=}CH{-}CH_3$

Figure 8-8: The dehydrohalogenation of a side chain.

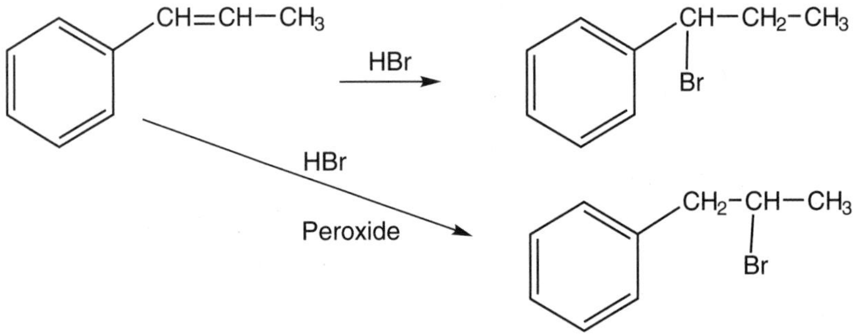

$CH{=}CH{-}CH_3$ → (HBr) → $CH(Br){-}CH_2{-}CH_3$

$CH{=}CH{-}CH_3$ → (HBr, Peroxide) → $CH_2{-}CH(Br){-}CH_3$

Figure 8-9: Markovnikov and anti-Markovnikov additions.

Figure 8-10 shows another pair of reactions for the halogenation of an aromatic compound. The reaction of the side chain is a free-radical substitution. Figure 8-11 shows the mechanism of this free-radical substitution.

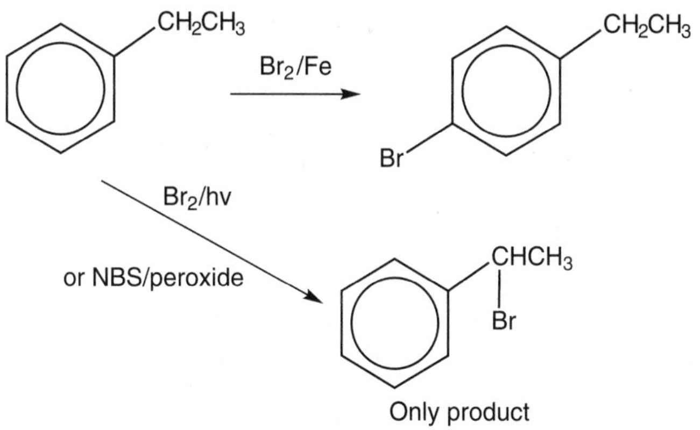

Figure 8-10: Two halogenation reactions.

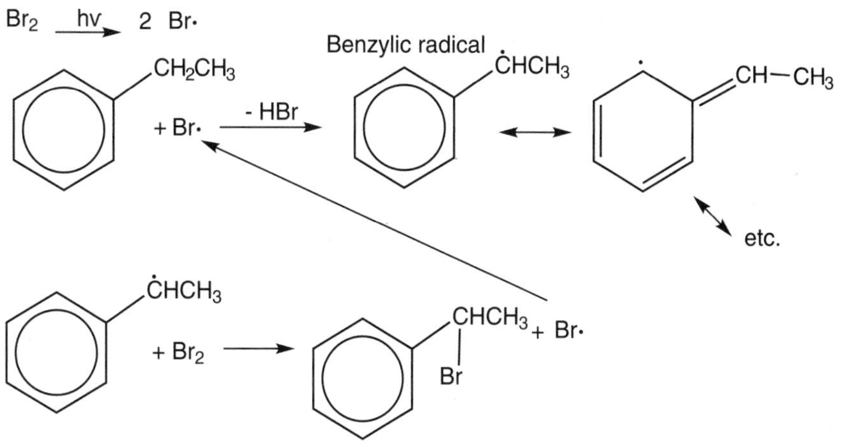

Figure 8-11: The free-radical substitution mechanisms for the reaction in Figure 8-10.

Part III

Carbonyls: Good Alcohols Gone Bad

"Yeah, those are the Carboxylic brothers, and you know what their alcohol absorption rate is."

In this part . . .

Carbonyls are a very large category of organic compounds that includes aldehydes, ketones, enols and enolates, carboxylic acids, esters, amides, and a slew of others. In the chapters in this part, you look at the structure, reactivity, and spectroscopy of all of these compounds, with special attention to aldehydes and ketones, enols and enolates, and carboxylic acids. You study a lot of reactions, including different ways of synthesizing these substances as well as the myriad reactions that these compounds undergo. Finally, you examine some chemical tests that help identify these compounds and some spectroscopic clues that can also be used for identification.

Chapter 9

Comprehending Carbonyls

In This Chapter

- Considering carbonyls
- Pondering polarity
- Reviewing resonance
- Clarifying carbonyl reactivity
- Shedding light on spectroscopy

The carbonyl group is part of a wide variety of functional groups that are important not only to organic chemistry but also to biochemistry. In fact, the carbonyl group is the basis of all the classes of organic compounds discussed in the next three chapters of this book. Figuring out the basics of the carbonyl group sets a good foundation and makes understanding the classes of organic compounds much easier (and isn't making it easier the whole point?).

In this chapter you see some classes of organic compounds that contain the carbonyl group. We then investigate polarity and resonance as they relate to the carbonyl group, and you have the opportunity to examine some reactions involving it. Finally, you discover the specifics of the different types of spectroscopy associated with the carbonyl group.

Carbonyl Basics

The carbon atom has sp^2 hybridization, which means the geometry around it is trigonal planar. The planar geometry leaves the carbon atom open to attack from either above or below the plane of the triangle.

A carbonyl group is a very stable group. This stability means that although many reactions alter what's attached to the group, few reactions actually change the C=O, and reactions that destroy the carbonyl group require a great deal of energy. Conversely, reactions that form a carbonyl group are energetically favorable.

Figure 9-1 shows the carbonyl group. ***Note:*** This name isn't part of the formal nomenclature of organic compounds, but just a simple name for a commonly seen group.

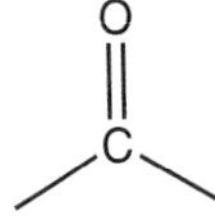

Figure 9-1: The carbonyl group.

Remember two important facts when examining the chemistry of the carbonyl group. First, the carbonyl group is very electrophilic, and this promotes attack by a nucleophile. (Remember, the carbon atom is the target of these attacks.) Second, after the attack, there's a strong drive for the carbonyl group to reform.

Considering compounds containing the carbonyl group

Functional groups containing a carbonyl differ in what's attached to the two open bonds on the carbon atom. Hydrogen atoms, alkyl groups, hydroxyl groups, and so on can be attached. Depending on what's attached, the compound is given a different functional group name and has a unique set of properties. The following sections discuss these different functional groups.

Aldehyde and ketones

Aldehydes and ketones are examples of carbonyl compounds. They differ in that either one or two alkyl groups are attached to the carbonyl carbon. See Figure 9-2, and check out Chapter 10 for more discussion of aldehydes and ketones.

O
R—C—H
Aldehyde
(R or H attached)

O
R—C—R'
Ketone

Figure 9-2: Aldehyde and ketone structure.

The similarity between an R (an alkyl group) and an H (plain old hydrogen) makes aldehydes and ketones similar in reactivities to each other.

You often see aldehydes represented as R-CHO. Don't confuse this condensed form with an alcohol, which is represented as R-OH.

Carboxylic acids

Attaching an -OH to a carbonyl function yields a carboxylic acid (shown in Figure 9-3 and covered in more detail in Chapter 12). This is a distinct group whose properties are not simply the sum of the properties of a carbonyl group and an alcohol group.

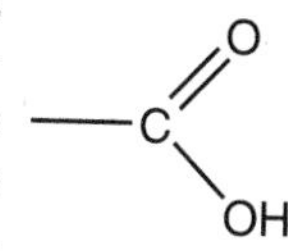

Figure 9-3: The structure of a carboxylic acid.

Don't confuse a carboxylic acid with a ketone or an alcohol. Carboxylic acids have entirely different properties and reactivities than either ketones or alcohols. In particular, the proton (H^+) on the oxygen in a carboxylic acid is unusually acidic (hence the name!), for reasons we talk about later in this chapter in the section "Reactivity of the Carbonyl Group."

The carboxylic acids, like all Brønsted-Lowry acids, can lose a hydrogen ion. The result is a carboxylate ion, shown in Figure 9-4.

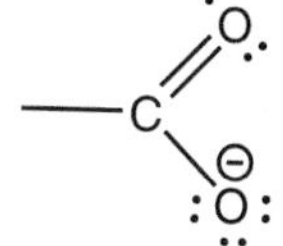

Figure 9-4: The carboxylate ion.

Acyl group

Another common combination is the acyl group, which is a carbonyl group with one alkyl group (R) attached. Compounds containing the acyl group get a closer examination in upcoming chapters, especially Chapter 12. Figure 9-5 shows an acyl group.

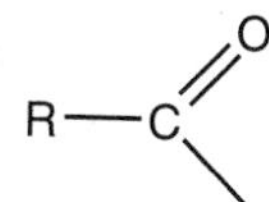

Figure 9-5: The acyl group.

Acyl chlorides

The acyl group appears in other compounds such as the acid chlorides, which are also known as acyl chlorides (see Figure 9-6). To find out about the synthesis and reactions of these types of compounds, check out Chapter 12.

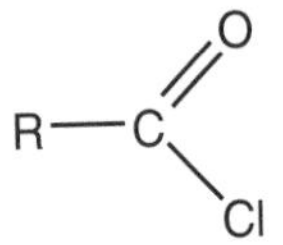

R—C(=O)—Cl

Figure 9-6: The acyl chloride structure.

Acid anhydride

Two carboxylic acid groups may combine to form an acid anhydride with the general structure shown in Figure 9-7. We discuss this combination in more detail, including synthesis and reactions, in Chapter 12.

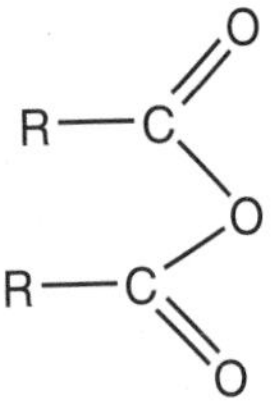

R—C(=O)—O—C(=O)—R

Figure 9-7: The acid anhydride structure.

Esters

A carboxylic acid may react with an alcohol to form an ester (see Figure 9-8). These esters are used quite a bit in the flavor and perfume industry. You can see their synthesis and reactions in much more detail in Chapter 12.

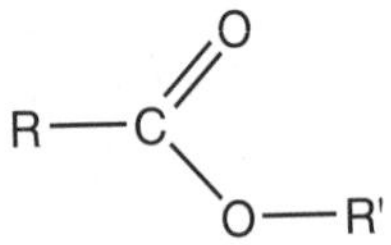

R—C(=O)—O—R'

Figure 9-8: The structure of an ester.

Amides

The interaction of a carboxylic acid with ammonia or an amine may form an amide. The amide derived from ammonia has an -NH_2 group attached to the carbonyl group. As you can see in Figure 9-9, primary and secondary amines yield amides with the nitrogen attached to one or two alkyl groups. (Tertiary amines don't combine to form amides.)

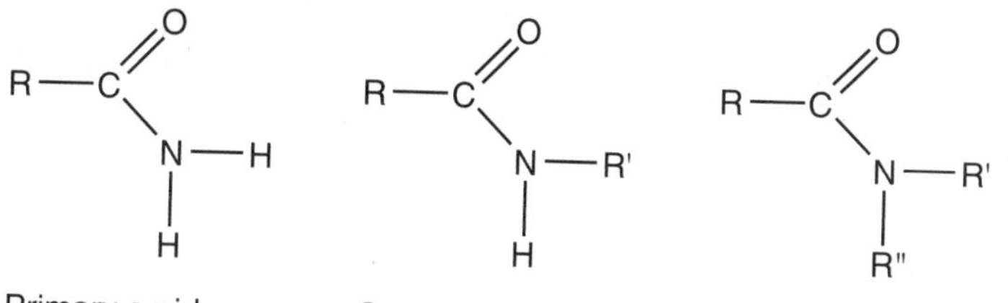

Primary amide Secondary amide Tertiary amide

Figure 9-9: Primary, secondary, and tertiary amides.

Getting to know the acidic carbonyl

Carboxylic acids aren't the only acidic carbonyl compounds. Acidity can occur in other ways; for example, hydrogen atoms attached to a carbon adjacent to the carbonyl group (the α hydrogen atoms only, not the β hydrogens) are acidic (see Figure 9-10).

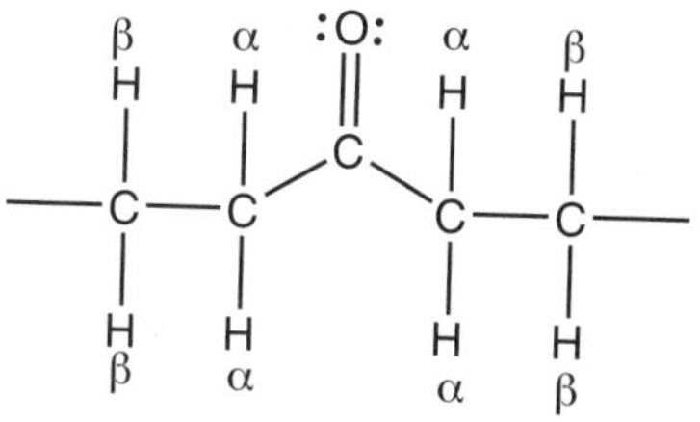

Figure 9-10: The α and β hydrogens of a carbonyl.

The acidic nature of the α hydrogen atoms is due to stabilization of the anion formed as a result of the loss of a hydrogen ion. See Figure 9-11.

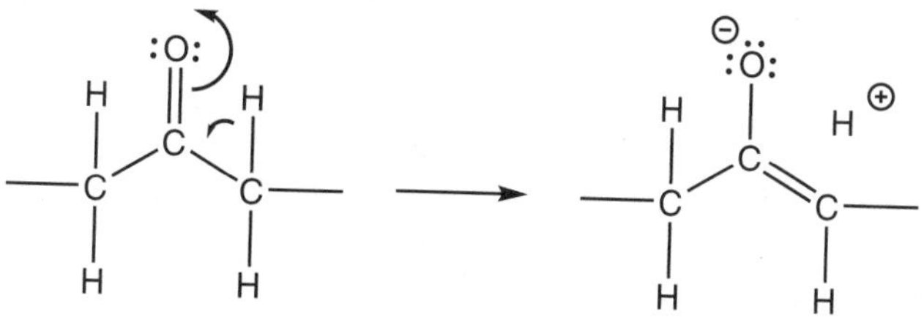

Figure 9-11: Resonance stabilization of an anion due to the loss of a hydrogen ion.

Polarity of Carbonyls

The electronegativity difference between carbon and oxygen causes the carbonyl group to be polar, with one part becoming negatively charged while another part becomes positively charged (see Figure 9-12). This is an induction effect. Resonance (see the next section) enhances this polarity. Other groups attached to the carbonyl group may also increase the polarity of the group.

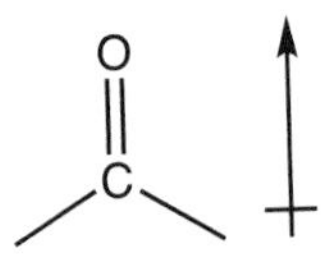

Figure 9-12: Polarity of the carbonyl group.

The electronegativity difference makes the oxygen atom δ– and the carbon atom δ+. This makes the oxygen atom a nucleophile and the carbon atom an electrophile, which sets the stage for nucleophilic attack on the carbon atom, examples of which are discussed in Chapter 10.

The polarity of the carbonyl group leads to dipole-dipole intermolecular forces, which, in general, increase the melting and boiling points of carbonyl compounds above that of comparably sized hydrocarbons. The alcohols, with their ability to hydrogen bond, tend to have higher melting and boiling points. Alcohols and carbonyl compounds can hydrogen bond to water molecules, thereby increasing the water-solubility of these compounds.

Additional groups attached to the carbonyl — for example, the -OH in the carboxylic acids — also contribute to the intermolecular forces. The carboxylic acids have higher melting and boiling points than nonpolar molecules of similar molecular masses due to the addition of hydrogen bonding to the intermolecular forces present. Converting a carboxylic acid to an ester decreases the intermolecular forces due to the loss of hydrogen bonding; converting a carboxylic acid to a carboxylate ion increases the strength of the intermolecular forces because it creates the possibility of ion-dipole interactions as well as ionic bonding.

Like the carboxylic acids, the primary and secondary amides have a hydrogen bonding contribution to their intermolecular forces.

Resonance in Carbonyls

Resonance enhances the polarity of the carbon-oxygen bond in the carbonyl group. The effect of resonance is shown in Figure 9-13. Shifting the electron from the double bond to the oxygen gives the oxygen a negative charge and the carbon a positive charge. This charge separation is important in the reactivity and polarity of the carbonyl group.

Figure 9-13: Resonance enhancing the polarity of the carbonyl group.

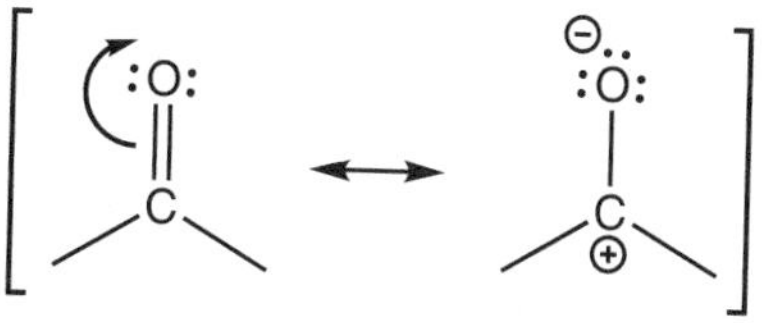

Make sure you keep in mind the charge distribution that's present in the right-hand structure while you're studying any reactions involving a carbonyl compound.

In amides, the lone electron pair on the nitrogen atom promotes resonance stabilization of the carbonyl region (see Figure 9-14). This stabilization is important not only to amides, but also to the secondary structure of proteins.

Figure 9-14: Stabilization of the carbonyl region in an amide by resonance.

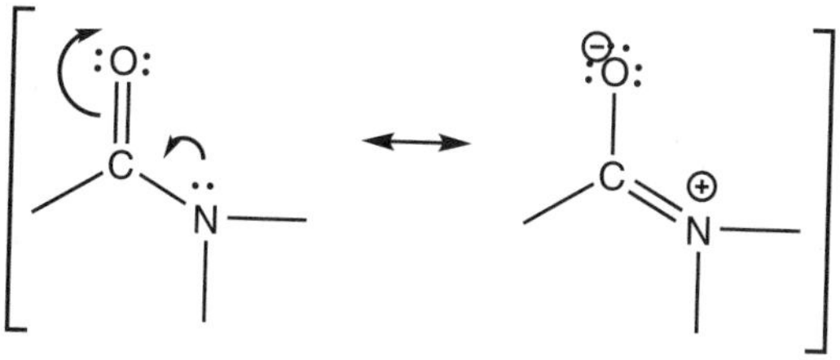

Resonance further increases the nucleophilic nature of the oxygen atom and the electrophilic nature of the carbon atom. In many cases, the nucleophilicity of the oxygen leads to its protonation (adding an H^+) in acidic media.

The increase in polarity due to resonance contributes to the strength of the intermolecular forces present in the compound.

Reactivity of the Carbonyl Group

A simple nucleophilic attack on a carbonyl group is shown in Figure 9-15. This process is reviewed in Chapter 2 and is further illustrated in the discussion of reactions in Chapter 10.

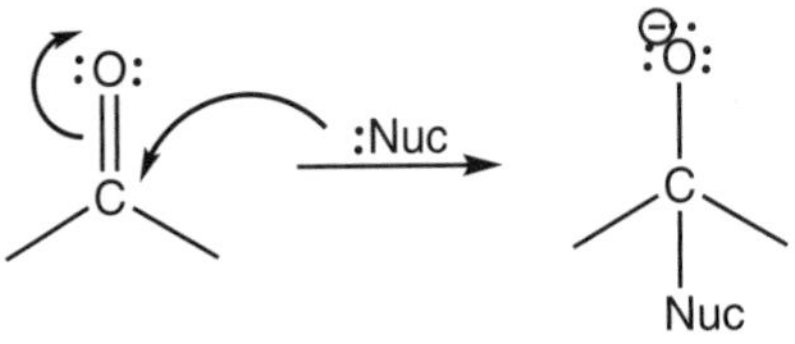

Figure 9-15: Nucleophilic attack on the carbonyl group.

The reactivity of the carbonyl group is enhanced by resonance, which stabilizes the carboxylate ion (see Figure 9-16). This increased stability of the carboxylate ion means that it's easier for a hydrogen ion to leave the carboxylic acid. Thus the resonance is one factor in accounting for the acidity of carboxylic acids.

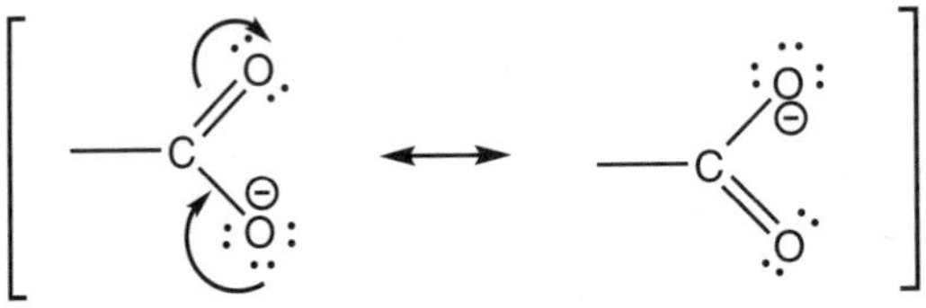

Figure 9-16: Resonance stabilizing the carboxylate ion.

Spectroscopy of Carbonyls

Unsurprisingly, organic compounds containing carbonyl groups exhibit many of the features of other organic compounds. For example, they normally have C-H and C-C stretches in the appropriate region of their infrared spectra. In this section we focus on the spectral properties of the carbonyl group and properties induced by the mere presence of the carbonyl group.

Infrared spectroscopy

The C=O in all carbonyl groups has a very intense band in the 1,700 cm^{-1} region of its infrared spectrum. In many cases this band is the most prominent feature of the spectrum. Table 9-1 shows some typical carbonyl stretches. (See Chapter 5 for a discussion of IR stretches and peaks.)

Table 9-1 Carbonyl Stretches in Various Functional Groups

Carbonyl Stretch	*cm^{-1}*
Aldehyde	1,740–1,720
Ketone	1,725–1,680
Carboxylic acid	1,725–1,700
Ester	1,750–1,735
Acid anhydride	1,850–1,725 (two bands)
Acid halide	1,800
Amides	1,650

Conjugation of the carbonyl group tends to shift the stretch to a lower frequency. Resonance may occur when a double bond or nitrogen atom is adjacent to the carbonyl group.

The C-H stretch for the aldehyde hydrogen tends to be a weak band in the 2,960–2,700 cm^{-1} region of the spectrum.

Carboxylic acids also have the characteristic broad O-H stretch around 3,600–3,000 cm^{-1}. See Figure 9-17.

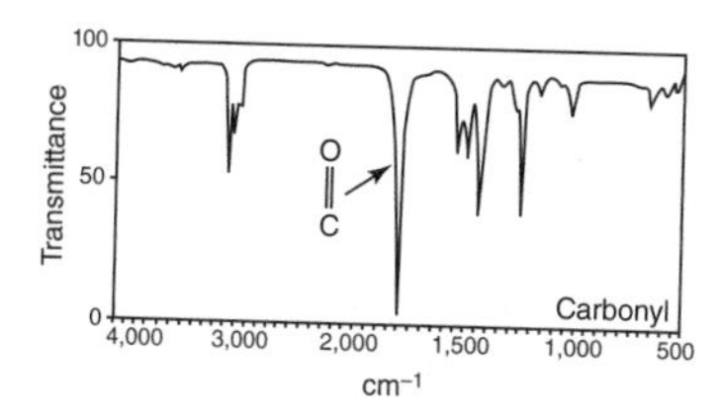

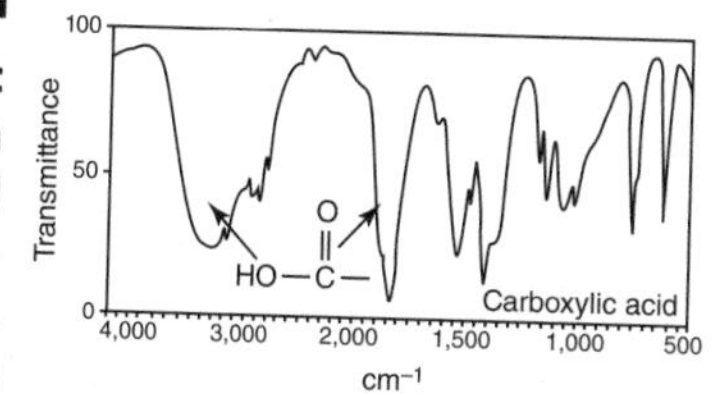

Figure 9-17: IR spectra of carbonyl and carboxylic acid.

Ultraviolet-visible (electronic) spectroscopy

In general, if no conjugation occurs, the UV-visible absorption of carbonyl compounds is below 215 nanometers (nm) and, for this reason, isn't very useful. However, in a few cases, a second absorption may be present. You can see some examples of the additional absorption band in Table 9-2.

Table 9-2 UV-Visible Absorptions in Various Functional Groups

Compound	*nanometers (nm)*
Aldehyde	290–295
Ketone	275–285
Acid chloride	≈235

Nuclear magnetic resonance (NMR) spectroscopy

The two important types of NMR spectroscopy are proton (^1H) NMR and ^{13}C NMR. Even though the oxygen can't be directly observed at work in these techniques, the presence of the electronegative oxygen atoms influences them both.

Proton NMR

In the proton NMR, the presence of the electronegative oxygen tends to shift the position of the chemical shift downfield. This can be seen in Table 9-3 and in the proton NMR spectra of propanal (Figure 9-18).

Table 9-3 Proton NMR Shifts

Proton NMR	*Parts-per-million (ppm)*
Aldehyde	9.5–10.5
Carboxylic acid	10–12

^{13}C NMR

In carbon-13 NMR, the presence of the oxygen also influences the chemical shift of the carbon atoms. Table 9-4 shows the chemical shifts of the carbonyl carbon atom for several classes. This can also be seen in the NMR spectrum of butyric acid in Figure 9-19.

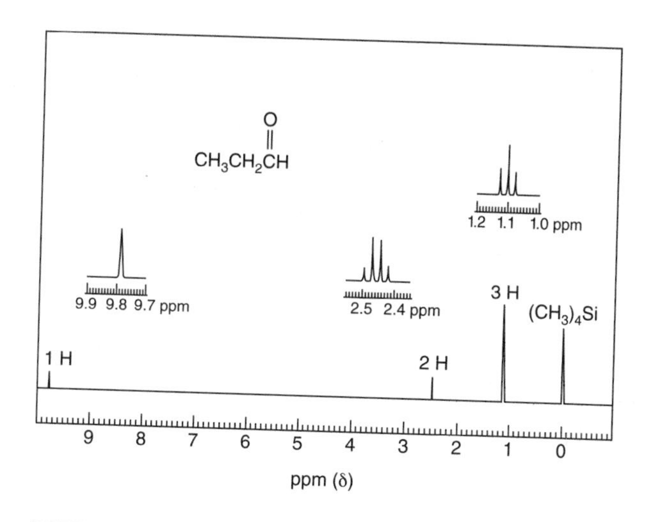

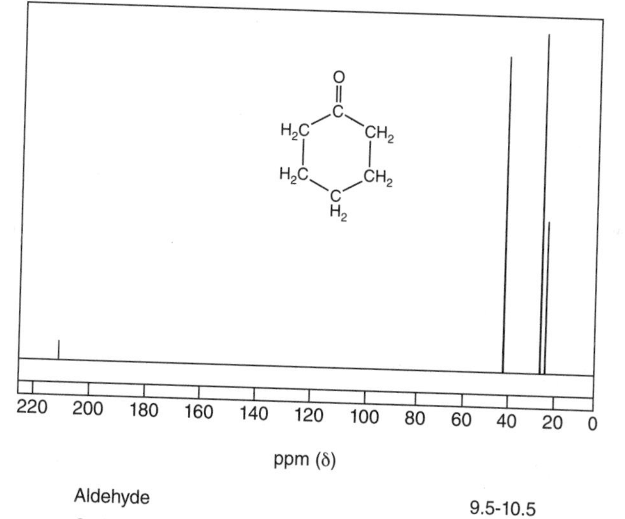

Figure 9-18: Proton NMR spectra of propanal and carbon-13 spectra of cyclohexanone.

Aldehyde 9.5-10.5

Carboxylic acid 10-12

Table 9-4 **^{13}C Shifts for the Carbonyl Carbon**

^{13}C NMR	*Parts-per-million (ppm)*
Aldehyde	190–200
Ketone	200–220
Carboxylic acid	177–185
Ester	170–175

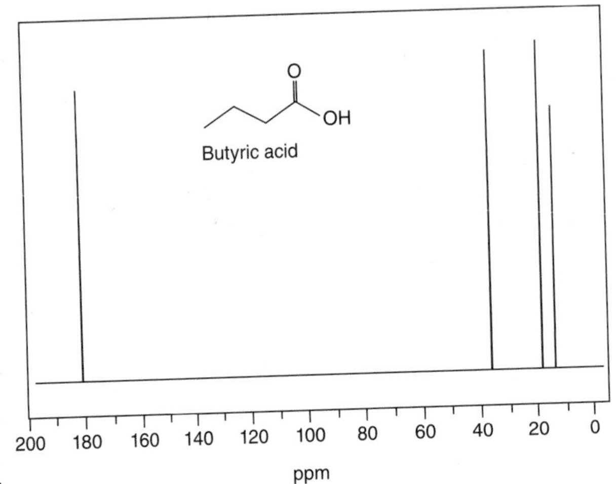

Figure 9-19: Carbon-13 spectrum of butyric acid.

Mass spectroscopy

Carbonyl compounds exhibit the normal fragmentation that happens with organic compounds, but they also display some other important features. Smaller aldehydes usually have a prominent CHO^+ peak (m/e 29). For ketones, fragmentation may occur on either side of the carbonyl group, which can influence the position of the group. The base peak is often CH_3CO^+ (m/e 43) for methyl ketones, and aliphatic and aromatic acids may have a $COOH^+$ peak (m/e 45).

The loss of CO_2 appears in the spectra of many dicarboxylic acids and substituted carboxylic acids. Aromatic acids often exhibit a prominent OH loss followed by CO loss. For example, in acid anhydrides, a break may occur on either side of the connecting oxygen atom, and this is followed by the loss of CO.

This break on either side is important for anhydrides with different R groups. The cleavage of the N-R bond is important for amides with N-substituted alkyl groups. A primary amide usually has a strong peak at m/e 44, which corresponds to $[O=C=NH_2]^+$.

Don't forget to consider the nitrogen rule when interpreting the mass spectra of amides. (See Chapter 5 to review the nitrogen rule.)

Table 9-5 and Figure 9-20 show partial mass spectral data for representative carbonyl compounds. You can analyze each of the spectra to try to explain each of the peaks listed.

Table 9-5 Mass Spectroscopy (m/e) Data for Figure 9-20

Compound A		*Compound B*		*Compound C*	
m/e	*%*	*m/e*	*%*	*m/e*	*%*
43	100	77	28	44	100
58	90	105	100	58	25
71	15	148	5	72	5
100	5			86	30
				115	10

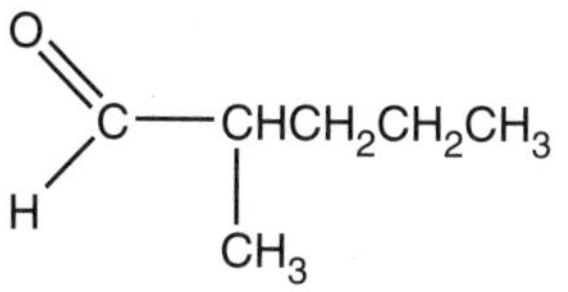

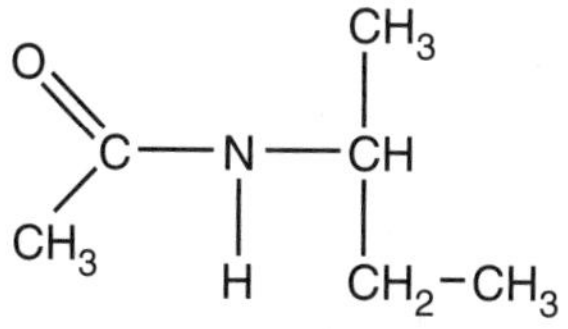

Figure 9-20: Carbonyl compounds (see Table 9-5 for mass spectra data).

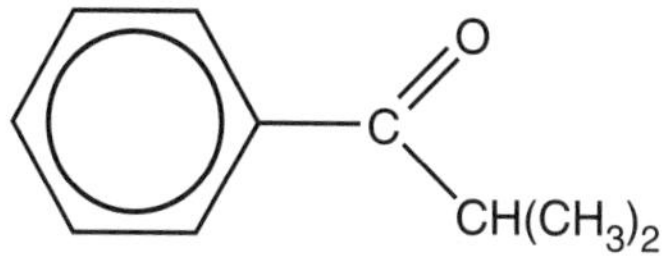

Take a look at Figure 9-21. It has the IR, proton NMR, and carbon-13 NMR of phenylethanal. Practice picking out the various peaks, bands, and splitting.

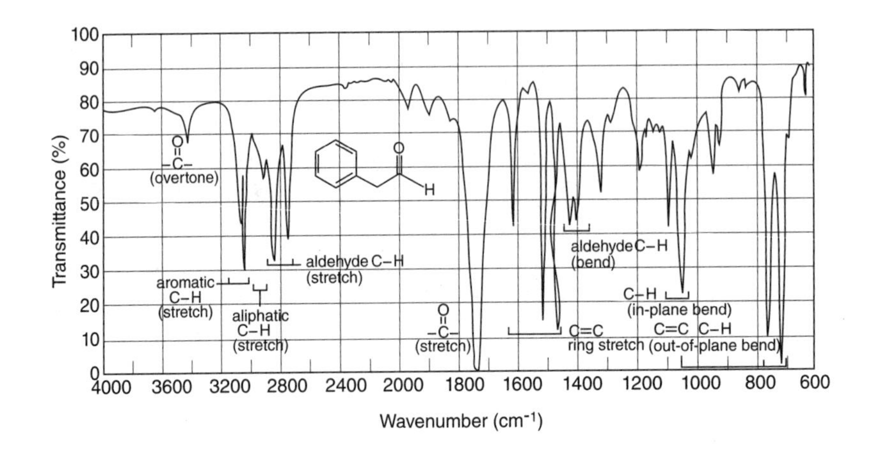

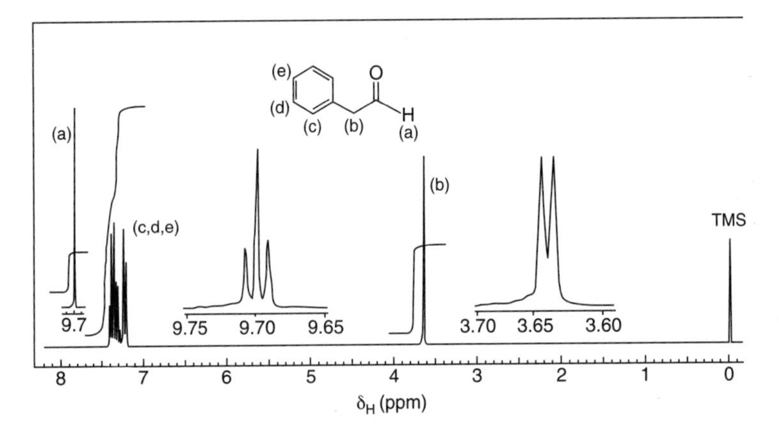

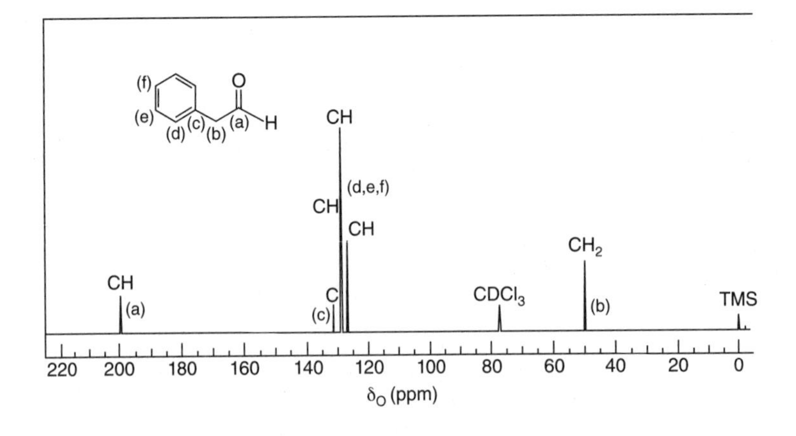

Figure 9-21: The IR, proton NMR, and carbon-13 NMR spectra of phenylethanal.

Chapter 10

Aldehydes and Ketones

In This Chapter

- Examining the structure and physical properties of aldehydes and ketones
- Finding out how aldehydes and ketones are formed
- Mastering the reactions of aldehydes and ketones
- Reviewing their spectroscopy

In this chapter, we focus on two types of compounds containing a carbonyl group, aldehydes and ketones (see Figure 10-1). The simplest aldehyde is methanal (formaldehyde), CH_2O, and the simplest ketone is propanone (acetone), C_3H_6O.

O
||
R—C—H
Aldehyde
(R or H attached)

O
||
R—C—R'
Ketone

Figure 10-1: The general structure of aldehydes and ketones.

Meeting Alcohol's Relatives: Structure and Nomenclature

The basic nomenclature for aldehydes and ketones follows that of other organic compounds. The following steps are the keys:

1. **Find the longest chain containing the carbonyl group to determine the parent name.**

   Replace the final *-e* of the hydrocarbon with an *-al* for aldehydes or an *-one* for ketones.

2. **Number the longest chain so the carbonyl has the lowest number.**

3. **Identify and name all substituents attached to the longest chain.**
4. **Place the names of the substituents alphabetically in front of the name of the longest chain.**
5. **For a ketone, add a number to indicate the position of the carbonyl group.**

   No number is necessary for aldehydes since the carbonyl group is always at carbon number one.

Some nomenclature examples are found in Figure 10-2.

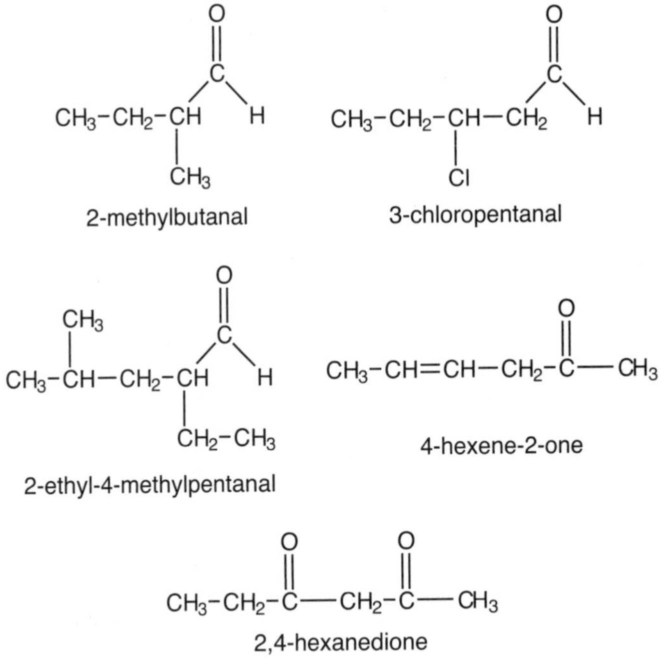

$CH_3-CH_2-CH(CH_3)-C(=O)H$

2-methylbutanal

$CH_3-CH_2-CH(Cl)-CH_2-C(=O)H$

3-chloropentanal

$CH_3-CH(CH_3)-CH_2-CH(CH_2-CH_3)-C(=O)H$

2-ethyl-4-methylpentanal

$CH_3-CH=CH-CH_2-C(=O)-CH_3$

4-hexene-2-one

$CH_3-CH_2-C(=O)-CH_2-C(=O)-CH_3$

2,4-hexanedione

Figure 10-2: The structure and nomenclature of some aldehydes and ketones.

As seen in most categories of organic compounds, the simpler members of the family also have common names. Some of the common names and systematic names are in Figure 10-3.

Formaldehyde (methanal)

Acetaldehyde (ethanal)

Benzaldehyde

CH_3-C-CH_3

Acetone (propanone)

Benzophenone

Figure 10-3: Common and systematic names of some simpler compounds.

Defining Physical Properties of Aldehydes and Ketones

The carbonyl group is a polar group (see Chapter 9 to review polarity). The polarity of the group leads to stronger intermolecular forces than in nonpolar substances such as the hydrocarbons. The dipole-dipole forces present are weaker than the hydrogen bonding present in alcohols. Therefore, the melting and boiling points of carbonyl compounds are higher than the hydrocarbons and lower than the alcohols. Aldehydes and ketones have similar melting and boiling points, but are above those of ethers.

When comparing melting points, boiling points, and solubility, you must use compounds of similar masses.

Aldehydes and ketones with six or fewer carbon atoms are soluble in water. The presence of the oxygen atom, which may be protonated, means they're dissolvable in concentrated sulfuric acid.

Creating Aldehydes and Ketones with Synthesis Reactions

Aldehydes and ketones can be synthesized in a number of ways. In most cases but not all, either an oxidation or a reduction is necessary. Oxidation may be a problem in the preparation of an aldehyde, because if care isn't taken, a carboxylic acid may be formed instead.

Oxidation reactions

The oxidation of a primary alcohol produces an aldehyde, while the oxidation of a secondary alcohol produces a ketone. (Tertiary alcohols don't undergo simple oxidation.) If you aren't careful, the oxidation of a primary alcohol may pass by the aldehyde to form a carboxylic acid. To isolate an aldehyde, either the oxidizing agent must be weak or you have to separate the aldehyde from the oxidizing agent before further oxidation can occur. One way to separate the aldehyde is to distill the aldehyde from the alcohol, which has a higher boiling point. An example of the oxidation of a primary alcohol is shown in Figure 10-4, and two examples of the oxidation of a secondary alcohol are shown in Figures 10-5 and 10-6.

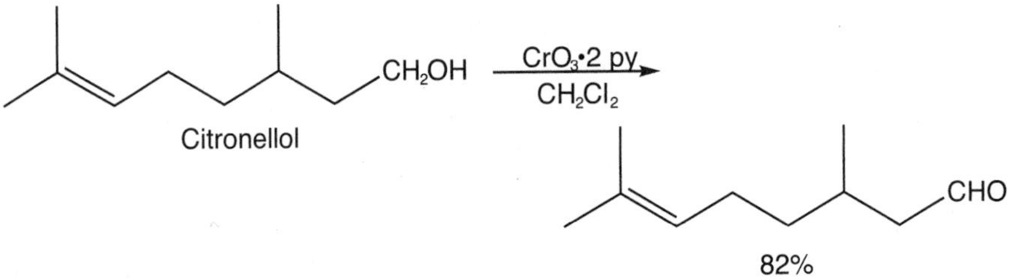

CH_2OH $\xrightarrow[CH_2Cl_2]{CrO_3\text{•}2\ py}$ CHO

Citronellol

82%

Figure 10-4: Forming an aldehyde through oxidation of a primary alcohol.

OH

$C(CH_3)_3$

$\xrightarrow[H_2O/HOAc/\Delta]{Na_2Cr_2O_7}$

O

$C(CH_3)_3$

3-tert-butylcyclohexanol

3-tert-butylhexanone

91%

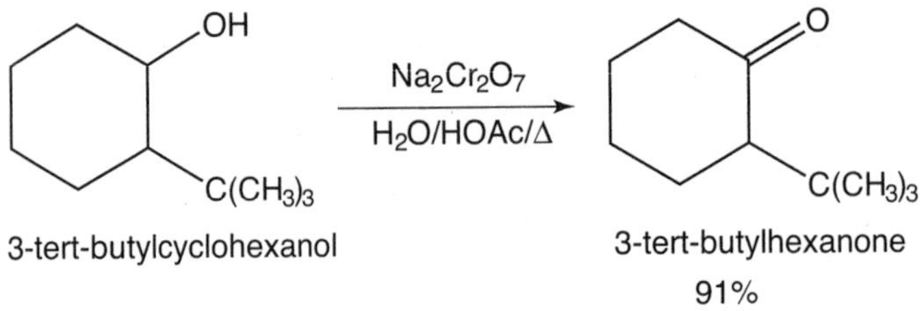

Figure 10-5: An example of ketone formation via oxidation of a secondary alcohol.

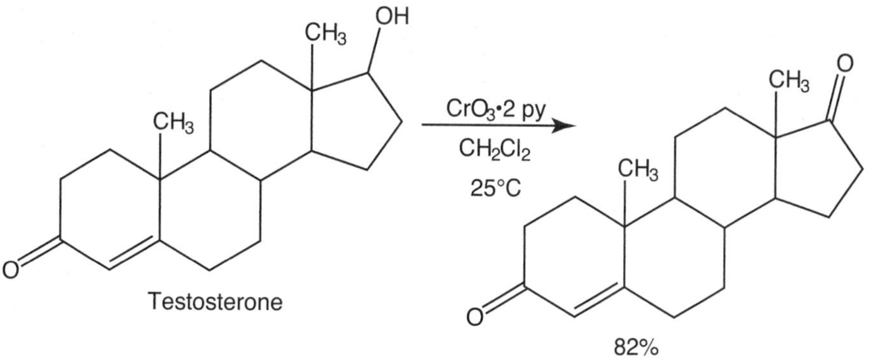

Figure 10-6: Another example of oxidation of a secondary alcohol to form a ketone.

To form ketones or aldehydes, a wide variety of oxidizing agents work, including air. The most common/useful oxidants contain chromium or manganese. Ozone, for ozonolysis, is also a useful oxidant to form aldehydes and ketones from alkenes.

Different organic chemistry instructors emphasize different oxidizing agents. Make sure you know the ones your instructor uses.

You cannot oxidize a tertiary alcohol other than by burning it.

The oxidation of an alkene with ozone followed by treatment with zinc in the presence of acid gives aldehydes and/or ketones. The reaction breaks the carbon-carbon double bond and changes each of carbon atoms of the C=C to a carbonyl group. Figure 10-7 shows an example of the ozonolysis of an alkene.

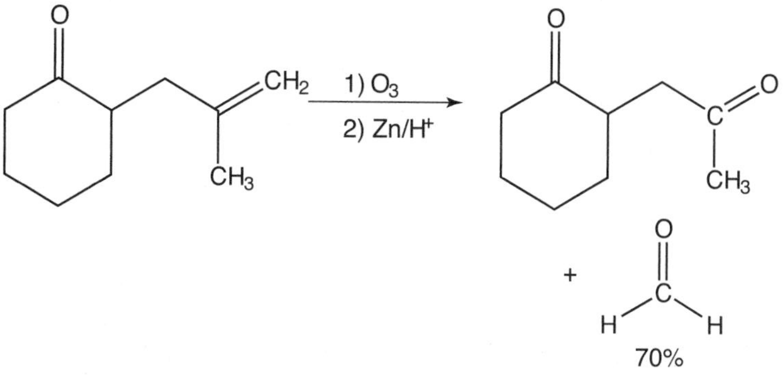

Figure 10-7: The ozonolysis of an alkene to form a ketone, and, in this case, an aldehyde.

On an exam you may be asked to determine the structural formula of the starting alkene given the ozonolysis products. A useful technique is to work backward from the products of ozonolysis . By cutting off the oxygens and then combining the two pieces, you get the starting alkene.

Reduction reactions

The stability of the carbonyl group limits the number of reagents that are sufficiently strong reducing agents to force a reduction. The hydride ion, $:H^-$, in various forms is a very strong reducing agent; in addition, it's a potential nucleophile and a strong base. However, the basic character of the hydride ion limits its usefulness as a nucleophile. For this reason, forming a complex with the hydride ion and boron or aluminum in order to minimize its basic character is helpful. Compounds such as sodium borohydride ($NaBH_4$) and lithium aluminum hydride ($LiAlH_4$) are examples of good reducing agents containing complexed hydride ion.

Lithium aluminum hydride, LAH, is a stronger reducing agent than sodium borohydride. By using some reducing agents, an aldehyde or a ketone can be reduced back to an alcohol, but in this section our emphasis is upon the reduction of a compound to form an aldehyde or a ketone.

An acid chloride can be reduced to form an aldehyde. If an acid chloride isn't available, an acid chloride can be formed from a carboxylic acid by refluxing the carboxylic acid with thionyl chloride ($SOCl_2$). Many reducing agents work, and a commonly used reducing agent is lithium tri-*tert*-butoxyaluminum hydride, $LiAlH[OC(CH_3)_3]_3$, at low temperature. The actual reducing agent is the hydride ion, $:H^-$. (This complexed hydride is a weaker reducing agent than either sodium borohydride or LAH.) Figure 10-8 shows an example of this reduction reaction. Shown parenthetically in this figure is the conversion of a carboxylic acid to an acid chloride by refluxing with thionyl chloride.

Many reactions are run at –78 degrees Celsius because this temperature can be reached by using dry ice to cool the reaction mixture.

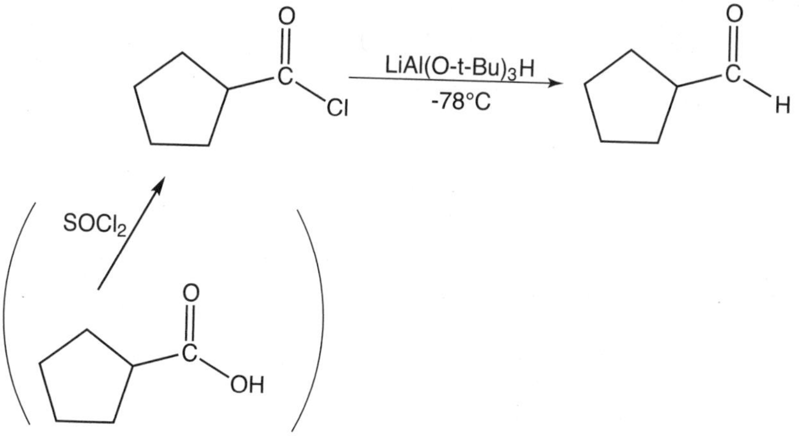

Figure 10-8: Reduction of an acid chloride to form an aldehyde, and conversion of carboxylic acid to acid chloride.

Other reactions

In addition to the methods previously described in this chapter, there are numerous other ways to make aldehydes and ketones, depending on the starting materials. These include using alkynes, doing a Friedel-Crafts acylation of an acid chloride and an aromatic compound, using organic nitriles, and the use of carboxylic acid. We examine each of these in the following sections.

Beginning with an alkyne

An alkyne can be converted to either an aldehyde or a ketone. To form an aldehyde, you begin with a terminal alkyne. Figures 10-9 and 10-11 show the formation of a ketone and an aldehyde from an alkyne. The reaction in Figure 10-11 always gives a ketone, while the reaction in Figure 10-9 gives an aldehyde from a terminal alkyne and a ketone from any other alkyne. The sia in Figure 10-10 is a siamyl group ($(CH_3)_2CH(CH_3)CH-$). We discuss enols and tautomerization in Chapter 11.

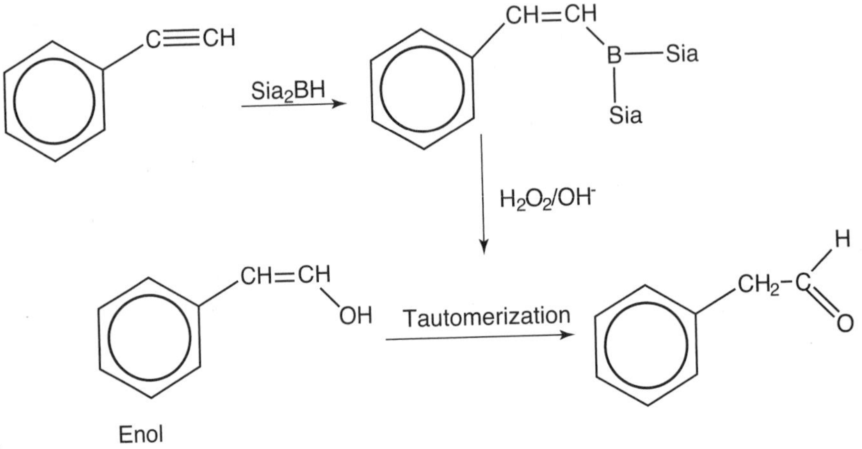
C≡CH
Sia2BH
CH=CH
B—Sia
Sia
H2O2/OH-
CH=CH
OH
Tautomerization
CH2-C
H
O
Enol

Figure 10-9: The conversion of an alkyne to an aldehyde (or ketone).

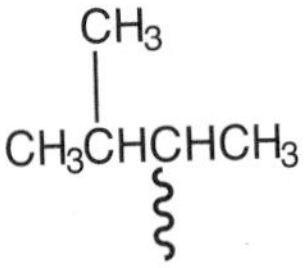
CH3
CH3CHCHCH3

Figure 10-10: The siamyl (sia) group.

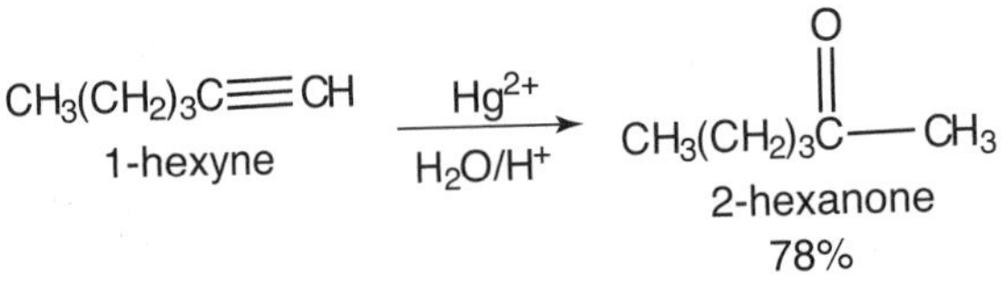
CH3(CH2)3C≡CH
1-hexyne
Hg2+
H2O/H+
O
CH3(CH2)3C—CH3
2-hexanone
78%

Figure 10-11: The conversion of an alkyne to a ketone.

Utilizing Friedel-Crafts acylation

A ketone can also be formed with a Friedel-Crafts acylation. The process requires an acid chloride and an aromatic compound. An aldehyde can't be formed by this procedure because the appropriate acid chloride, formyl chloride (HCOCl), is unstable and decomposes to carbon monoxide and hydrogen chloride. Figure 10-12 illustrates the preparation of acetophenone from benzene and acetyl chloride.

Creating ketones two ways with organic nitriles

Organic nitriles react with either Grignard reagents or organolithium compounds to form ketones. Organolithium compounds tend to be more reactive than the Grignard reagents, and, for this reason, are useful when reacting with "stubborn" nitriles. Figure 10-13 illustrates the conversion of a nitrile to a ketone with a Grignard reagent.

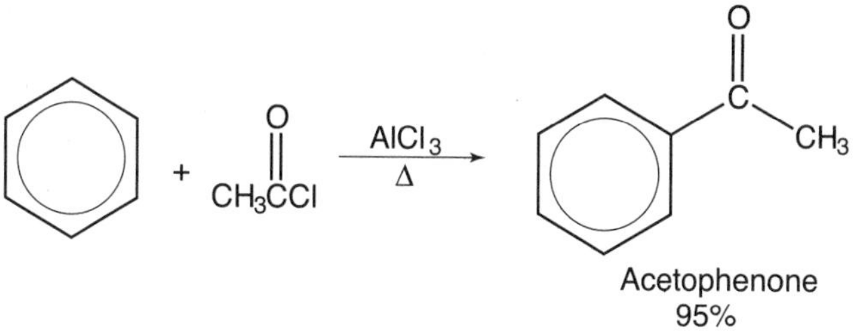

Figure 10-12: The preparation of acetophenone from benzene and acetyl chloride.

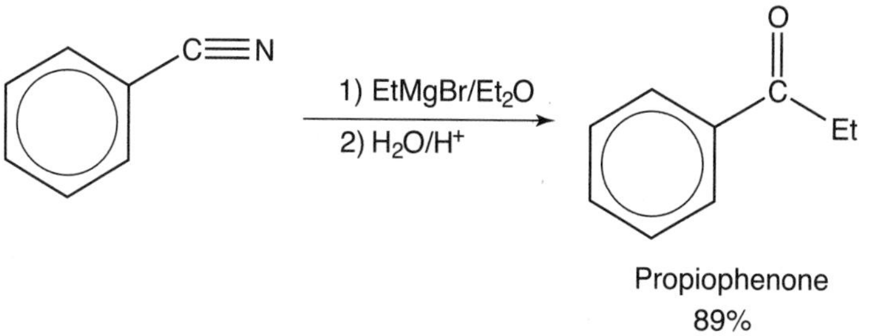

$$C_6H_6 + CH_3CCl(=O) \xrightarrow[\Delta]{AlCl_3} C_6H_5C(=O)CH_3$$

Acetophenone
95%

Figure 10-13: Using a Grignard reagent to convert a nitrile to a ketone.

$$C_6H_5C{\equiv}N \xrightarrow[\text{2) } H_2O/H^+]{\text{1) } EtMgBr/Et_2O} C_6H_5C(=O)Et$$

Propiophenone
89%

Figure 10-14 shows a partial mechanism for the conversion of a nitrile to a ketone by the reaction with a Grignard reagent.

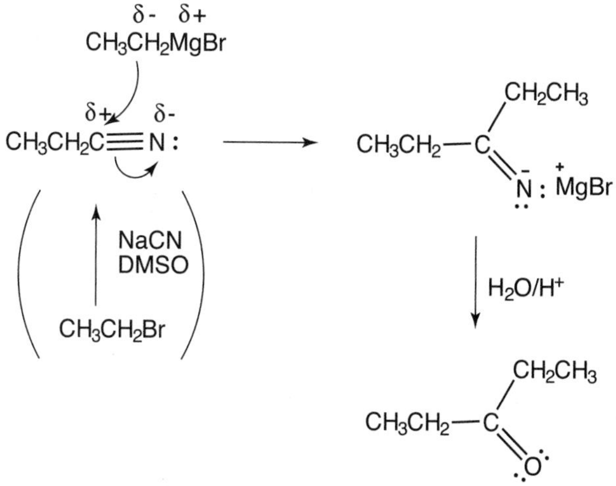

Figure 10-14: A partial mechanism for the conversion of an alkyl halide to a nitrile, which reacts to form a ketone with a Grignard reagent.

$$CH_3CH_2Br \xrightarrow{NaCN,\ DMSO} CH_3CH_2C{\equiv}N: \xrightarrow{\overset{\delta-\ \delta+}{CH_3CH_2MgBr}} CH_3CH_2{-}C(CH_2CH_3){=}\ddot{N}^{-}:\ \overset{+}{M}gBr \xrightarrow{H_2O/H^+} CH_3CH_2{-}C(CH_2CH_3){=}\ddot{O}:$$

Forming from carboxylic acid

Finally, a carboxylic acid can be converted to a ketone with an organolithium reagent. The reaction requires two moles of organolithium per mole of carboxylic acid, because one mole must react with the acid hydrogen and the second mole attacks the carbonyl group. Figure 10-15 illustrates the mechanism for this reaction.

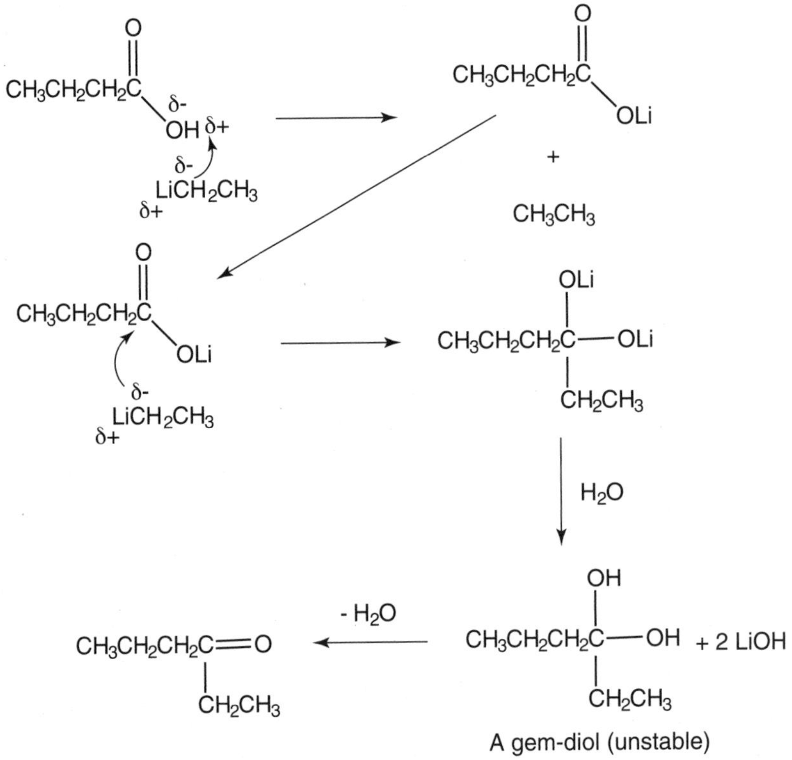

$CH_3CH_2CH_2C(=O)OH$ + $LiCH_2CH_3$ → $CH_3CH_2CH_2C(=O)OLi$ + CH_3CH_3

$CH_3CH_2CH_2C(=O)OLi$ + $LiCH_2CH_3$ → $CH_3CH_2CH_2C(OLi)_2CH_2CH_3$

$\xrightarrow{H_2O}$ $CH_3CH_2CH_2C(OH)_2CH_2CH_3$ + 2 LiOH (A gem-diol (unstable))

$\xrightarrow{-H_2O}$ $CH_3CH_2CH_2C(=O)CH_2CH_3$

Figure 10-15: The mechanism for the reaction of an organolithium compound with a carboxylic acid.

Taking Them a Step Further: Reactions of Aldehydes and Ketones

When considering the reaction of carbonyl groups, remember the polarity of the carbon-oxygen bond, the hybridization of the carbon atom *(sp^2)*, and the bond angles of the planar group. Figure 10-16 summarizes these features.

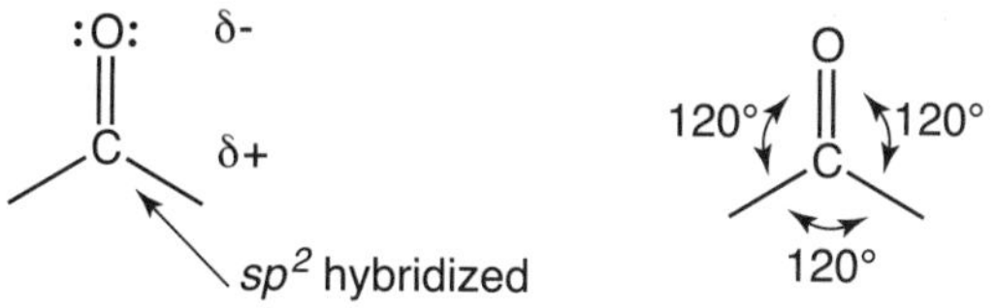

Figure 10-16: Important features to keep in mind when considering the reaction of carbonyl groups.

Strong reducing agents like sodium borohydride and lithium aluminum hydride are capable of reducing aldehydes to primary alcohols and ketones to secondary alcohols. The general reaction is the reverse of the reactions used to form aldehydes and ketones by the oxidation of primary and secondary alcohols, respectively (to review, see the earlier section "Oxidation reactions"). However, the mechanisms for reduction are different.

Nucleophilic attack of aldehydes and ketones

Figures 10-17 and 10-18 summarize the important reaction features of carbonyl chemistry. Remember that both of these processes are reversible, and note that the carbocation resonance structure is susceptible to nucleophilic attack.

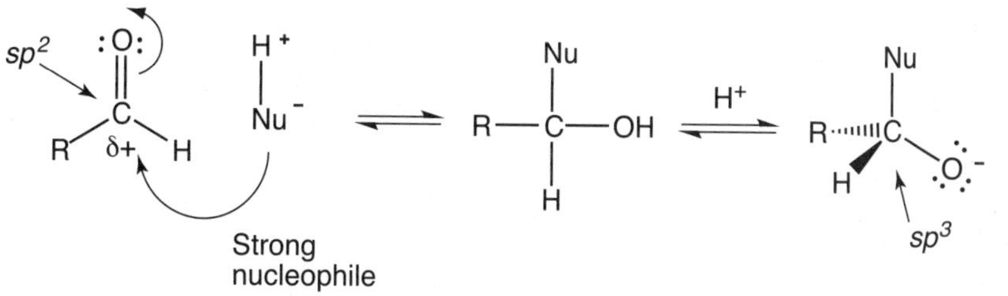

Figure 10-17: Nucleophilic attack upon a carbonyl group.

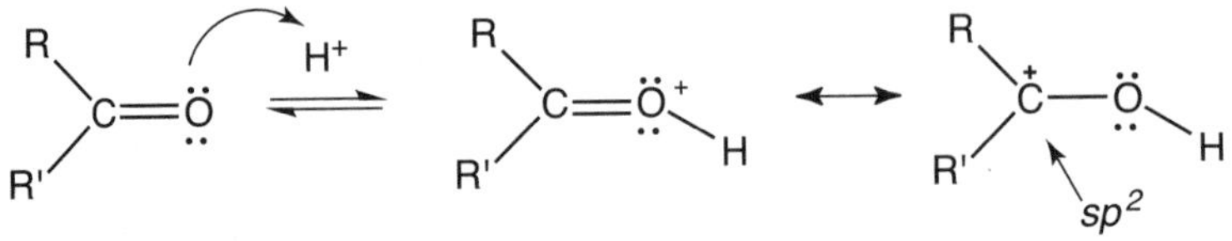

Figure 10-18: The behavior of a carbonyl group in acidic media.

Oxygen-containing nucleophiles

In the presence of acid, an alcohol may react with a carbonyl group to produce a hemiacetal or an acetal. This is an example of alcohol addition, and the process begins with the mechanism shown in Figure 10-18. Figure 10-19 shows the general structures of hemiacetals and acetals.

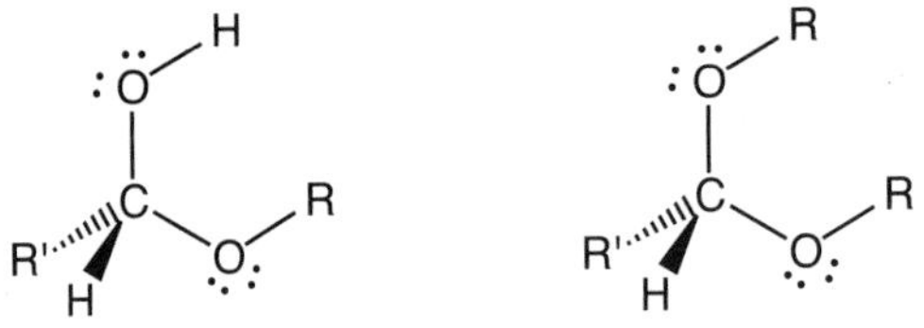

Figure 10-19: The general structures of hemiacetals and acetals.

Hemiacetal

Acetal

In general, hemiacetals are too unstable to isolate; however, they readily form when an aldehyde dissolves in an alcohol. Cyclic hemiacetals are more stable if the ring consists of five or six atoms. Many carbohydrates form hemiacetals or acetals through an internal reaction of an alcohol with the carbonyl group present in the same molecule. (Check out Chapter 16 for more on this topic.) Figure 10-20 illustrates the general mechanism for forming a hemiacetal and an acetal. Note that all steps in Figure 10-20 are reversible.

The mechanism in Figure 10-20 provides the foundation for many other mechanisms in Organic Chemistry II. Understanding this mechanism backwards and forwards is very useful to you.

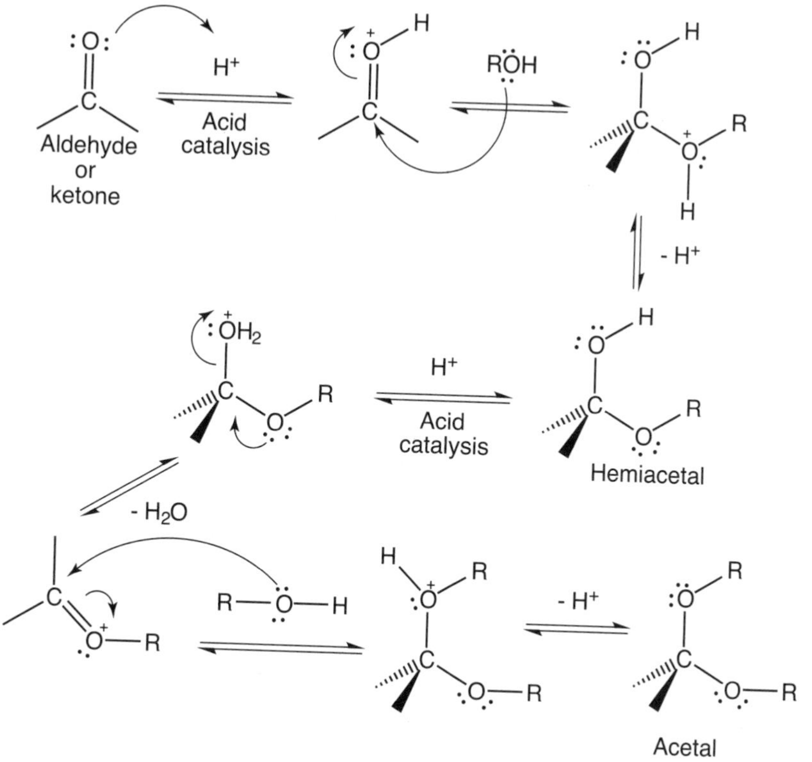

Figure 10-20: The general mechanism for forming a hemiacetal and an acetal.

If you look in older texts, you may still see the terms hemiketal and ketal. At one time, four terms were used for the products of alcohols with carbonyl groups: hemiacetal, acetal, hemiketal, and ketal. A hemiketal is now a type of hemiacetal and a ketal is now a type of acetal. Originally, acetals and hemiacetals came from aldehydes and ketals and hemiketals came from ketones. The structures of hemiacetals and acetals contained a C-H bond, but ketals and hemiketals did not.

In synthesis, an acetal can be used to protect a functional group. An example of protection is shown in Figure 10-21. In this case, the formation of the cyclic acetal protects the aldehyde group from oxidation by the permanganate. The cyclic acetal forms through the use of a 1,2-dihydroxy compound (a glycol). After the oxidation of the double bond, acidification of the product regenerates the aldehyde group.

The alcohol in hemiacetal/acetal formation can be replaced with a thiol (R-SH). In addition, a glycol can be replaced with a dithiol, and then you can follow a procedure similar to the one outlined in Figure 10-21. This procedure leads to an easy method for reducing a carbonyl (Figure 10-22). The reaction with a thiol is in the presence of the Lewis acid boron trifluoride, BF_3.

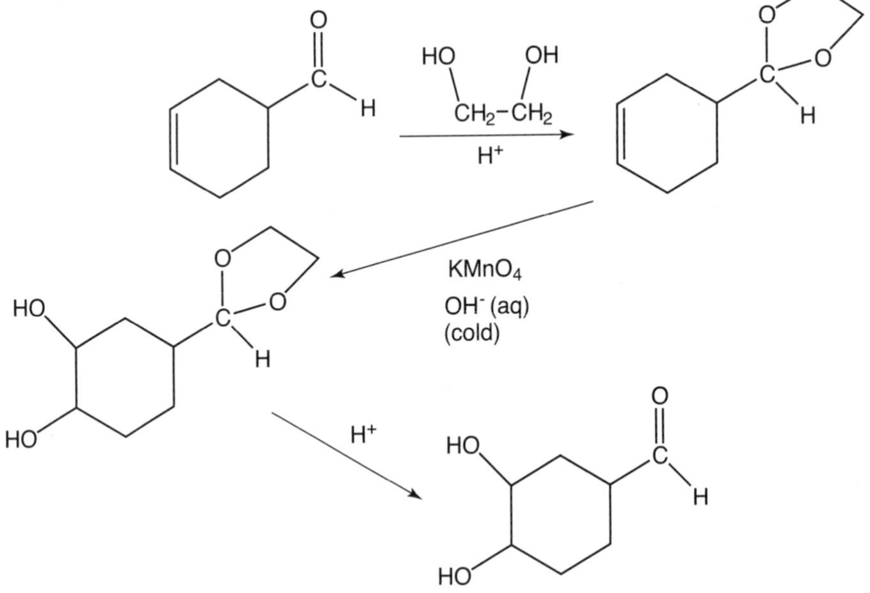

HO OH
CH_2-CH_2
H^+
$KMnO_4$
OH^- (aq)
(cold)
H^+

Figure 10-21: The use of an acetal to protect an aldehyde group from oxidation.

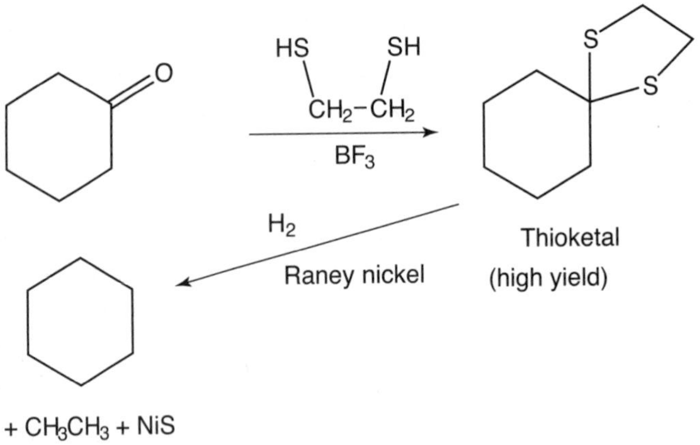

HS SH
CH_2-CH_2
BF_3
Thioketal
(high yield)
H_2
Raney nickel
$+ CH_3CH_3 + NiS$

Figure 10-22: A dithiol provides an easier method to reduce a carbonyl group.

Nitrogen-containing nucleophiles

Many different nitrogen-containing nucleophiles can attack a carbonyl group. In this section you consider primary amines (RNH_2), secondary amines (R_2NH), hydroxylamine (NH_2OH), and hydrazine (NH_2NH_2).

Primary amines add to aldehydes and ketones to form imines, $R\text{-}N{=}CR_2$. Secondary amines react to form enamines, $R'_2C{=}CR(NR_2)$. Hydroxylamine reacts to form an oxime, $R_2C{=}NOH$. Hydrazine reacts to form a hydrazone, $R_2C{=}NNH_2$.

Primary amines ($R\text{-}NH_2$)

Figure 10-23 shows the reaction of a primary amine with an aldehyde, the product of which is an imine. With the exception of the last step, the mechanism for the formation of an imine is similar to the mechanism for the formation of a hemiacetal/acetal. In the formation of an acetal, the final step involves the positive charge on the protonated hemiacetal attacking a second alcohol to form an acetal. However, in the corresponding step in the formation of an imine, the loss of a hydrogen ion is easier than a second attack.

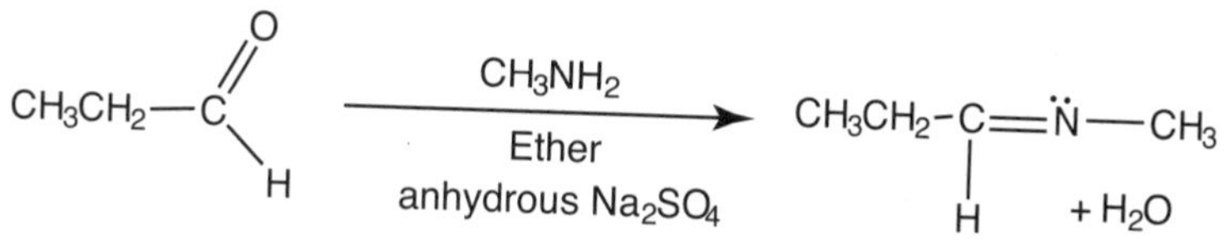

$$CH_3CH_2\text{—}C(=O)H \xrightarrow[\text{Ether, anhydrous } Na_2SO_4]{CH_3NH_2} CH_3CH_2\text{-}C(H){=}\ddot{N}\text{—}CH_3 + H_2O$$

Figure 10-23: The reaction of methylamine with propanal to form an imine.

The C=N bond can exist in both E and Z isomers. This isn't important for symmetrical ketones, but it is important for aldehydes and unsymmetrical ketones. The same is also true for the oximes in the later section on hydroxylamines.

Secondary amines (R_2NH)

Figure 10-24 shows the reaction of a secondary amine with a ketone to form an enamine. With the exception of the last step, the mechanism for the formation of an enamine is similar to both the mechanism for the formation of a hemiacetal/acetal and the mechanism for the formation of an imine. In this case, because losing a hydrogen ion from a carbon adjacent to the C=N carbon occurs more easily than a second attack, an enamine forms. The hydrogen loss is on the side that yields the more stable alkene.

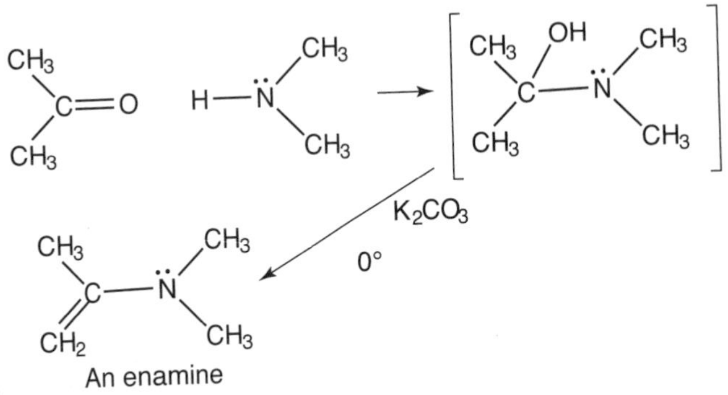

Figure 10-24: The reaction of dimethylamine with propanone (acetone).

The more stable alkene is, the more substituted alkene.

Hydroxylamine (NH_2OH)

Figure 10-25 shows the reaction of hydroxylamine with a carbonyl to form an oxime. Because pure hydroxylamine isn't very stable, in this reaction the source of the hydroxylamine is hydroxylamine hydrochloride, NH_2OH•HCl. The mechanism of this reaction is very similar to the mechanism leading to the formation of an imine.

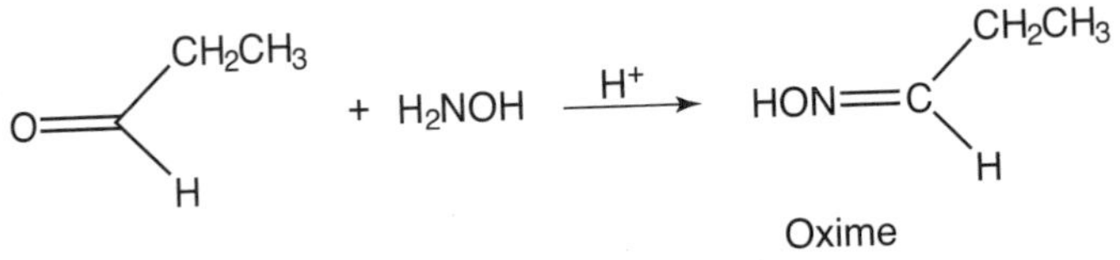

Figure 10-25: The formation of an oxime by the reaction of hydroxylamine with propanal.

Hydrazine (NH_2NH_2)

The formation of a hydrazone is illustrated in Figure 10-26. Hydrazones form when a carbonyl reacts with hydrazine. The mechanism for the formation of a hydrazone is also similar to the formation of an imine and an oxime. Again, the presence of the C=N means that both E and Z isomers can form.

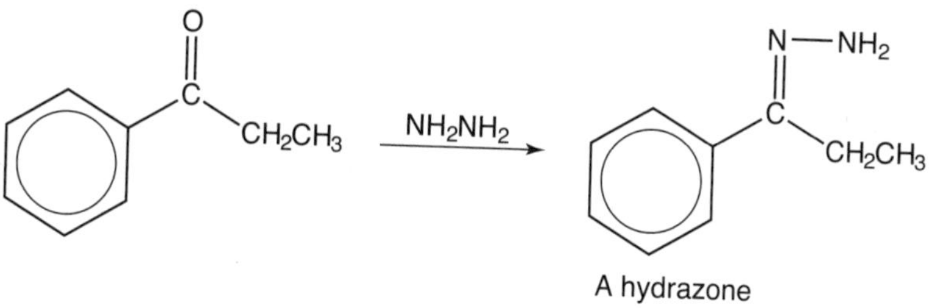

Figure 10-26: The formation of a hydrazone by the reaction of a ketone with hydrazine.

At one time, hydrozones were important in the identification of ketones. However, with advances in spectroscopic methods, using NMR data to identify the compound is easier.

Hydrazones are important intermediates in a Wolff-Kishner reduction, a procedure for reducing a carbonyl group. An example of a Wolff-Kishner reduction appears in Figure 10-27, and the mechanism is in Figure 10-28.

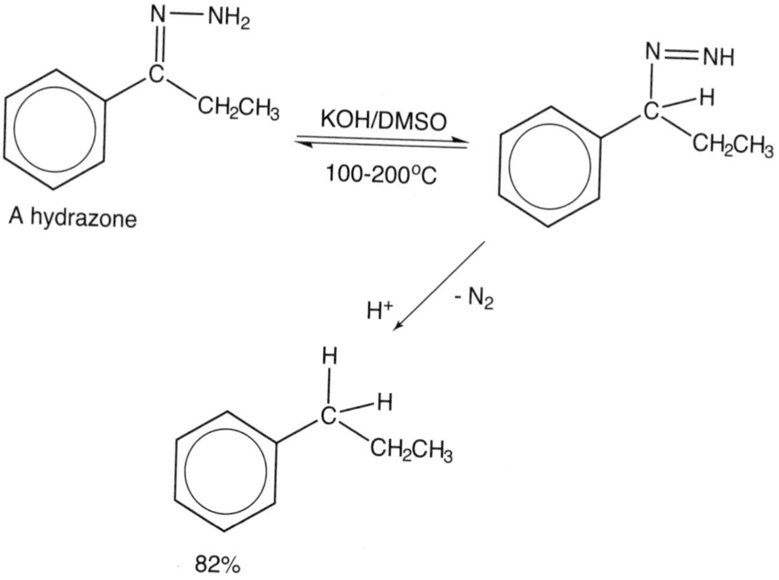

Figure 10-27: The Wolff-Kishner reduction of a hydrazone.

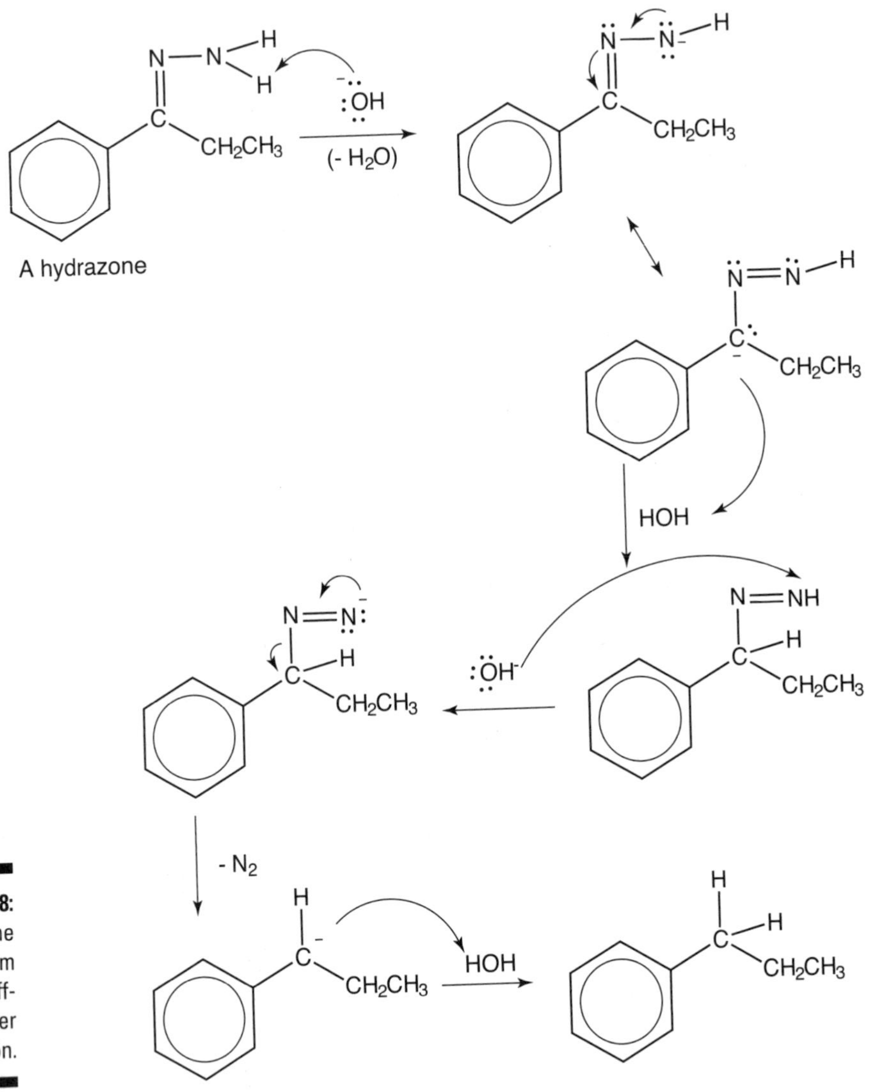

Figure 10-28: The mechanism of a Wolff-Kishner reduction.

The formation of the carbanion near the end of the mechanism would seem to make the yield low because forming a carbanion is difficult. However, the loss of the very stable nitrogen molecule, N_2, promotes the reaction.

Carbon-containing nucleophiles

Ylides (pronounced *il*-ids) are important compounds containing a negative carbon atom adjacent to a positive heteroatom. The two important types of ylides are those that contain phosphorus and those that contain sulfur.

The negative carbon atom serves as the nucleophile. Phosphorus ylides are important to the Wittig reaction, which converts a carbonyl to an alkene.

The Wittig reaction begins with the preparation of a phosphorus ylide (a Wittig reagent). Figure 10-29 illustrates to formation of a typical phosphorus ylide by an S_N2 mechanism, in this case, methylenetriphenylphosphine. The reagent is a hybrid of the two resonance forms illustrated in the figure. The mechanism in Figure 10-30 shows how the ylide formed in Figure 10-29 attacks a carbonyl.

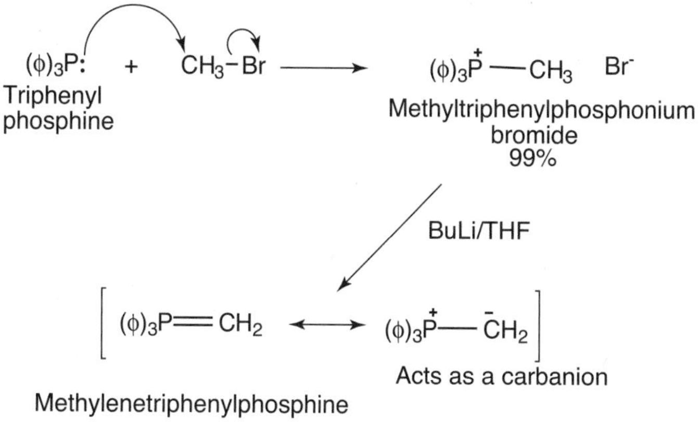

$(\phi)_3P:$ + CH_3-Br → $(\phi)_3\overset{+}{P}-CH_3$ Br^-

Triphenyl phosphine

Methyltriphenylphosphonium bromide 99%

BuLi/THF

$[(\phi)_3P{=}CH_2 \longleftrightarrow (\phi)_3\overset{+}{P}-\overset{-}{C}H_2]$

Methylenetriphenylphosphine

Acts as a carbanion

Figure 10-29: The formation of a phosphorus ylide.

$(\phi)_3\overset{+}{P}-\overset{-}{C}H_2$

O

$(\phi)_3$

P

CH_2

O

CH_2

$O{=}P(\phi)_3$ +

Triphenylphosphine oxide

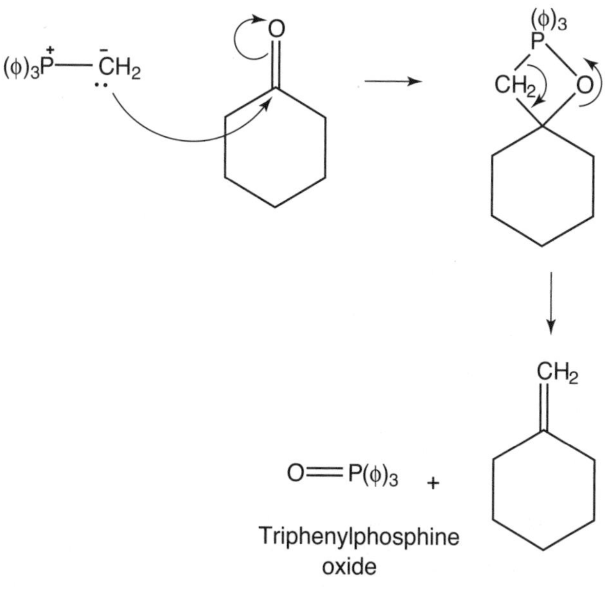

Figure 10-30: The mechanism of methylenetriphenylphosphine attacking a carbonyl group.

Sulfur ylides behave similarly to phosphorus ylides, but the final products are different. Figure 10-31 shows the mechanism for the preparation of a sulfur ylide and the reaction of the sulfur ylide with a carbonyl group. Notice that the mechanism for the formation of the sulfur ylide is similar to the formation of a phosphorus ylide. However, the last step in the sulfur ylide mechanism is an internal S_N2 reaction, which eliminates the original thioether (dimethyl sulfide). The reaction of a sulfur ylide with a ketone yields epoxides, whereas the product of a phosphorus ylide with a ketone is an alkene.

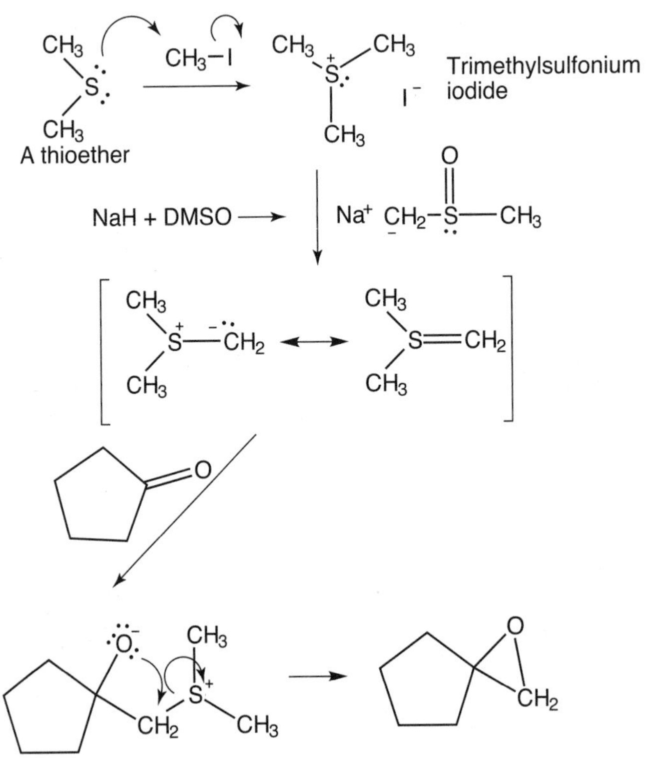

CH_3 CH_3–I CH_3 CH_3 Trimethylsulfonium iodide I^- CH_3 A thioether CH_3 NaH + DMSO → Na^+ CH_2–S–CH_3 CH_3 CH_2 CH_3 S=CH_2 CH_3 CH_3 O O CH_3 CH_2 CH_3 O CH_2

Figure 10-31: The mechanism for the formation of a sulfur ylide and attack by the sulfur ylide on a carbonyl group.

Oxidation of aldehydes and ketones

Aldehydes easily oxidize to carboxylic acids or to carboxylates. In fact, preventing the oxidation of an aldehyde is difficult. Ketones oxidize with difficulty, since a change in the backbone must first take place.

Aldehydes

Aldehydes are easy to oxidize: They slowly oxidize in the presence of air; thus, in the laboratory, many old open bottles of aldehydes are acidic. In practice, the oxidation of an aldehyde may employ several reagents. Figure 10-32 shows one common reagent mixture (the Jones reagent — CrO_3/H_2SO_4/acetone). Another common procedure is a two-step reaction where basic potassium permanganate oxidizes the aldehyde to a carboxylate ion and the second step involves the acidification of the product to form the carboxylic acid.

$$CH_3(CH_2)_4CHO \xrightarrow[\text{Acetone, 0°C}]{CrO_3/H_2SO_4} CH_3(CH_2)_4COOH \quad 85\%$$

Figure 10-32: The oxidation of an aldehyde to a carboxylic acid.

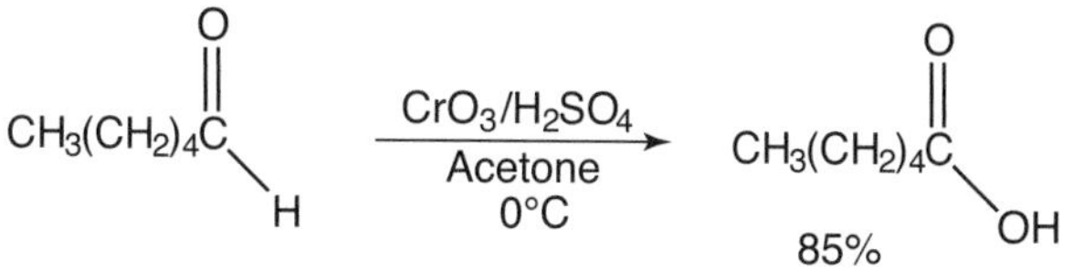

For years, Tollen's Reagent ($Ag^+(NH_3)_2OH^-$) was used in the identification of aldehydes. Aldehydes reacted with Tollen's reagent to deposit silver metal on the walls of the reaction vessel, forming a mirror. An example of a positive Tollen's test is in Figure 10-33.

Figure 10-33: The reaction of an aldehyde with Tollen's reagent.

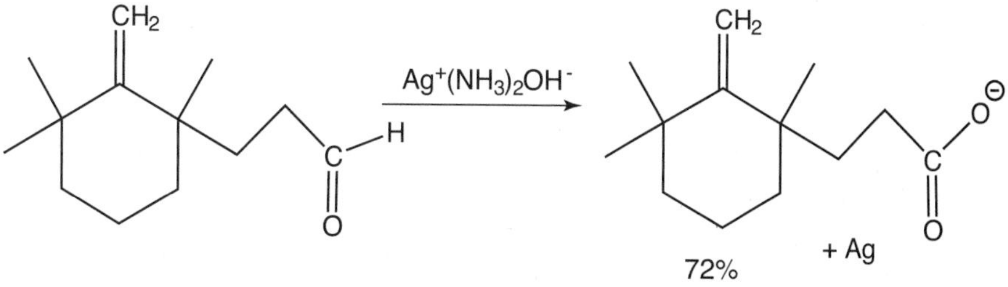

Ketones

Aldehydes are easy to oxidize, but ketones are more challenging. The two important oxidation reactions of ketones are the oxidation with a strong oxidant and the iodoform test.

Strong oxidants such a hot basic potassium permanganate oxidize a ketone with an alteration of the carbon backbone. Figure 10-34 illustrates the oxidation of a ketone with permanganate followed by acidification to produce a carboxylic acid. The oxidation cleaves the carbon-carbon on one side of the carbonyl group.

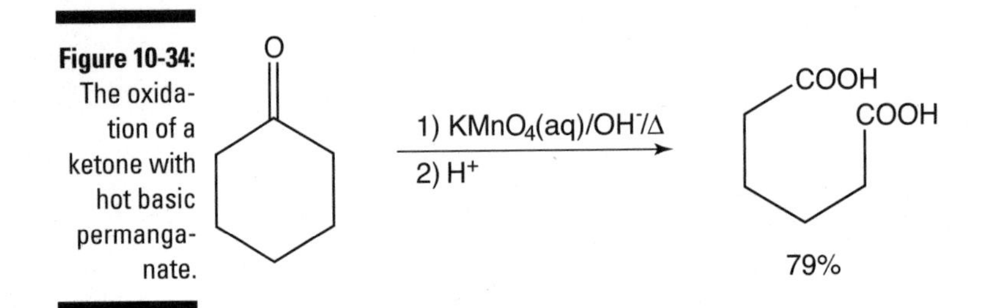

Figure 10-34: The oxidation of a ketone with hot basic permanganate.

For years the iodoform test was a laboratory method for the identification of a methyl ketone (a ketone where one of the R groups is a methyl group). A positive test produced the compound iodoform. Iodoform, CHI_3, is a yellow precipitate with a characteristic odor. The oxidation utilizes sodium hypoiodite, which is generated *in situ* by the reaction of iodine with sodium hydroxide. Figure 10-35 shows an example of the iodoform test.

NaOH + I_2 → NaOI

R–C(=O)–CH_3 → R–C(=O)–O^- + CHI_3

Figure 10-35: The iodoform test.

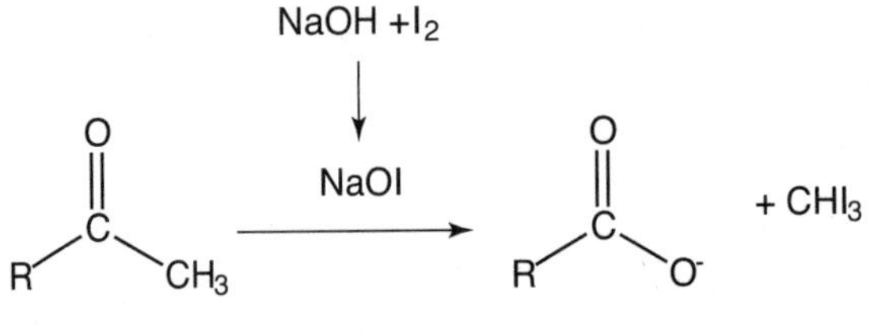

The Baeyer-Villiger reaction

The Baeyer-Villiger reaction uses a peroxyacid, or peracid (RCO_3H), which is an oxidizing agent and a nucleophile. An example of a peracid is mCPBA (meta-chloroperbenzoic acid), shown in Figure 10-36. The peroxyacid inserts an oxygen atom next to the carbonyl to form a carboxylic acid (from an aldehyde) or an ester (from a ketone). Both aldehydes and ketones are susceptible to a Baeyer-Villiger reaction — see Figure 10-37.

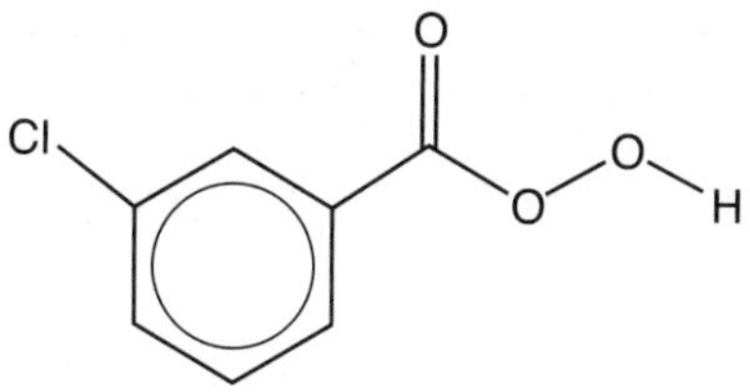

Figure 10-36: The structure of mCPBA.

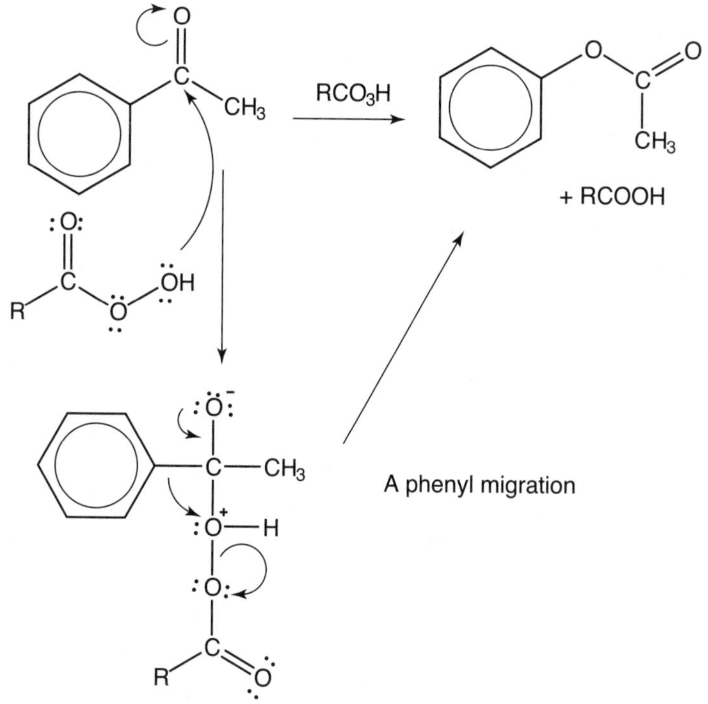

Figure 10-37: The mechanism of a Baeyer-Villiger reaction.

During the reaction, a group migrates. The mechanism is unusual in that the migration is effectively the movement of a carbanion. An aldehyde or an unsymmetrical ketone has two possible (different) groups that could migrate, but the group that actually migrates is the one with the higher migratory aptitude. The relative ranking of migratory aptitude is: H > phenyl > 3° > 2° > 1° > Me.

Checking Out Spectroscopy Specs

In this section we take a quick look at the characteristic spectra of aldehydes and ketones.

The following list describes the characteristic absorption of aldehydes and ketones in the infrared region:

- Both aldehydes and ketones exhibit a very strong characteristic absorption at $\approx$1,700 cm^{-1}.
- Conjugation shifts the band to lower frequency.
- The hydrogen attached to the carbonyl group of aldehydes gives two bands at 2,900–2,820 and 2,775–2,700 cm^{-1}.
- A 2,720 cm^{-1} shoulder indicates a saturated aldehyde.

The NMR spectra of aldehydes and ketones contain the following characteristics:

- The proton NMR of the aldehyde hydrogen appears in the region of δ = 9–10. This high shift is due to the inductive effect (electronegative oxygen).
- In ^{13}C-NMR, the carbonyl carbon of aldehydes appears in the δ = 190–200 region and for ketones in the δ = 200–220 region.

The following bands appear in the UV-vis spectra of aldehydes and ketones:

- A weak band for aldehydes and ketones is between 270–300 nanometers (nm).
- In both cases, conjugation shifts the band to 300–350 nm.

Aldehydes and ketones commonly exhibit the following characteristics in their mass spectra:

- Smaller aldehydes usually have a large m/e 29 peaks corresponding to the CHO^+ ion.
- Methyl ketones usually have a large m/e 43 peak corresponding to the CH_3CO^+ ion.
- Ketones usually fragment on either side of the carbonyl group and help in positioning the carbonyl group.

Chapter 11

Enols and Enolates

In This Chapter

- Exploring enols and enolates: Common intermediates
- Considering synthesis of enols and enolates
- Looking at how enols and enolates react

Aldehydes and ketones are the starting materials for a large number of reactions, and these reactions often involve intermediates known as enols and enolates. The mechanisms in this chapter are similar to each other and to mechanisms seen in the last chapter. Many organic chemistry students have trouble because they treat every mechanism as a member of a large group of independent entities and not different aspects of a small number of related entities. So as you go through this chapter, examine each of these mechanisms and focus on the fact that many of these mechanisms are different examples of the same mechanism.

Getting to Know Enols and Enolates

Before we look at enols and enolates, we need to examine some aspects of carbonyl groups not covered in Chapter 10.

A key feature of many carbonyl groups is that a hydrogen atom is attached to the α-carbon (the carbon atom next to the carbonyl group). The K_a, acid dissociation constant, for the α-hydrogen is about the same as for an alcohol (10^{-19} to 10^{-20} or pK_a = 19 to 20). The acidity of this hydrogen atom is partly because of the electron-withdrawing power of the oxygen and partly because of the resonance stabilization of the conjugate base resulting from the loss of H^+. The resonance stabilization is the more important factor. Figure 11-1 shows the result of hydrogen ion loss and the two resonance structures contributing to the resonance hybrid, which is the enolate ion.

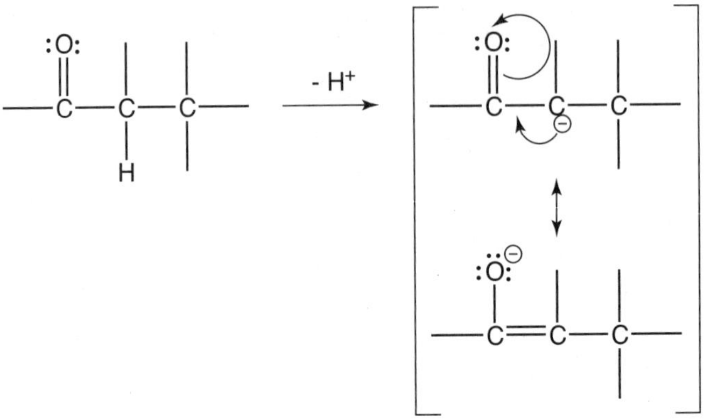

Figure 11-1: The acidity of the α-hydrogen atom and the resonance structures of the conjugate base, the enolate ion.

Enough already: Structure of enols and enolates

The enolate ion formed in Figure 11-1 is a conjugate base that reacts with a hydrogen ion. However, the hydrogen ion can attack in two possible places. Figure 11-2 shows what forms when a hydrogen ion attacks at each of these two sites. Attack at the α-carbon yields the keto form (the original carbonyl compound). Attack at the oxygen atom produces the enol form (an alcohol adjacent to a carbon-carbon double bond). The presence of a negative charge on the enolate ion means that it acts as a nucleophile. In general, the carbon acts as the nucleophile, but in special cases the oxygen may be the nucleophile.

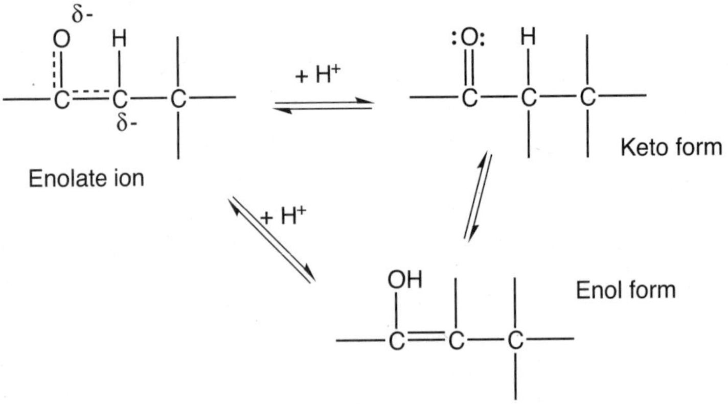

Enolate ion

Keto form

Enol form

Figure 11-2: The attack of an enolate ion by a hydrogen ion.

I thought I saw a tautomer

The keto and enol forms seen in Figure 11-2 are readily interconvertable isomers called *tautomers.* The interconversion of these two forms is tautomerization. Normally the keto form predominates because its carbon-oxygen double bond is more stable than the enol form's carbon-carbon double bond.

The small value for the K_a of the α-hydrogen means that only very strong bases can readily remove the hydrogen ion. However, in α-β dicarbonyl compounds, the enol form predominates because an internal hydrogen bond stabilizes the enol form. Figure 11-3 shows an α-β dicarbonyl compound. Because of the influence of the second carbonyl group, the α-hydrogen becomes more acidic than a simple carbonyl. A stable hydrogen-bonded and resonance-stabilized species can form, so acidity is enhanced (see Figure 11-4).

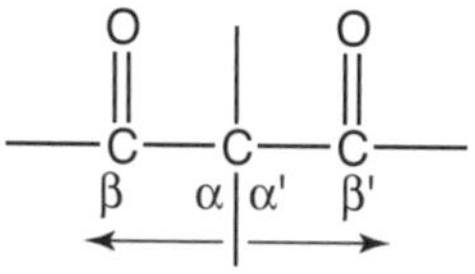

Figure 11-3: An α-β dicarbonyl compound.

The formation of five- and six-membered rings tends to increase the stability of a species, as shown in Figure 11-4.

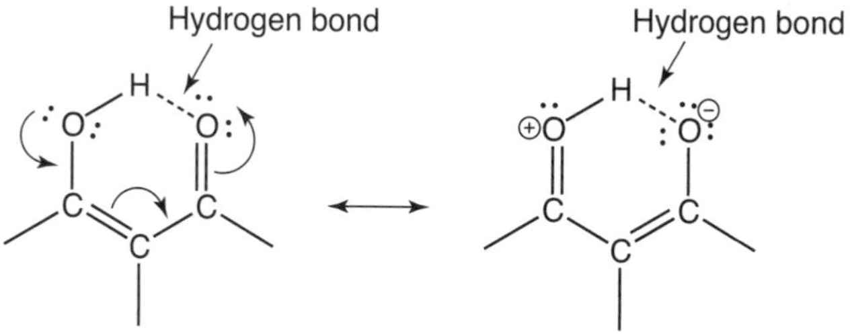

Figure 11-4: The stabilization of an α-β dicarbonyl compound through hydrogen bonding and resonance.

Tautomerization doesn't occur without an α-hydrogen. Most ketones do contain one or more α-hydrogen atoms, so they undergo tautomerization. These ketones exist in equilibrium with the enol form. In most cases, the ketone form predominates at equilibrium, but in a few cases the enol is particularly stable and it predominates.

Keto-enol-tautomerization is not resonance. The ketone and enol forms are different compounds that are in equilibrium.

Studying the Synthesis of Enols and Enolates

Tautomerization can be induced through the addition of an acid or base. (See the previous section for details on tautomerism.) We begin here by investigating the racemization (the formation of both enantiomers) of the compound shown in Figure 11-5, and we use this reaction to investigate both the acid and base mechanisms. The acid-catalyzed mechanism is shown in Figure 11-6 and the base-catalyzed mechanism is shown in Figure 11-7.

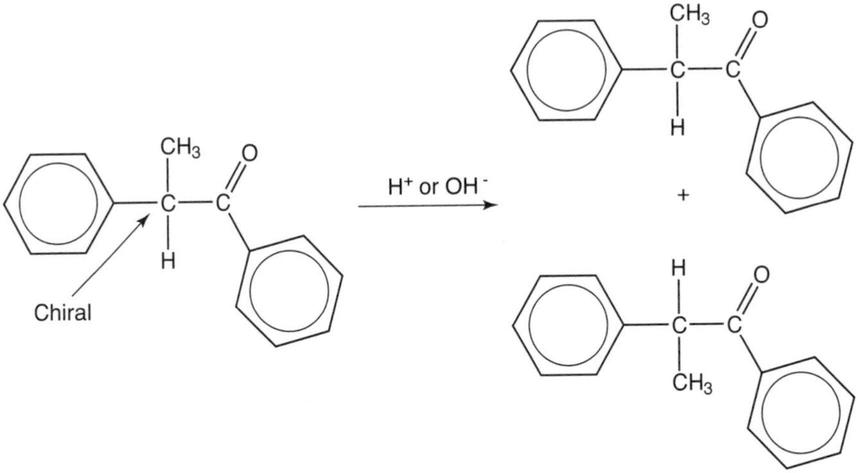

Figure 11-5: The racemization of a ketone through the action of acid or base.

In the acid-catalyzed mechanism (Figure 11-6), the protonation of the intermediate cation may produce either enantiomer because the hydrogen ion may attack the carbon from either side. ("One if by land, two if by sea!") Equal amounts of each enantiomer result and give a racemic mixture.

In the base-catalyzed racemization (Figure 11-7), the carbon is also vulnerable to attack from either side. Attack from one side gives one enantiomer, while attack from the other side gives the other enantiomer. (The attack results in deprotonation.) Since the probability for attack from either side is equal, a racemic mixture again results.

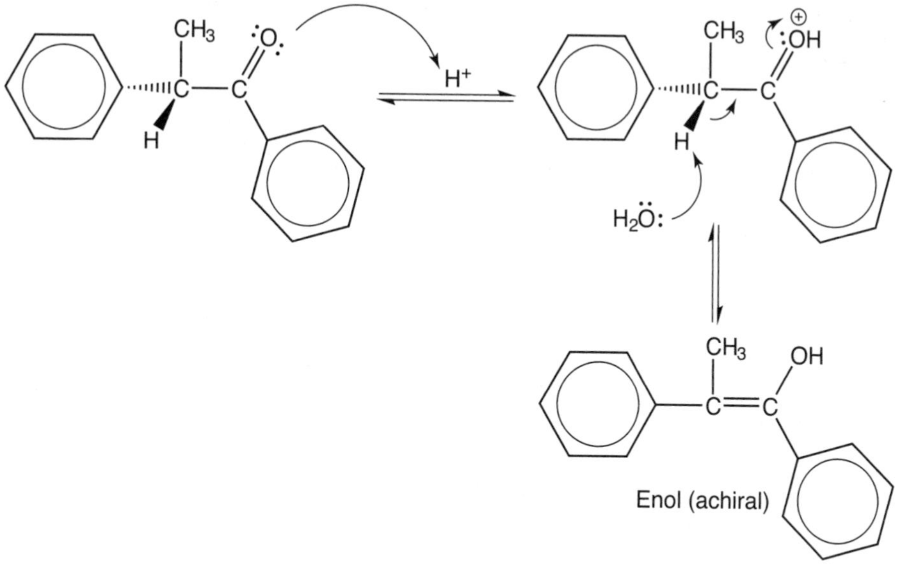

Enol (achiral)

Figure 11-6: The beginning of the mechanism for the acid-catalyzed racemization.

Enolate (achiral)

Enol (achiral)

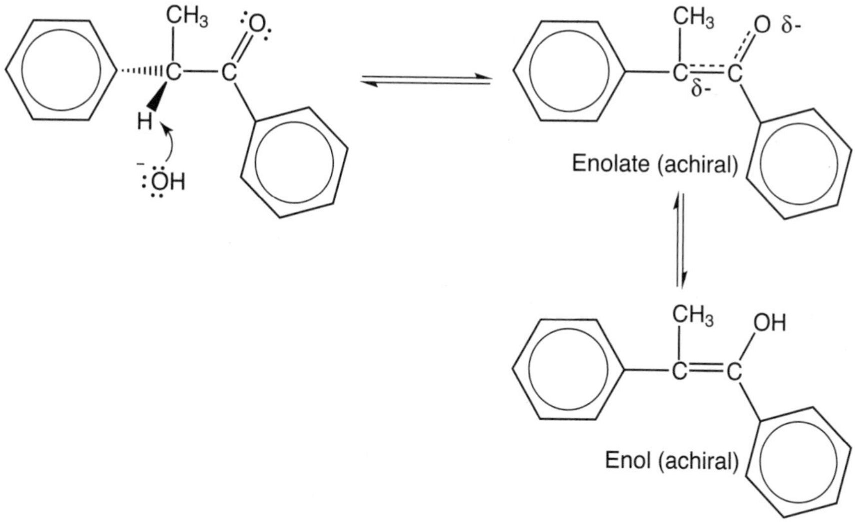

Figure 11-7: The beginning of the mechanism for the base catalyzed racemization.

In each of these mechanisms, the steps involve the gain or loss of a hydrogen ion. The enolate ion, shown in Figure 11-7, is an important nucleophile and is important to many other mechanisms. Under acidic conditions, the mechanism begins with the gain of a hydrogen ion, but under basic conditions, the mechanism begins with the loss of a hydrogen ion.

A mechanism can be under acidic *or* basic conditions; it can't be under acidic *and* basic conditions. That is to say, H^+ (acid) and OH^- (base) are never in the same mechanism.

To produce an enolate, the α-hydrogen ion must be removed. Strong bases, such as hydroxide (OH^-) and alkoxide (OR^-) ions, only form a small amount of enolate because these bases aren't sufficiently strong. However, forming an enolate is easier if you use one of these bases and an α-β dicarbonyl compound.

Thinking Through Reactions of Enols and Enolates

Enols and enolates undergo several types of reactions that are important in organic chemistry synthesis. In this section we take a look at the major reactions that these compounds undergo.

Haloform reactions

In Chapter 10, you see the iodoform test as a means of identifying methyl ketones. Here we reexamine this reaction in light of enols and enolates.

Haloform reactions involve the interaction of methyl ketones with alkaline halogen (X_2 where X = Cl, Br, or I) solution to give haloform (CHX_3) plus a carboxylate ion. The process begins with attack on the α-hydrogen of the methyl group followed by attack by a halogen. The addition of each halogen atom makes the remaining hydrogen atoms on the same carbon more acidic (due to electron withdrawal by the halogen). The process is useful in the synthesis of a carboxylate (carboxylic acid) with one less carbon atom.

Any compound that oxidizes to a methyl ketone also gives a haloform reaction, because halogens are also oxidizing agents. For example, the compound shown in Figure 11-8 reacts.

```
      |    H
      |    |
    —C—C—Me
      |    |
      |    OH
```

Figure 11-8: This compound undergoes a haloform reaction.

Figure 11-9 illustrates the mechanism for the haloform reaction. The mechanism involves a repeated series of base attacks (removal of an α-hydrogen) followed by the reaction with the halogen until all three α-hydrogen atoms are replaced. Then the base attacks the carbonyl carbon to induce the loss of a carbanion ($:CX_3^-$). The highly reactive carbanion quickly attacks and removes the hydrogen from the carboxylic acid group.

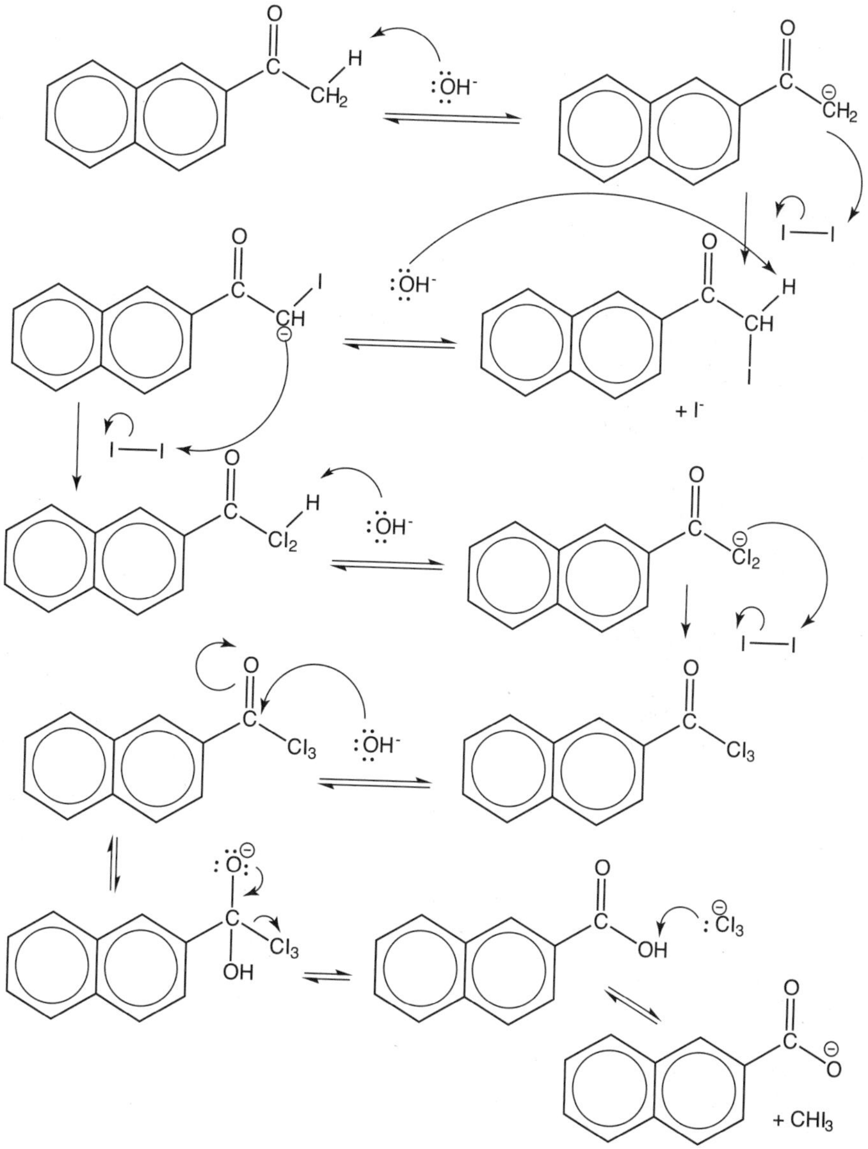

Figure 11-9: The mechanism of the haloform reaction.

The haloform reaction is a useful method of preparing a carboxylic acid (carboxylate ion) with one less carbon. It is one of the very few cases where carbanion loss occurs. It's only possible because the three halogen atoms are capable of stabilizing the negative charge.

A ketone can be halogenated even when it isn't a methyl ketone. This process can be either acid or base catalyzed. The general mechanism is shown in Figure 11-10.

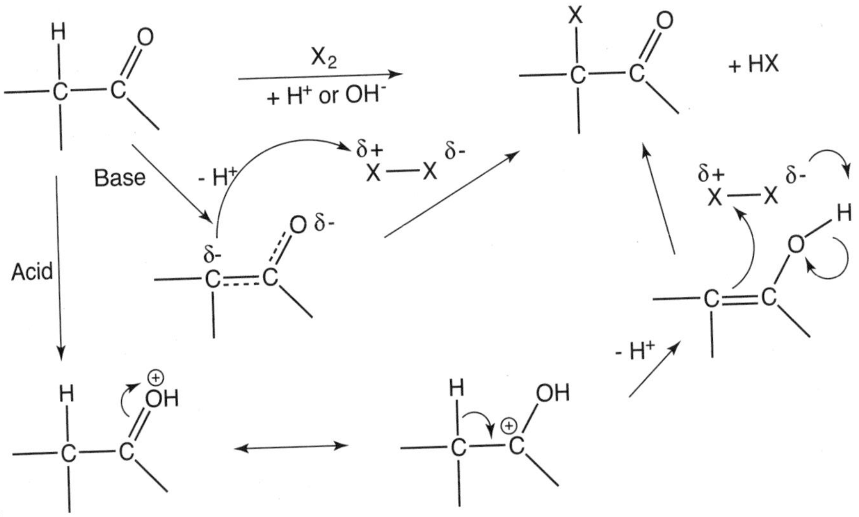

Figure 11-10: Mechanisms for the halogenation of a ketone.

Aldol reactions and condensations

The reaction of a carbonyl (aldehyde or ketone) with a base produces an enolate ion (a nucleophile). This nucleophile attacks any electrophile. What happens when you add a base to a carbonyl with no electrophile present? It turns out that a reaction still occurs because the carbonyl group itself is an electrophile. As the enolate forms, it can attack the carbonyl group of another aldehyde or ketone molecule. This is an aldol reaction or aldol condensation, also called an aldol addition.

Ketones are less reactive towards the nucleophile. In Organic Chemistry I, you saw that alkyl groups are electron donating. In ketones, the presence of the two alkyl groups attached to the carbonyl do a better job at compensating for the δ+ on the carbon atom than do one alkyl group and a hydrogen atom in an aldehyde. For this reason, aldehydes are more reactive than ketones.

In the aldol reaction, aldehydes or ketones with α-hydrogen atoms react in the presence of dilute base to give an aldol (a β-hydroxyaldehyde or ketone). An aldol contains an aldehyde (or ketone) and an alcohol group in the β-position. Figure 11-11 shows the general reaction. The aldol formed in this reaction is 2-methyl-3-hydroxypentanal, which forms with an 86 percent yield. (That's really good — when we were taking organic, we would have killed at times for a 5-percent yield!)

$$CH_3CH_2C(=O)H + CH_3CH_2C(=O)H \xrightarrow{OH^-} CH_3CH_2CH(OH)CH(CH_3)C(=O)H$$

Figure 11-11: An example of an aldol reaction.

The product of an aldol reaction of RR'CO is $(RR'CO)_2$. R' is an alkyl group, and R may be an H (aldehyde) or an alkyl (ketone).

Figure 11-12 shows the mechanism for the aldol reaction. Notice that in the last step the hydroxide ion formed is a weaker base than the alkoxide ion.

Forming a stronger base from a weaker base is very highly unlikely.

The aldol formed by the aldol reaction, especially if heated, can react further. The heating causes dehydration (loss of H_2O), and the overall reaction involving an aldol reaction followed by dehydration is the aldol condensation. The product of an aldol condensation, favored by the presence of extended conjugation, is an α,β-unsaturated aldehyde (an enal) or ketone. The mechanism for dehydration (Figure 11-13) begins where the mechanism of the aldol reaction (Figure 11-12) ends. This process works better if extended conjugation results. The aldol reaction and condensation are reversible.

The product of an aldol condensation of RR'CO is $(RR'CO)_2 - H_2O$.

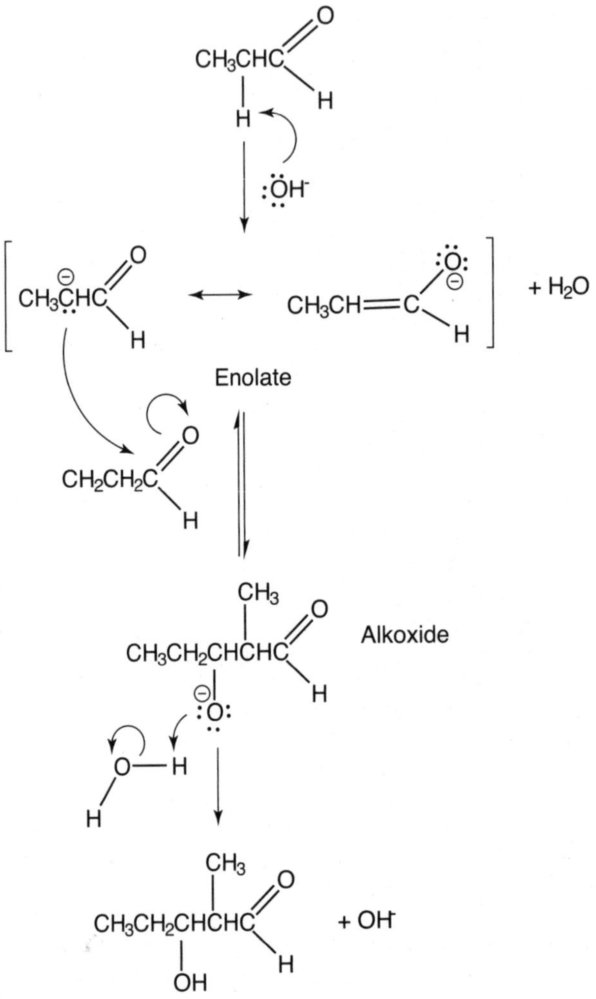

Enolate

Alkoxide

$CH_3CHC(=O)H$ + $:\ddot{O}H^-$ → [$CH_3\ddot{C}^{\ominus}HC(=O)H$ ↔ $CH_3CH{=}C(O^{\ominus})H$] + H_2O

Enolate + $CH_2CH_2C(=O)H$ ⇌ $CH_3CH_2CH(O^{\ominus})CH(CH_3)C(=O)H$ (Alkoxide)

Alkoxide + H_2O → $CH_3CH_2CH(OH)CH(CH_3)C(=O)H$ + OH^-

Figure 11-12: The mechanism of the aldol reaction.

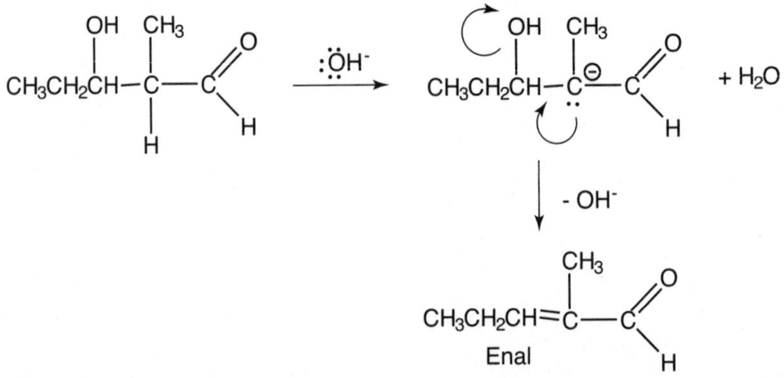

$CH_3CH_2CH(OH)-CH(CH_3)-C(=O)H$ $\xrightarrow{:\ddot{O}H^-}$ $CH_3CH_2CH(OH)-\ddot{C}^{\ominus}(CH_3)-C(=O)H$ + H_2O

$\xrightarrow{-OH^-}$ $CH_3CH_2CH{=}C(CH_3)-C(=O)H$

Enal

Figure 11-13: The mechanism for the dehydration of the product of an aldol reaction.

Crossed aldol condensations

An aldol reaction/condensation occurs when the enolate ion from an aldehyde or ketone attacks a molecule of the parent compound. If, however, two different carbonyl compounds are present, a crossed aldol reaction/condensation occurs.

Crossed aldol condensation reactions may present a problem because multiple products can be formed. For example, a mixture of carbonyl A and carbonyl B can give two different enolates, each of which can then attack either an A or a B molecule. Therefore, four products are possible (A_2, A + B, B + A, and B_2).

The formation of multiple products limits the practicality of crossed aldol condensations. Following are the two ways to increase the practicality of a crossed aldol condensation:

- One way is to begin with one aldehyde (A) and slowly add the other aldehyde (B). This decreases the chances of two of the products, B_2 and B + A, from forming (leaving A_2 and A + B).
- The other method is to choose a compound that doesn't have any α-hydrogen atoms, because a compound with no α-hydrogen cannot form an enolate. For example, if B has no α-hydrogen, then only two products are possible (A_2 and A + B). Examples of aldehydes with no α-hydrogen atoms are formaldehyde and benzaldehyde.

Aldol cyclization

A molecule that contains two carbonyl groups may undergo an internal aldol condensation. Ideally, either two or three carbon atoms should be between the carbonyl groups. (The carbonyl groups need to be at positions one and four or one and five relative to each other.) In these cases, a five- or six-membered ring forms, both of which are stable and help facilitate the reaction. (A seven-membered ring can be formed this way, but other sizes don't form easily.) Figures 11-14 and 11-15 show examples of aldol cyclization reactions. These reactions lead to the formation of a five- and six-membered ring, respectively. The mechanism for the preparation of jasmone (Figure 11-16) illustrates the general mechanism for this process.

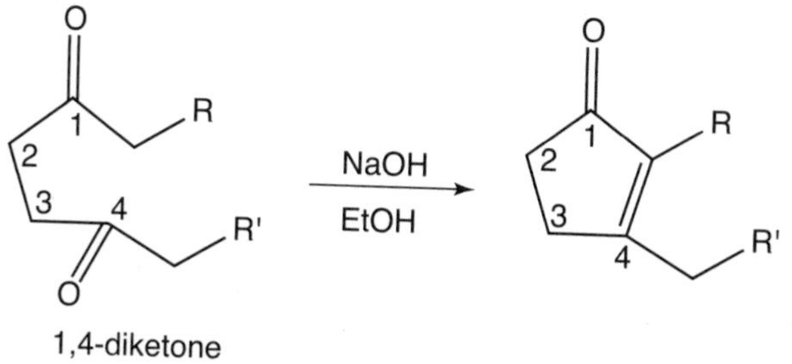

Figure 11-14: An aldol-cyclization reaction leading to the formation of a five-membered ring.

O
1
R
2
3
4
R'
O
1,4-diketone
NaOH
EtOH
O
1
R
2
3
4
R'

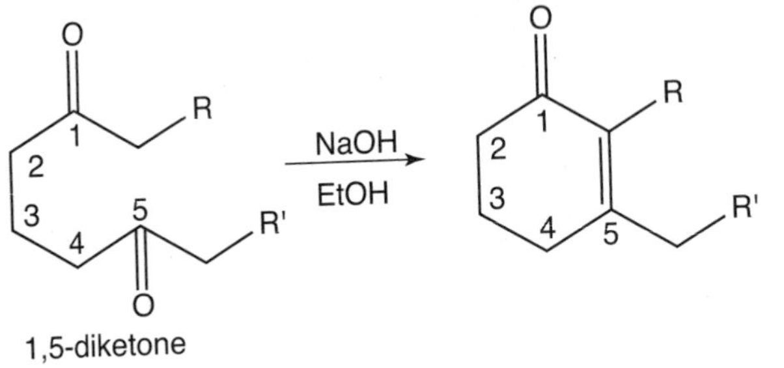

Figure 11-15: An aldol-cyclization reaction leading to the formation of a six-membered ring.

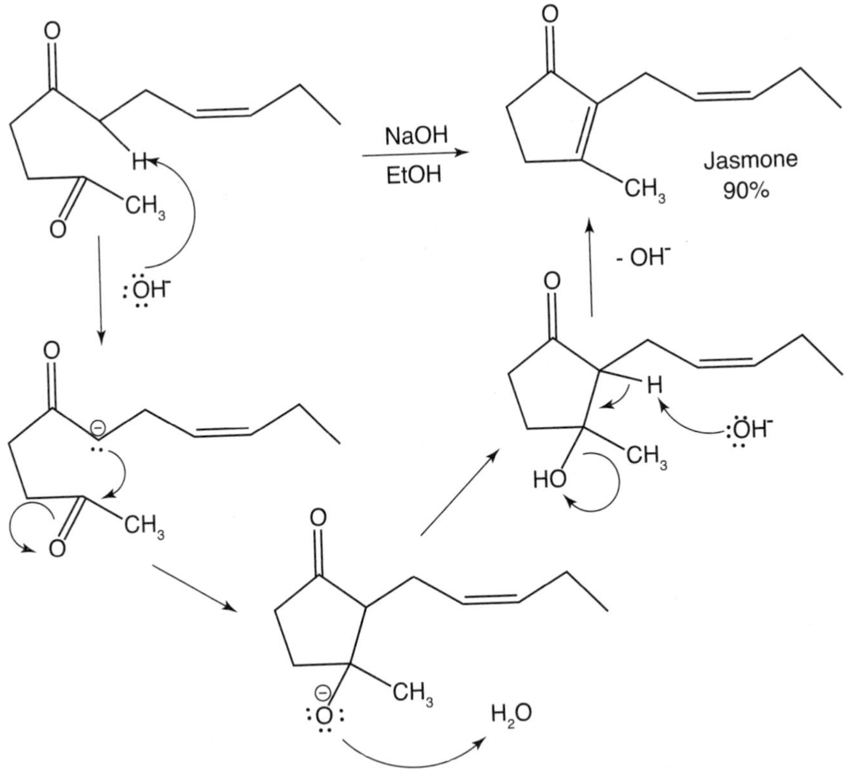

Figure 11-16: The mechanism for the formation of jasmone.

Addition reactions to unsaturated aldehydes and ketones

The α,β-unsaturated aldehydes or ketones prepared earlier in this chapter are useful as starting materials for a number of reactions. In this section, you investigate some of these reactions.

Claisen-Schmidt reaction

The Claisen-Schmidt reaction (Figure 11-17) produces an α,β-unsaturated aldehyde or ketone, the general structure of which is shown in Figure 11-18. The Claisen-Schmidt reaction is a crossed aldol condensation.

Figure 11-17: An example of a Claisen-Schmidt reaction for the formation of an α,β-unsaturated ketone.

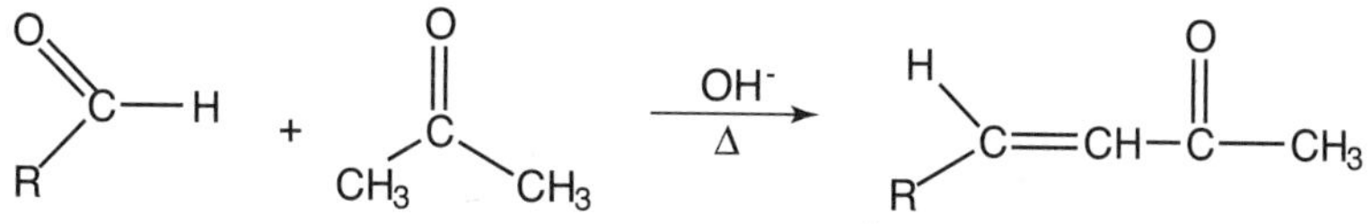

$$RCHO + CH_3COCH_3 \xrightarrow[\Delta]{OH^-} RCH{=}CH{-}CO{-}CH_3$$

Figure 11-18: The general structure of an α,β-unsaturated aldehyde or ketone.

$$C_{\beta}{=}CH_{\alpha}{-}C(=O)$$

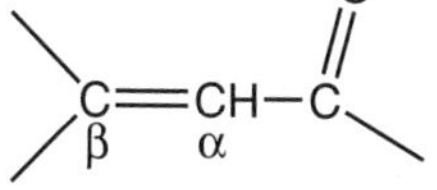

The addition reactions to an α,β-unsaturated aldehyde or ketone may be a simple addition of a nucleophile or conjugate addition. Conjugate addition involves tautomerization (see the earlier section "I thought I saw a tautomer" for details). The general structure of the product of simple addition is shown in Figure 11-19, while the general structure of the product of conjugate addition is in Figure 11-20. The mechanism for conjugate addition is in Figure 11-21.

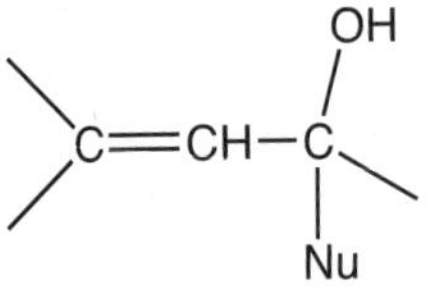

Figure 11-19: The general structure of the product of simple addition to an α,β-unsaturated aldehyde or ketone.

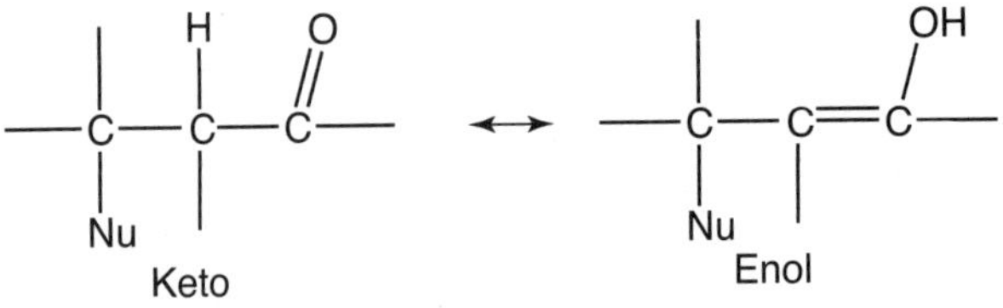

Figure 11-20: The general structure of the product of conjugate addition to an α,β-unsaturated aldehyde or ketone.

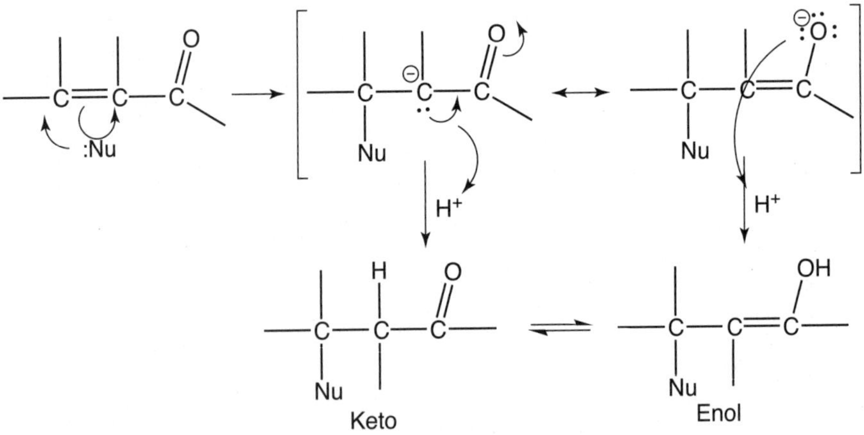

Figure 11-21: The mechanism for conjugate addition to an α,β-unsaturated aldehyde or ketone.

Conjugate addition occurs because there are two sites on the electrophile where a nucleophile can attack. The structure of the resonance hybrid and the two resonance structures contributing to the hybrid are shown in Figure 11-22. The presence of this resonance is apparent in the infrared spectrum because the carbonyl stretch shifts to a longer wavenumber.

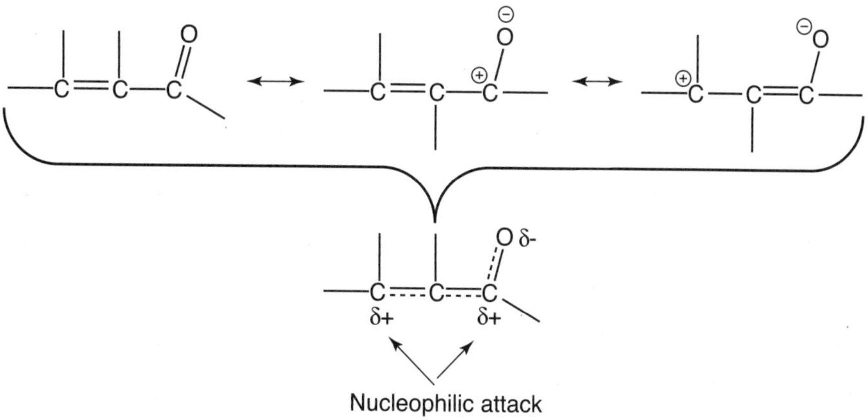

Figure 11-22: The resonance hybrid and its contributing resonance structures resulting from nucleophilic attack in a conjugate addition reaction.

The simple addition and the conjugate addition reactions compete with each other. Figure 11-23 shows an example of a reaction with both the simple addition and the conjugate addition products. In this reaction, an increase in the size of the alkyl group on the Grignard reagent leads to an increase in the yield of the conjugate addition product, while substitution on the C=C leads to an increase in the yield of the simple addition product.

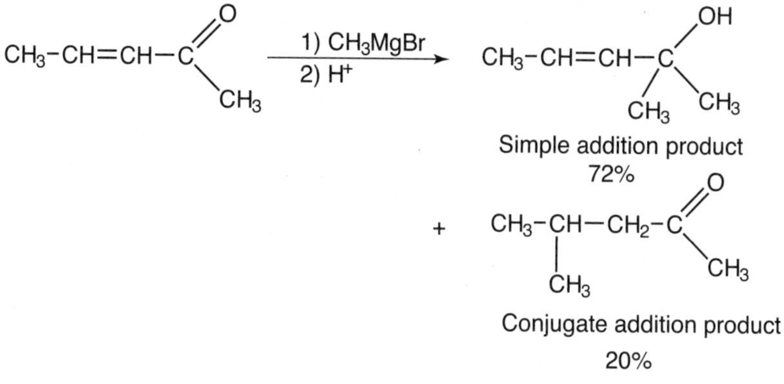

Figure 11-23: The formation of a simple addition product and a conjugate addition product.

Michael addition

The Michael addition is an enolate ion addition to an α,β unsaturated carbonyl. This reaction takes advantage of the increased acidity of a hydrogen atom that's α to two carbonyl groups. This enolate ion is very stable, so it's less reactive than normal enolates. The more-stable enolate leads to a greater control of the reaction so that only one or two products form instead of multiple products from a less stable (and therefore more reactive) enolate. An example of this type of reaction is in Figure 11-24 with the mechanism in Figure 11-25.

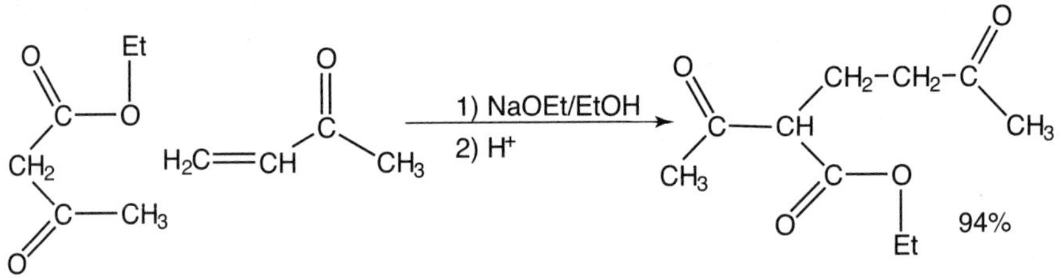

Figure 11-24: An example of a Michael addition.

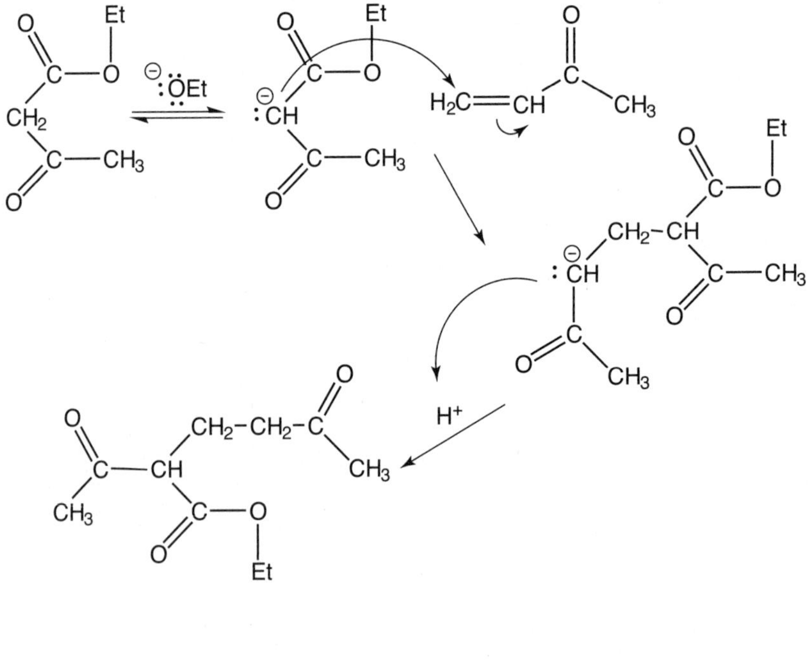

Figure 11-25: The mechanism for the Michael addition reaction in Figure 11-24.

The addition is to the alkene carbon atom furthest from the carbonyl group. This is position four, so this is a 1,4-addition or a conjugate addition. (If the addition had been to the carbonyl carbon, this would be a 1,2-addition.) The 1 in 1,4-addition or 1,2-addition refers to the addition of a hydrogen to the carbonyl oxygen atom to form an enol.

The very stable enolate ion is a Michael donor, all of which react like the enolate in Figure 11-25. Some important Michael donors are shown in Figure 11-26. The α,β unsaturated carbonyl is a Michael acceptor. Figure 11-27 shows some important Michael acceptors, which behave like the α,β unsaturated carbonyl in Figure 11-25.

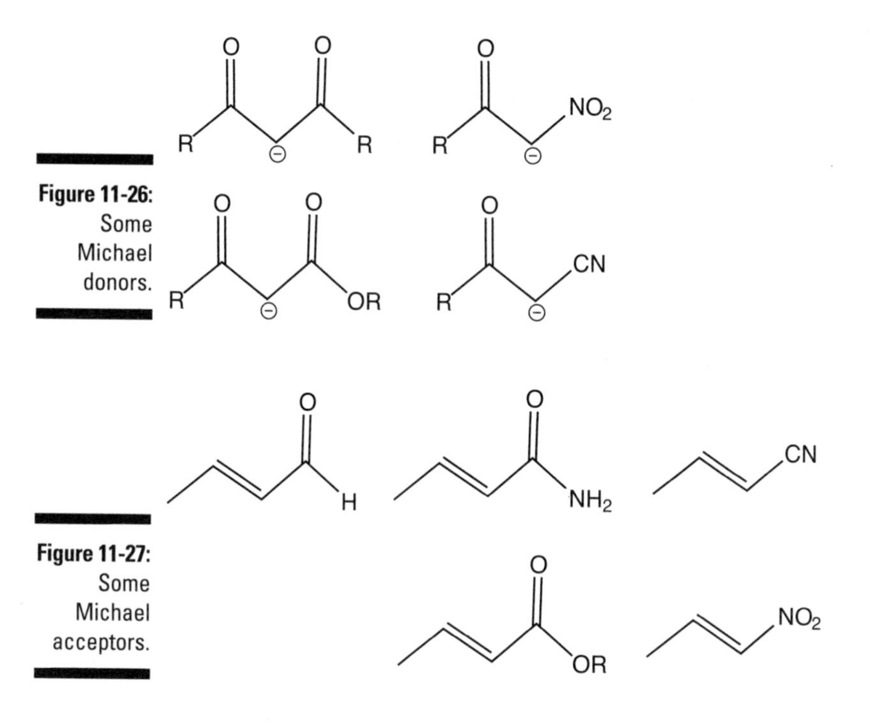

Figure 11-26: Some Michael donors.

Figure 11-27: Some Michael acceptors.

Figure 11-28 shows the mechanism for another Michael addition.

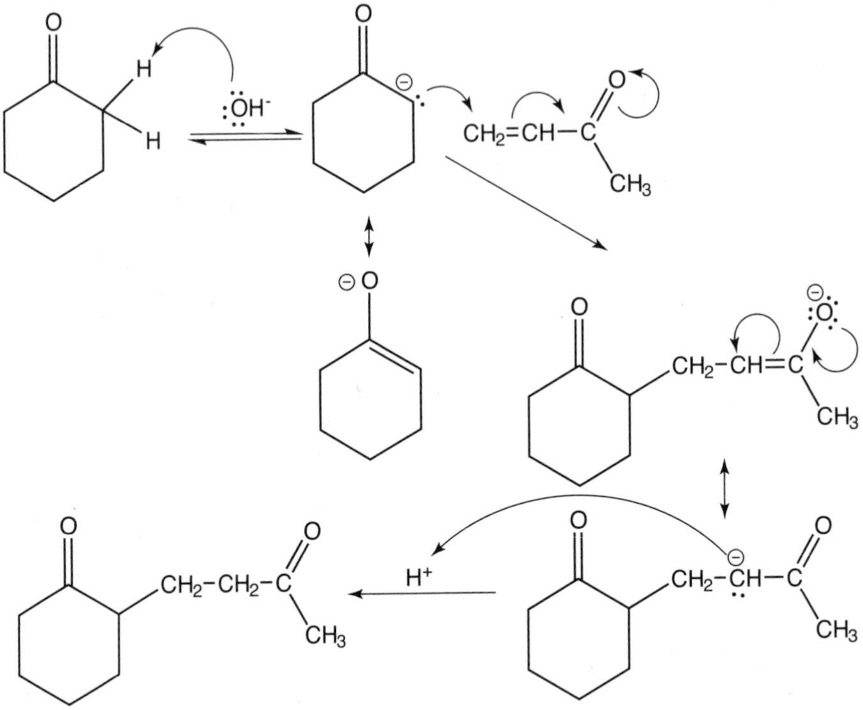

Figure 11-28: The mechanism for a Michael addition reaction.

Other enolate-related reactions

A number of species, such as nitroalkanes and nitriles, have an acidic α-hydrogen atom. These compounds can lose a hydrogen ion to produce an anion that is analogous to and reacts like an enolate ion.

Nitroalkanes

The loss of the acidic hydrogen from a nitroalkane produces a resonance-stabilized anion. This anion (a nucleophile) is capable of attacking a carbonyl compound. Synthetically, the advantage of using a nitroalkane is that reducing the nitro group to an amine group is easy. Figure 11-29 shows the formation and use of the anion from a nitroalkane followed by reduction.

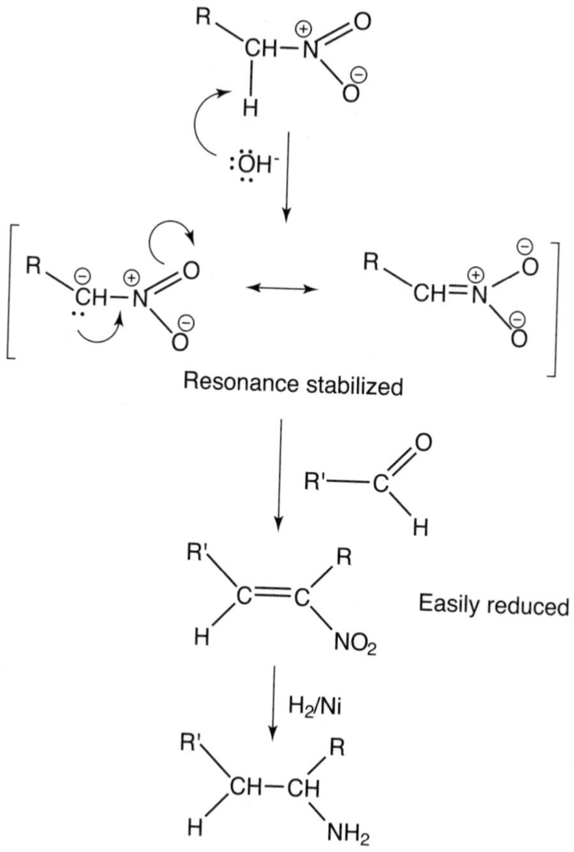

Resonance stabilized

H_2/Ni

Easily reduced

Figure 11-29: The use of a nitroalkane as a substitute for an enolate ion.

Nitriles

The α-hydrogen atoms of nitriles are also acidic. Figure 11-30 shows the formation of the resonance-stabilized ion and the reaction of the anion with a carbonyl compound. These reactions are important in some synthesis reactions.

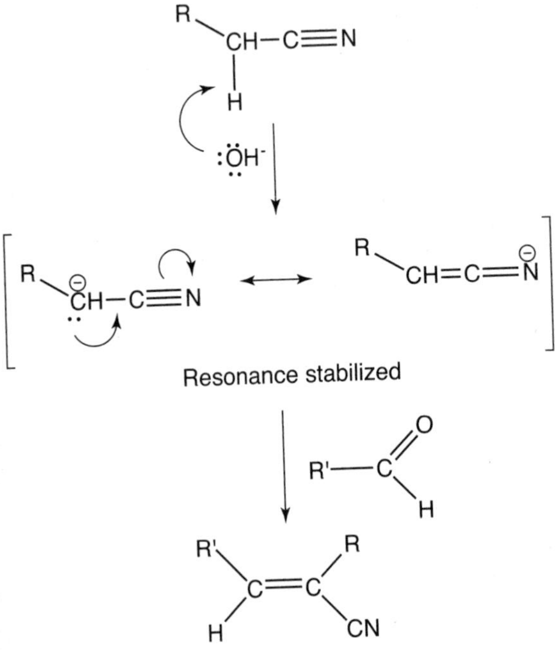

Resonance stabilized

Figure 11-30: The use of a nitrile as a substitute for an enolate ion.

Miscellaneous reactions

A number of other reactions involve, directly or indirectly, enols or enolates.

Other additions

These reactions mainly involve conjugate additions to carbonyl compounds by nucleophiles such as the cyanide ion, CN^-, and primary or secondary amines, RNH_2 or R_2NH. Figure 11-31 shows the conjugate addition by the cyanide ion, and Figure 11-32 shows the conjugate addition by a secondary amine.

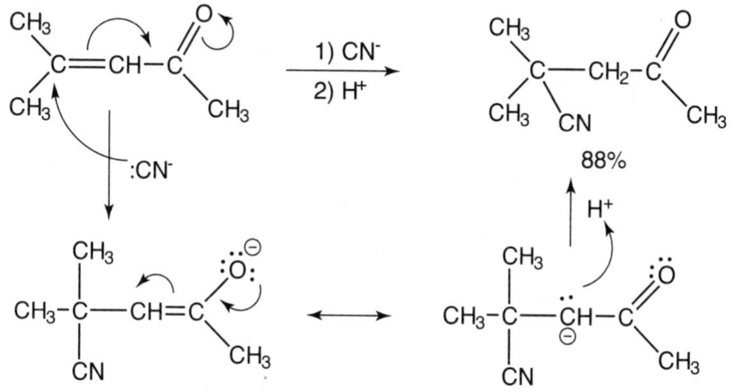

Figure 11-31: The conjugate addition of a cyanide ion.

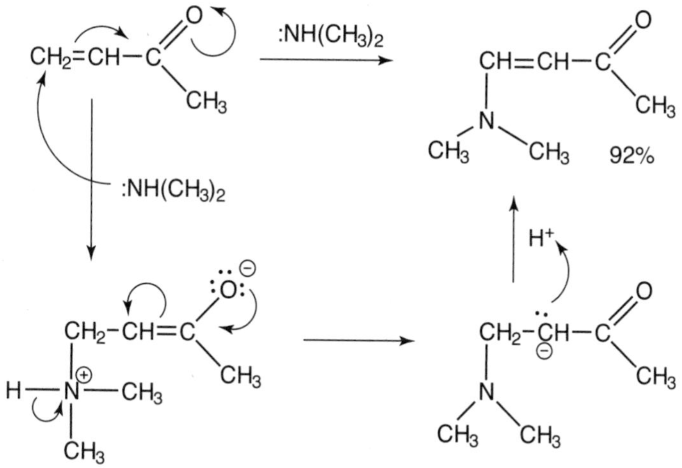

Figure 11-32: The conjugate addition of a secondary amine.

Cannizzaro reaction

The Cannizzaro reaction is a redox reaction, which requires a concentrated base and a carbonyl group with no α-hydrogen atoms. Normally, the oxidation converts an aldehyde to a carboxylate (carboxylic acid), while the reduction generates an alcohol. Figure 11-33 shows an example of a Cannizzaro reaction.

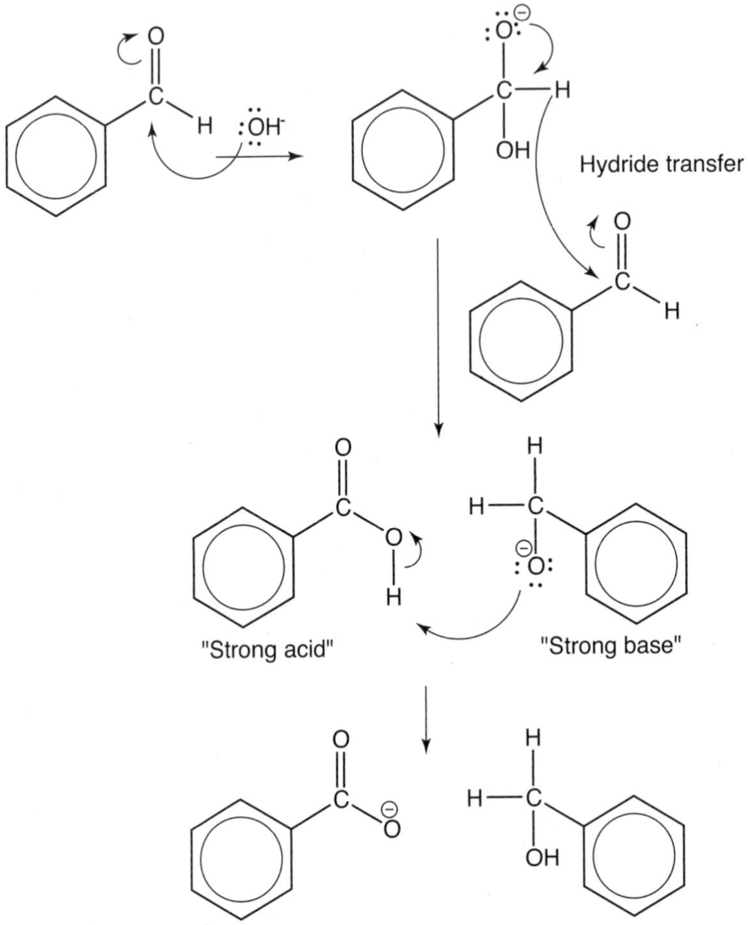

Figure 11-33: An example of a Cannizzaro reaction.

A crossed Cannizzaro reaction is similar to a normal Cannizzaro reaction; however, two different aldehydes are reacting. Normally one of the aldehydes is formaldehyde because there are fewer chances for side reactions. (It also has the advantage of being cheap.) The reactions in Figures 11-34 and 11-35 are crossed Cannizzaro reactions using an excess of formaldehyde. The excess of formaldehyde increases the probability of the initial attack by the hydroxide being on the formaldehyde instead of the other aldehyde. Figure 11-35 shows the synthesis of pentaerythritol.

Acid catalysis

In general, we use bases to initiate many of the reactions in this chapter. However, acids can catalyze some reactions. Figure 11-36 shows the formation of 4-methyl-3-penten-2-one by acid catalysis.

O
C
H + O
C
H H
:ÖH⁻
O
H O⁻
+ CH₂OH

Figure 11-34: A crossed Cannizzaro reaction with excess formaldehyde.

O
C
CH_3 H + O
C
H H
:ÖH⁻
CH_2–OH
HO—CH_2–C—CH_2–OH
CH_2–OH

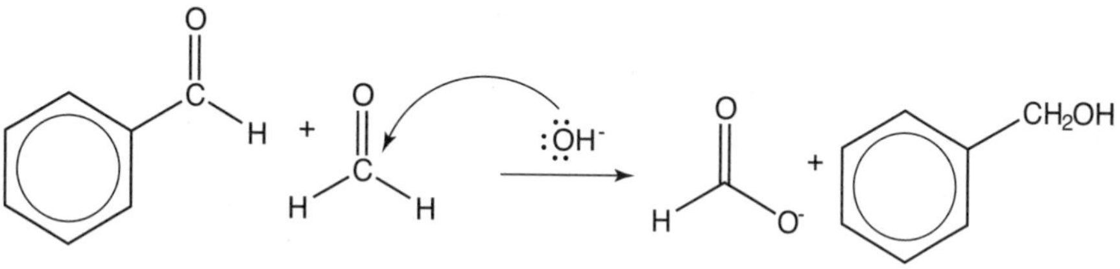

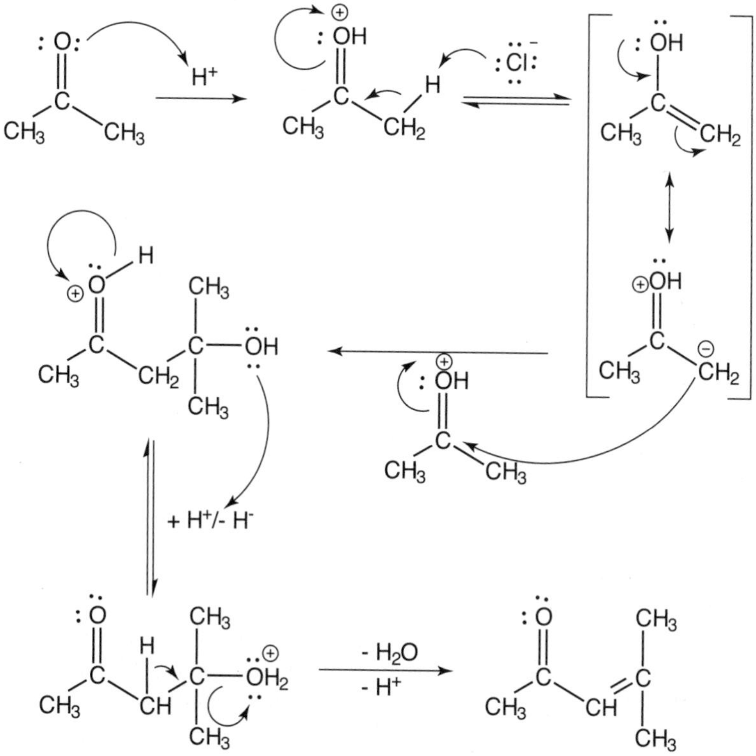

Figure 11-35: Another crossed Cannizzaro reaction with excess formaldehyde.

: O:
C
CH_3 CH_3
H^+
⊕
: OH
C
CH_3 CH_2
H
: Cl: ⁻
: OH
C
CH_3 CH_2
⊕OH
C
CH_3 ⊖ CH_2
⊕
: OH
C
CH_3 CH_3
H
⊕Ö
C
CH_3 CH_2
CH_3
C—ÖH
CH_3
+ H^+/- H^-
: Ö
C
CH_3 CH
H
CH_3
C—ÖH₂ ⊕
CH_3
- H_2O
- H^+
: Ö
C
CH_3 CH
CH_3
C
CH_3

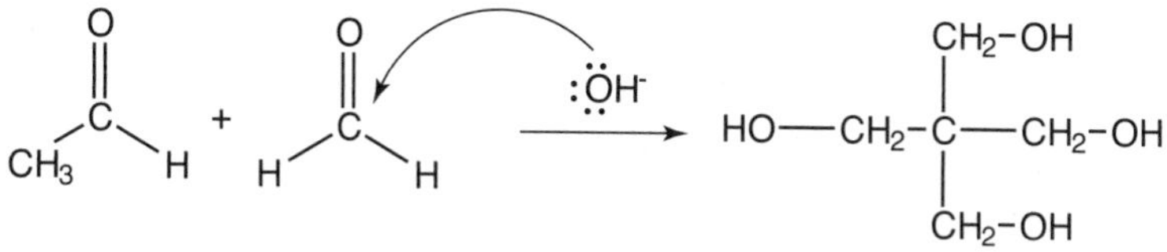

Figure 11-36: The formation of 4-methyl-3-penten-2-one by acid catalysis.

Robinson annulation

The Robinson annulation begins with a Michael addition followed by an aldol condensation (see the earlier section "Michael addition" for more info on that reaction). An example of a Robinson annulation is shown in Figure 11-37. Another example, the reaction of 2-methyl-1,3-cyclohexandione with methyl vinyl ketone, is given in Figure 11-38. The mechanism of the reaction through the formation of Compound 1 is in Figure 11-39, and mechanism for the reaction of Compound 1 to give the final product is in Figure 11-40.

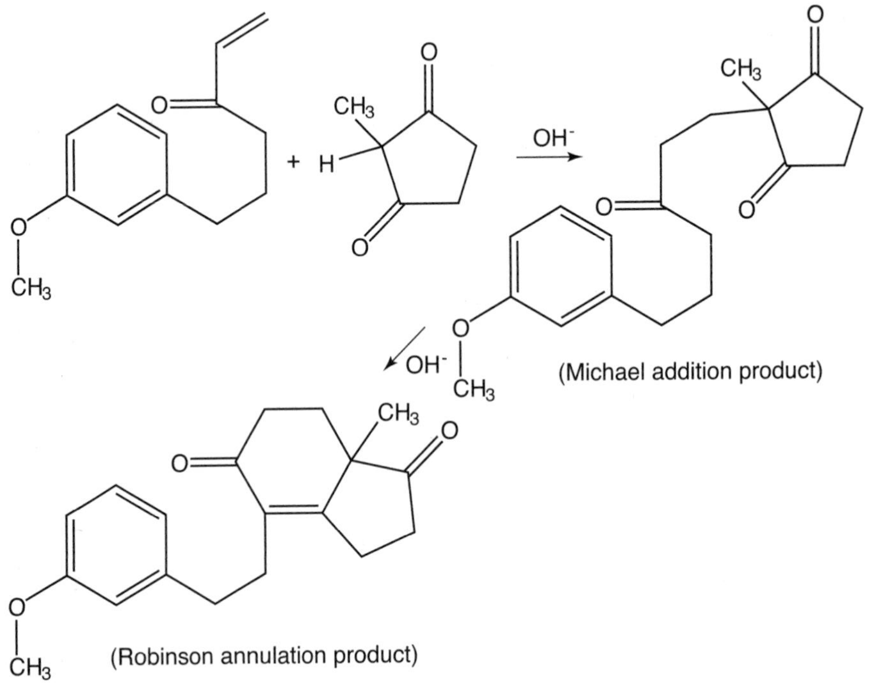

Figure 11-37: A Robinson annulation.

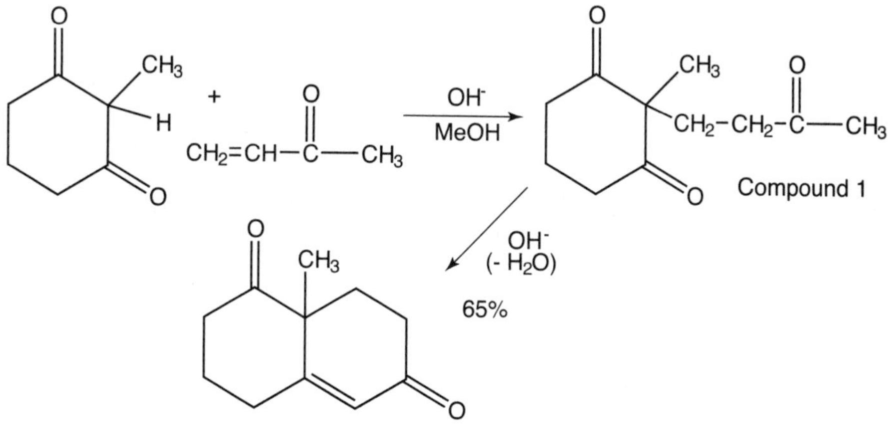

Figure 11-38: The Robinson annulation applied to the reaction of 2-methyl-1,3-cyclohexandione with methyl vinyl ketone.

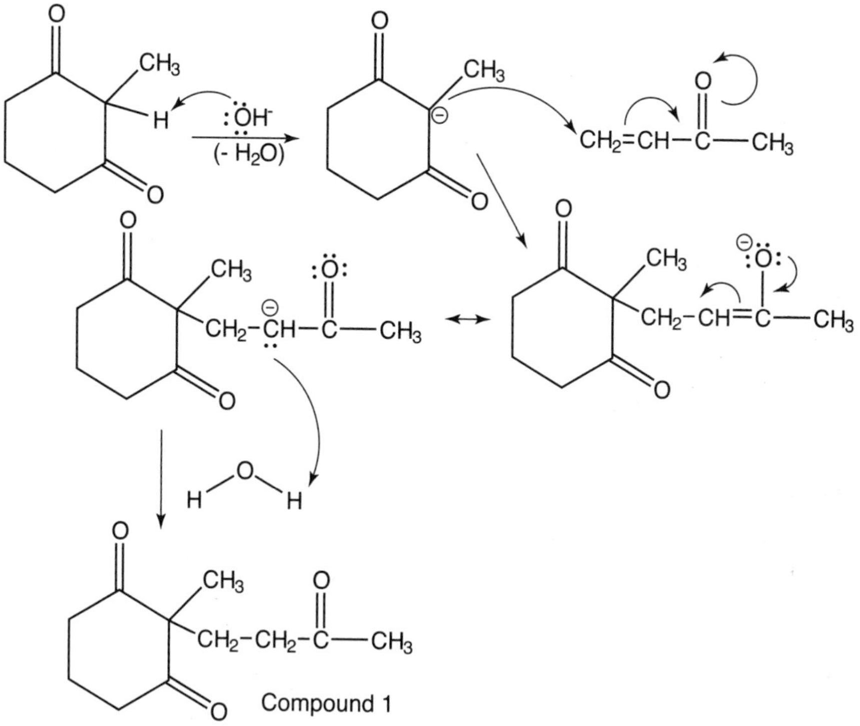

Figure 11-39: The beginning steps in the mechanism in Figure 11-38.

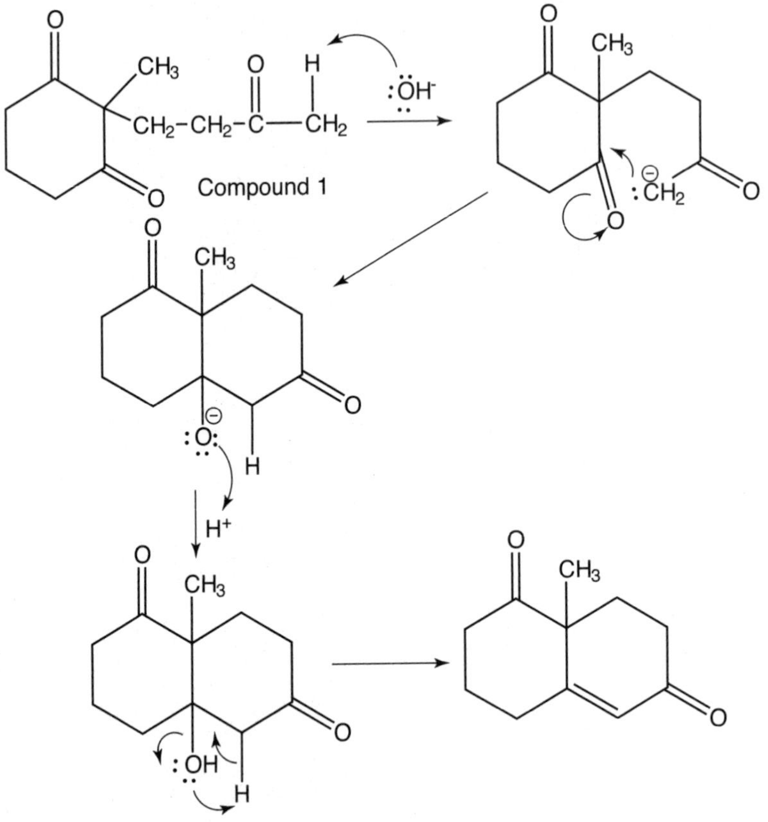

Figure 11-40: The final steps of the mechanism began in Figure 11-39.

Chapter 12

Carboxylic Acids and Their Derivatives

In This Chapter

- Tackling the names and forms of carboxylic acids and their derivatives
- Analyzing the physical properties of these compounds
- Assessing how they are synthesized
- Looking at their reactions
- Determining them with spectroscopy and chemical tests

This chapter continues your examination of compounds containing the carbonyl group; in this case carboxylic acids and their derivatives. You see many carbonyl compounds in Chapter 9, and in Chapter 10 you see the role of carbonyl groups in aldehydes and ketones. Many of the characteristics of the carbonyl group seen in those compounds are part of the chemistry of all carbonyl compounds. So remember while reading this chapter that you're looking at an extension of those characteristics and reactions and not a totally different set of compounds. (And if you skipped Chapter 10, you'll probably find it helpful to go back and read it first.)

Acid-base chemistry is important when studying carboxylic acids because, in general, they are significantly stronger acids than the α-hydrogen atoms seen in Chapter 9. The carboxylic acids are stronger acids because of the inductive effect of the carbonyl oxygen and the oxygen to which the hydrogen is bonded. These carboxylic acids are the most important organic acids. You find them in citrus fruits (citric acid), vinegar (acetic acid), aspirin (acetylsalicylic acid), and numerous other natural and synthetic compounds, as well on numerous organic exams. In this chapter you explore the structure, synthesis, and reactions of these acids and acids like them.

Seeing the Structure and Nomenclature of Carboxylic Acids and Derivatives

In Chapter 9 you see the basic structure of each of the carboxylic acids and carboxylic acid derivatives. In this chapter we focus on the carboxylic acids and related compounds, such as esters, acyl chlorides, and acid anhydrides, and we also include some information on amides (see Chapter 13 for an additional examination of amides). Before you can get into synthesis and reactions, though, you need to understand the structure and nomenclature of these compounds.

Structure

You see in Chapter 10 that aldehydes and ketones contain a carbonyl group attached to carbon or hydrogen atoms. In the case of carboxylic acids and their derivatives, a carbonyl group is attached to an electronegative element such as oxygen, chlorine, or nitrogen. The presence of these elements tends to increase the $\delta+$ charge on the carbonyl carbon, which makes the carbon atom more susceptible to nucleophilic attack.

The general formula for a carboxylic acid is RCOOH, where R may be hydrogen or any alkyl or aryl group. The derivatives vary slightly from that formula:

- In esters, the -OH is replaced with a -OR'.
- In acyl chlorides, the -OH is replaced with a -Cl.
- In acid anhydrides, two carboxylic acid molecules join (with the removal of a water molecule) to produce a molecule where an oxygen atom joins two carbonyl groups.

Nomenclature

In Chapter 9 you see how to identify the different types of compounds containing the carbonyl group. Now you take a look at the naming (nomenclature) of these compounds.

Finding out what carboxylic acids are called

When naming carboxylic acids, the final *-e* of the hydrocarbon is replaced with either *-ic* acid (common name) or *-oic* acid (IUPAC name). The carbonyl carbon on the acid is position one. When naming the salts, the *-ic* of the acid name is replaced with *-ate*.

Figure 12-1 shows the structures of some carboxylic acids, and their common names and IUPAC names are given in the following list. Figure 12-2 shows the structures and names of some of carboxylic acid salts.

- (I) formic acid or methanoic acid
- (II) acetic acid or ethanoic acid
- (III) propionic acid or propanoic acid
- (IV) butyric acid or butanoic acid
- (V) valeric acid or pentanoic acid
- (VI) 4-methylpentanoic acid

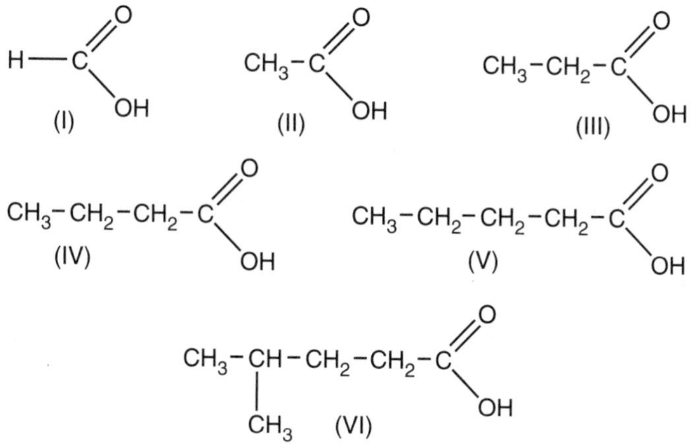

Figure 12-1: Some typical carboxylic acids.

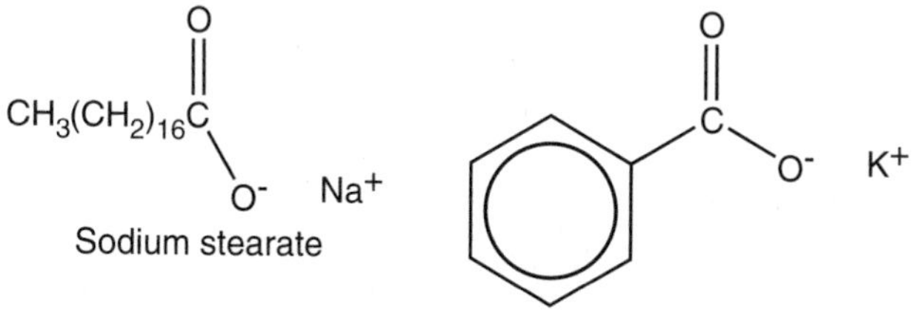

Figure 12-2: Two carboxylic acid salts.

Sodium stearate

Potassium benzoate

Designating dicarboxylic acids

Molecules may contain more than one carboxylic acid group. The dicarboxylic acids,which contain two carboxylic acid groups, are very important in areas such as organic synthesis. Many dicarboxylic acids have the general formula $HOOC\text{-}(CH_2)_n\text{-}COOH$. Table 12-1 lists how the names of the dicarboxylic acids relate to the value of *n*.

Table 12-1 Some Dicarboxylic Acids ($HOOC-(CH_2)_n-COOH$)

Value of n	*Dicarboxylic Acid*
$n = 0$	Oxalic acid
$n = 1$	Malonic acid
$n = 2$	Succinic acid
$n = 3$	Glutaric acid
$n = 4$	Adipic acid

A few important unsaturated dicarboxylic acids are shown in Figure 12-3. The position of the acid groups in a dicarboxylic acid is significant:

- If the two acid groups are ortho, the acid is phthalic (be sure to pronounce the *th!*).

  The phthalic acids are examples of aromatic dicarboxylic acids.
- If the two acid groups are meta, the acid is isophthalic.
- If the two carboxylic acid groups are para, the acid is terephthalic.

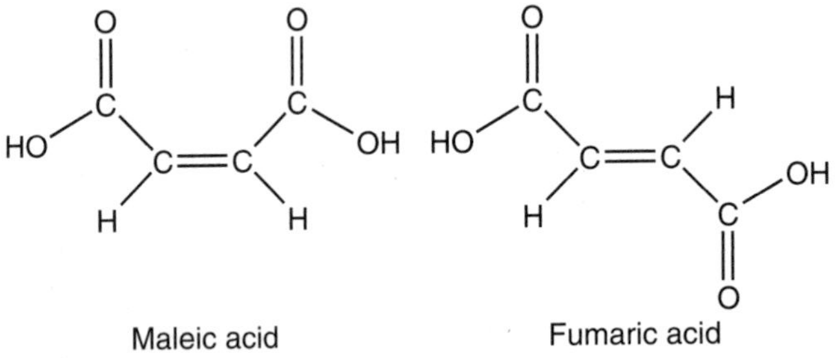

Maleic acid

Fumaric acid

Figure 12-3: Two unsaturated dicarboxylic acids.

Examining the nomenclature of esters

As you see later in this chapter in the section "Uniting acids and alcohols to make esters," esters come from an alcohol and an acid. The name of an ester reflects this origin. The alcohol name appears first (as an alkyl), and the acid name comes second, with the suffix *-ate* replacing the *-ic acid* part of the acid name. Two examples of ester structures and names are in Figure 12-4.

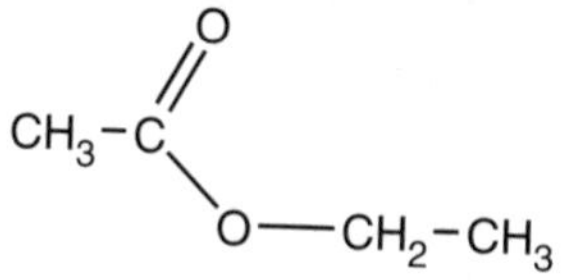

Ethyl acetate
or
ethyl ethanoate

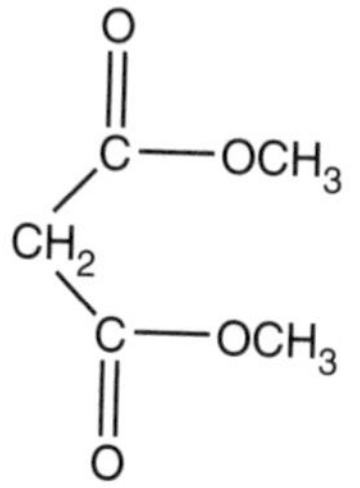

Dimethyl malonate

Figure 12-4: Examples of two esters with their names.

Naming acid anhydrides

Acid anhydrides form by joining two acids together. When naming, replace the word *acid* with the word *anhydride*. For example, two acetic acid molecules join to form acetic anhydride. Dicarboxylic acids may react internally to form an acid anhydride. See Figure 12-5 for some examples.

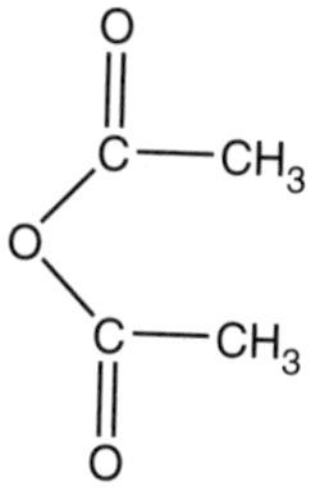

Acetic anhydride

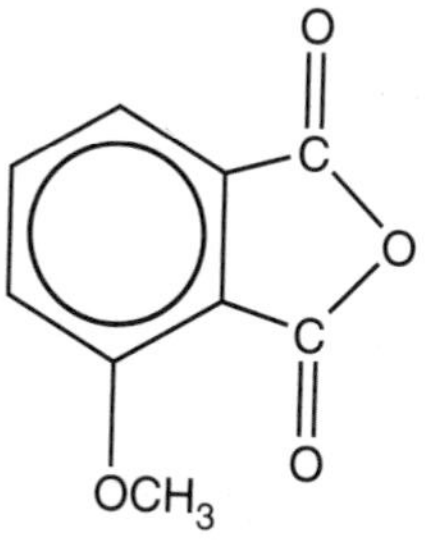

3-methoxyphthalic anhydride

Figure 12-5: Examples of two acid anhydrides with their names.

Labeling acyl chlorides

When naming an acyl chloride, you simply need to replace *-ic acid* with *-yl chloride*. See Figure 12-6 for three examples.

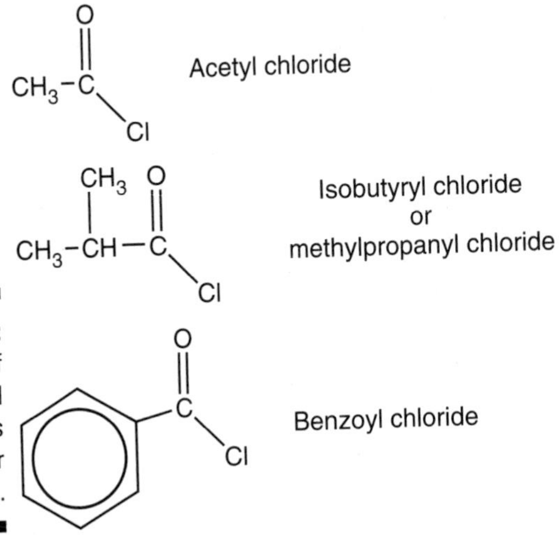

CH_3COCl — Acetyl chloride

$(CH_3)_2CHCOCl$ — Isobutyryl chloride or methylpropanyl chloride

C_6H_5COCl — Benzoyl chloride

Figure 12-6: Examples of three acyl chlorides with their names.

Clarifying amide nomenclature

When naming amides, replace the *-ic* (or *-oic*) *acid* with *amide*. Each R group attached to the nitrogen is represented by an N at the beginning of the name. See Figure 12-7 for some examples.

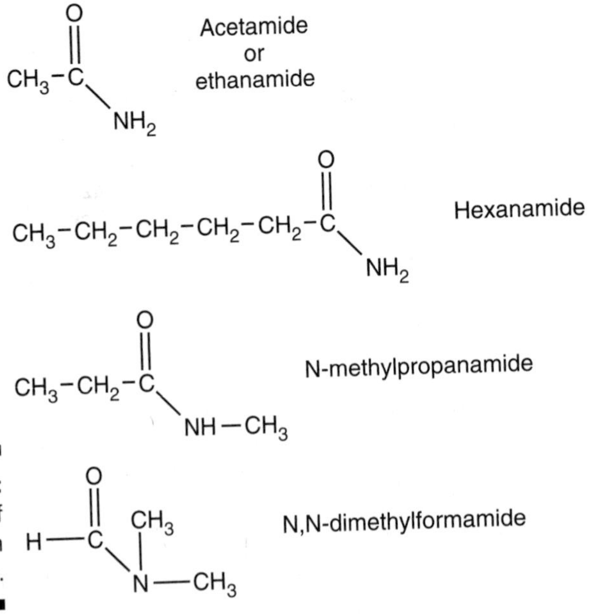

CH_3CONH_2 — Acetamide or ethanamide

$CH_3CH_2CH_2CH_2CH_2CONH_2$ — Hexanamide

$CH_3CH_2CONHCH_3$ — N-methylpropanamide

$HCON(CH_3)_2$ — N,N-dimethylformamide

Figure 12-7: Examples of amides with their names.

Checking Out Some Physical Properties of Carboxylic Acids and Derivatives

Physical properties of carboxylic acids and derivatives include solubility, melting point, boiling point, and a few other characteristics. In this section we examine each class and discuss the most important physical properties. (In the upcoming section "Considering the Acidity of Carboxylic Acids," we discuss the most important chemical property of carboxylic acids — acidity.)

Carboxylic acids

Carboxylic acids with six or fewer carbon atoms are soluble in water because of the polarity of the acid functional group and the ability of the acidic hydrogen atom to hydrogen bond. Carboxylic acids with more than six carbon atoms react with and dissolve in either aqueous sodium bicarbonate or aqueous sodium hydroxide solution.

A useful means of distinguishing between larger carboxylic acids and phenols that don't dissolve in water is that the phenols dissolve in aqueous sodium hydroxide but don't dissolve in aqueous sodium bicarbonate. Neither sodium hydroxide nor sodium bicarbonate affects the solubility of alcohols.

In general, the carboxylic acids have disagreeable odors, high melting points, and high boiling points. The high melting and boiling points are due to hydrogen bonding. In some cases the hydrogen bonding is sufficient to hold two carboxylic acids molecules together as a *dimer* (two molecules held together). When this occurs, the molecular weight (MW) appears to be about twice the weight of the acid. For example, a solution of benzoic acid (MW = 122 g/mol) in naphthalene has a molecular weight for the benzoic acid dimer of about 244 g/mol.

Esters

In general, esters have sweet odors. For this reason, many are useful in perfumes or as flavorings. The boiling points of esters are similar to those of aldehydes and ketones of comparable molar masses, which means that the boiling points are lower than comparable alcohols.

Amides

If one or two hydrogen atoms are attached to the nitrogen atom, hydrogen bonding can occur. The presence of hydrogen bonding increases the melting and boiling points.

Considering the Acidity of Carboxylic Acids

The carboxylic acids are the most important of the organic acids. (Notice that many of the mechanisms in this chapter have one or more steps involving H^+ transfer, a sure sign of acidity.) The K_a values (acid dissociation constants) are normally between 10^{-4} and 10^{-5}, indicating that an equilibrium has been established with only a small percentage of the weak acid in its dissociated form. As acids, they have a sour taste. (Vinegar is a 4- to 5-percent solution of acetic acid.) Two factors enhance the acid behavior of the carboxylic acids. The first factor is an inductive effect (see Figure 12-8), which is a result of the electron-withdrawing power of the two oxygen atoms. ***Note:*** The arrows in the figure indicate the electron-withdrawing power of the two oxygen atoms.

The other factor is the resonance stabilization of the carboxylate ion (see Figure 12-9). Remember that resonance stabilizes the molecular structure.

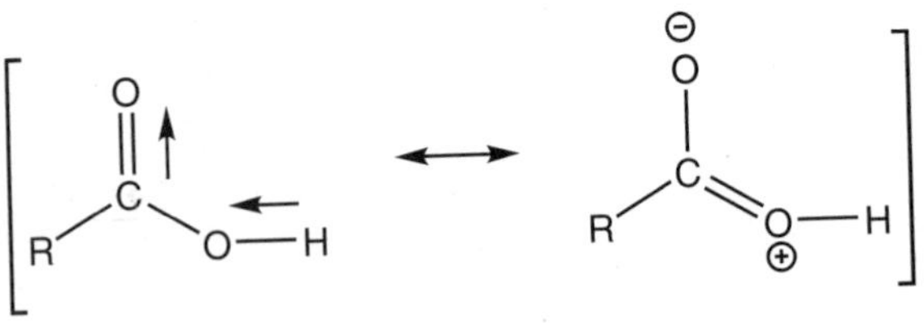

Figure 12-8: The inductive effect.

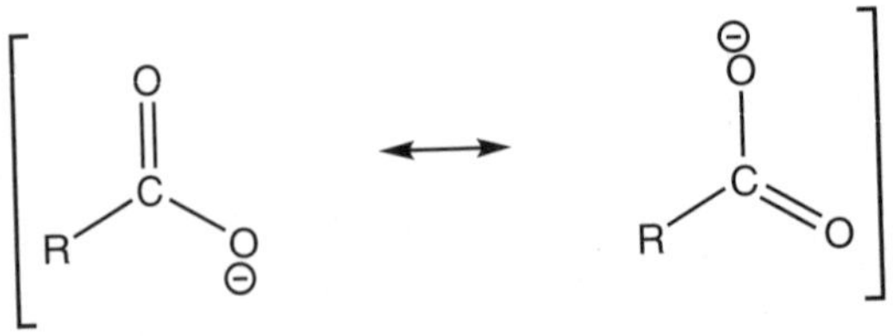

Figure 12-9: Resonance stabilization of the carboxylate ion.

The higher the K_a (lower the pK_a) value, the stronger the acid.

The acidity can be increased by adding electron-withdrawing groups to the R (electron donors have the opposite effect). For example, the acidity of acetic acid increases as chlorine atoms replace hydrogen atoms. Acetic acid has $K_a = 1.76 \times 10^{-5}$, chloroacetic acid has $K_a = 1.40 \times 10^{-3}$, dichloroacetic acid has $K_a = 3.32 \times 10^{-2}$, and trichloroacetic acid has $K_a = 2.00 \times 10^{-1}$.

The distance the electron-withdrawing group is from the carboxylic acid group is also important. For example, butanoic acid has $K_a = 1.5 \times 10^{-5}$, 4-chlorobutanoic acid has $K_a = 3 \times 10^{-5}$, 3-chlorobutanioic acid has $K_a = 8.9 \times 10^{-5}$, and 2-chlorobutanoic acid has $K_a = 1.4 \times 10^{-3}$. This shows that the chlorine is more effective the closer it gets to the carboxylic acid group.

For the aromatic carboxylic acids, substituents on the aromatic ring may also influence the acidity of the acid. Benzoic acid, for example, has $K_a = 4.3 \times 10^{-5}$. The placements of various activating groups on the ring decrease the value of the equilibrium constant, and deactivating groups increase the value of the equilibrium constant. Table 12-2 illustrates the influence of a number of para-substituents upon the acidity of benzoic acid.

Table 12-2 Comparison of K_a Values of Benzoic Acid to Para-Substituted Benzoic Acids

Benzoic acid	$K_a = 4.3 \times 10^{-5}$
Para-Substituted Benzoic Acid, Activating Groups	***K_a Value***
-OH	$K_a = 2.8 \times 10^{-5}$
$-OCH_3$	$K_a = 3.5 \times 10^{-5}$
$-CH_3$	$K_a = 4.3 \times 10^{-5}$
Para-Substituted Benzoic Acid, Deactivation Groups	***K_a Value***
-Br/-Cl	$K_a = 1.1 \times 10^{-4}$
-CHO	$K_a = 1.8 \times 10^{-4}$
-CN	$K_a = 2.8 \times 10^{-4}$
$-NO_2$	$K_a = 3.9 \times 10^{-4}$

The dicarboxylic acids have two K_a values with $K_{a1} >> K_a$. The second K_a value is lower because the loss of the first acidic hydrogen leaves an anion, which can back-donate electron density (inductive effect). The difference between the K_a values decreases as the value of *n* increases for the series $HOOC(CH_2)_nCOOH$.

Determining How Carboxylic Acids and Derivatives Are Synthesized

The syntheses of carboxylic acids and their derivatives are important reactions in organic chemistry. In this section you take a look at several ways to make these compounds.

Synthesizing carboxylic acids

A number of methods are used in the synthesis of carboxylic acids. Most of these methods involve the oxidation of some organic molecule, but other methods can be used. In this section we take a look at a few of these methods.

Oxidation of alkenes

The synthesis of carboxylic acids by the oxidation of alkenes is a two-step process. In the first step, a hot basic potassium permanganate ($KMnO_4$) solution oxidizes an alkene, and in the second step, the oxidized alkene is acidified. The process cleaves the carbon backbone at the carbon-carbon double bond to produce two smaller carboxylic acid molecules. For example, oleic acid ($CH_3(CH_2)_7CH{=}CH(CH_2)_7COOH$) yields of mixture of nonanoic acid ($CH_3(CH_2)_7COOH$) and nonadioic acid ($HOOC(CH_2)_7COOH$).

Oxidation of aldehydes and primary alcohols

The oxidation of either primary alcohols or aldehydes doesn't change the carbon backbone, so you end up with a carboxylic acid containing the same number of carbon atoms as the aldehyde or alcohol. Alcohols require considerably stronger oxidizing conditions than aldehydes do.

The oxidation of a secondary alcohol gives a ketone, and neither ketones nor tertiary alcohols readily oxidize.

The oxidants for alcohols include one of the following:

- Hot acidic potassium dichromate ($K_2Cr_2O_7$)
- Chromium trioxide (CrO_3) in sulfuric acid (H_2SO_4)(Jones reagent)
- Hot basic permanganate followed by acidification

An example of this type reaction is the conversion of 1-decanol ($CH_3(CH_2)_8CH_2OH$) to decanoic acid ($CH_3(CH_2)_8COOH$).

The oxidants for aldehydes include one of the following:

- Any reagent that can oxidize an alcohol
- Cold dilute potassium permanganate
- A number of silver compounds including $Ag(NH_3)_2OH$ and Ag_2O in base followed by acidification
- Air (over a long period of time)

An example of this type of reaction is the conversion of hexanal ($CH_3(CH_2)_4CHO$) to hexanoic acid ($CH_3(CH_2)_4COOH$).

Oxidation of alkyl benzenes

Strong oxidizing agents are capable of attacking alkyl benzenes if the carbon atom nearest the ring has at least one hydrogen atom attached. When this occurs, the oxidation removes all of the alkyl group except the carbon atom closest to the ring. Oxidizing agents include the following:

- Hot acidic potassium dichromate solution
- Hot (95 degrees Celsius) potassium permanganate solution followed by acidification

Figure 12-10 illustrates the reaction of *p*-nitrotoluene to form *p*-nitrobenzoic acid.

CH_3 NO_2 *p*-nitrotoluene

1) $KMnO_4/H_2O/95°$
2) H^+

O C OH NO_2 *p*-nitrobenzoic acid 88%

Figure 12-10: The oxidation of *p*-nitrotoluene to *p*-nitrobenzoic acid.

Oxidation of methyl ketones

In general, ketones don't undergo oxidation; however, methyl ketones undergo a haloform reaction. In a haloform reaction, the oxidation converts the methyl group to a haloform molecule (usually iodoform (CHI_3)), which leaves the carbon backbone one carbon atom shorter. The oxidant in a haloform reaction is sodium hypohalite (NaOX), which forms by the reaction of sodium hydroxide (NaOH) with a halogen (X_2, where X = Cl, Br, or I). Figure 12-11 illustrates the oxidation of a methyl ketone.

Figure 12-11: The oxidation of a methyl ketone.

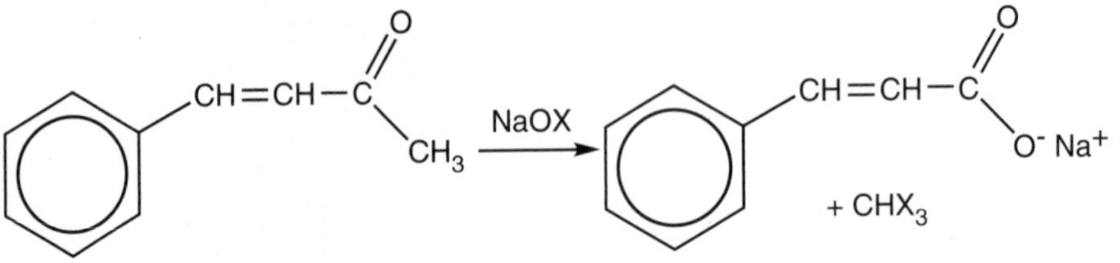

Hydrolysis of cyanohydrins and other nitriles

The basic hydrolysis (reaction with water) of a nitrile (R-CN) followed by acidification yields a carboxylic acid. In general, an S_N reaction (nucleophilic substitution) of an alkyl halide is used to generate the nitrile before hydrolysis. Figure 12-12 illustrates the formation of a carboxylic acid beginning with an alkyl halide.

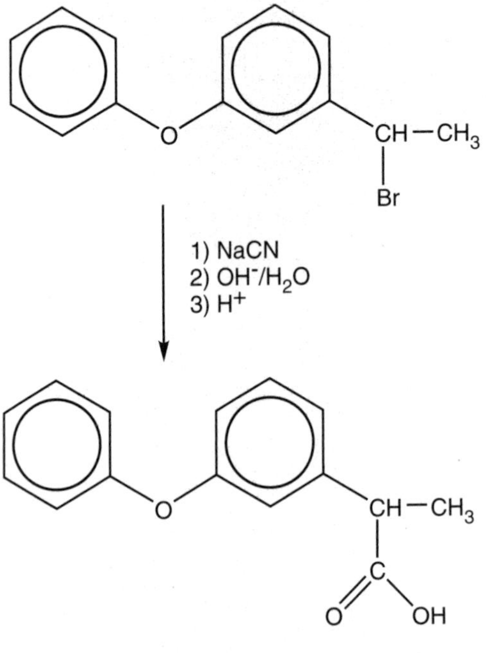

Figure 12-12: Formation of carboxylic acid from alkyl halide using the hydrolysis of a nitrile.

The product of the reaction in Figure 12-12 is fenoprofen, an antiarthritic agent.

Carbonation of Grignard reagents

After multiple steps, an organic halide can be converted to a carboxylic acid. The organic halide converts to a Grignard reagent, which reacts with carbon dioxide and then acidification forms the acid. Figures 12-13 and 12-14 illustrate the steps in this process.

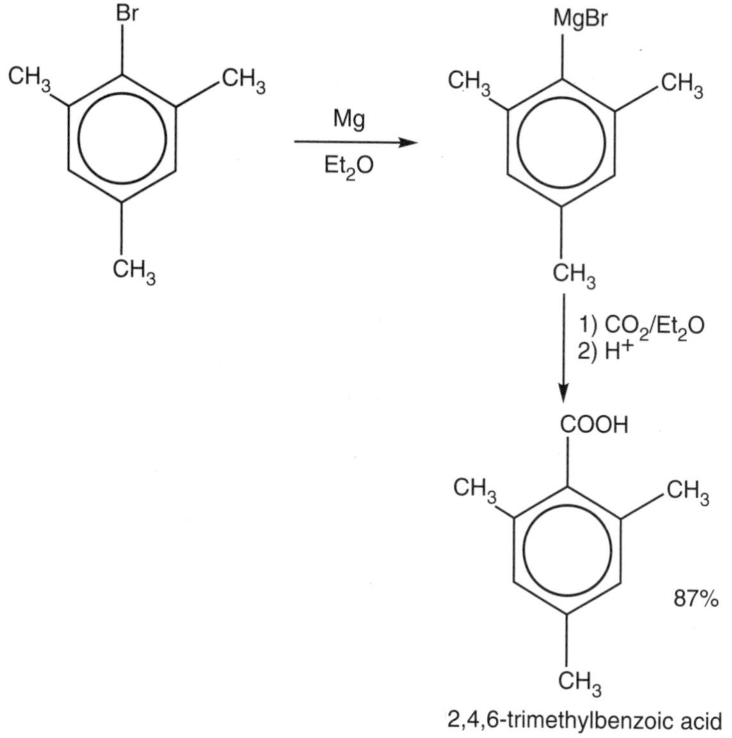

Br, CH_3, CH_3, CH_3 $\xrightarrow[Et_2O]{Mg}$ MgBr, CH_3, CH_3, CH_3

1) CO_2/Et_2O
2) H^+

COOH, CH_3, CH_3, CH_3

87%

2,4,6-trimethylbenzoic acid

Figure 12-13: The use of a Grignard reagent to form carboxylic acid.

$$CH_3CH_2CH_2CH_2Cl \xrightarrow[Et_2O]{Mg} CH_3CH_2CH_2CH_2MgCl$$

1) CO_2/Et_2O
2) H^+

$CH_3CH_2CH_2CH_2COOH$

73%

Pentanoic acid

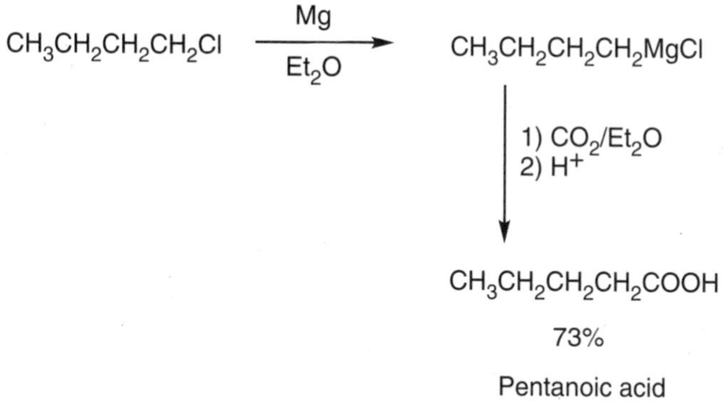

Figure 12-14: Another example using a Grignard reagent to form carboxylic acid.

Developing acyl halides with halogen

The formation of an acyl halide involves the reaction of a carboxylic acid with a halogen source. The common halogen sources are compounds like PX_3, PX_5, ClOCCOCl (oxalyl chloride), or SOX_2, where X is a halogen. The most commonly used acyl halides are the chlorides, and the simplistic reaction is RCOOH → RCOCl. Figure 12-15 illustrates the mechanism using thionyl chloride ($SOCl_2$) as the halogen source. One aid in the reaction is the formation of a transition state containing a six-membered ring. This reaction works because the $-SO_2Cl$ is a better leaving group than Cl^-.

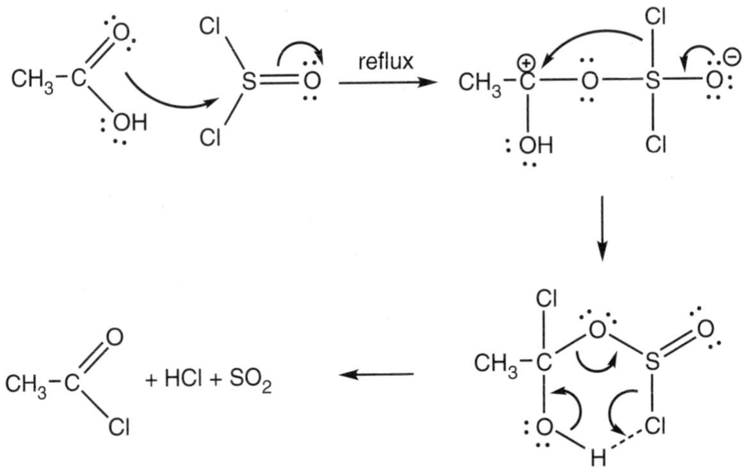

Figure 12-15: Formation of acyl chloride by the reaction of carboxylic acid with thionyl chloride.

Removing water to form acid anhydrides

The name *anhydride* means *without water,* so that makes it pretty clear that the general idea behind the formation of an acid anhydride is to remove water from a carboxylic acid. Both acyl chlorides and acid anhydrides are very effective at removing water. In some cases, heat can be used to remove water.

Sodium salt plus acid chloride

The reaction of a carboxylic acid with sodium hydroxide (NaOH) produces the sodium salt of the carboxylic acid. The sodium salt then reacts with an acid chloride to form the anhydride. Figure 12-16 illustrates the final step in this process. In this reaction, the carboxylate ion behaves as a nucleophile and attacks the carbonyl carbon atom of the acid chloride. The reaction of a carboxylic acid with sodium hydroxide also generates water, which, if not removed, reacts with the acid chloride and lowers the yield of the reaction.

This synthesis can produce either a *symmetric anhydride* (both acids the same) or an *asymmetric anhydride* (different acids).

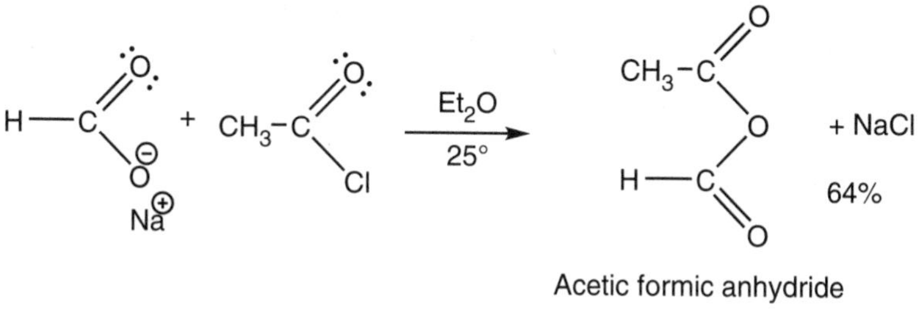

Figure 12-16: The reaction of sodium salt of carboxylic acid (sodium formate) with acid chloride.

$$H-C(=O)-O^{-}Na^{+} + CH_3-C(=O)-Cl \xrightarrow[25°]{Et_2O} CH_3-C(=O)-O-C(=O)-H + NaCl$$

64%

Acetic formic anhydride

Acid plus acid chloride plus pyridine

This process that happens when you combine an acid with an acid chloride with pyridine is similar to the reaction of a sodium salt with an acid chloride. The pyridine behaves as a base in place of the sodium hydroxide. The advantage of this process is that no water forms to react with the acid chloride. Figure 12-17 illustrates this reaction.

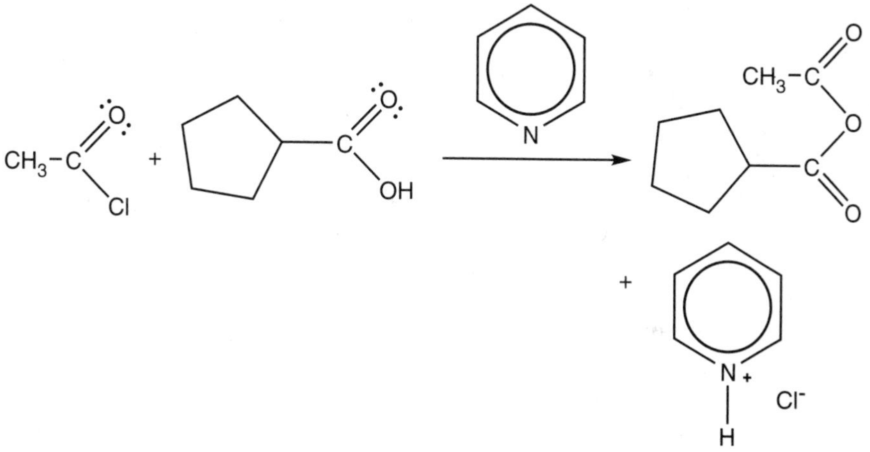

Figure 12-17: Acid anhydride formed by the reaction of acid chloride with carboxylic acid in the presence of pyridine.

$CH_3-C(=O)-Cl$ + cyclopentyl$-C(=O)-OH$ → (N) → $CH_3-C(=O)-O-C(=O)-$cyclopentyl + pyridinium $N^{+}-H$ Cl^{-}

Acetic anhydride plus acid

The dehydrating properties of an acid anhydride can be used to produce another acid anhydride. This is an equilibrium process. By heating the mixture, the more volatile acid vaporizes to shift the equilibrium toward the products. Acetic acid, from acetic anhydride, is useful because it's more volatile than most other carboxylic acids. Figure 12-18 illustrates this reaction.

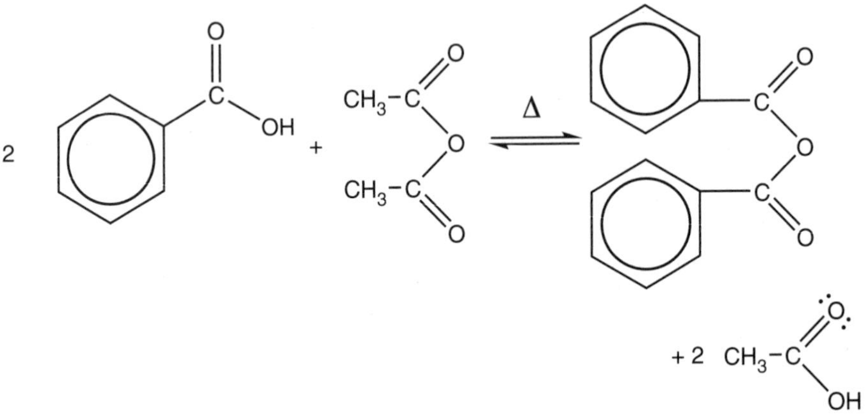

Figure 12-18: Formation of acid anhydride by the reaction of carboxylic acid with acetic anhydride.

Cyclic anhydrides

A cyclic anhydride can be formed from a dicarboxylic acid by heating if the anhydride that forms has a five- or six-membered ring. If the dicarboxylic acid contains a ring, only a *cis* isomer (not the *trans* isomer) reacts. Figure 12-19 illustrates this type of reaction. The black circles in the figure indicate that the *cis*-isomer is reacting to form a *cis* product.

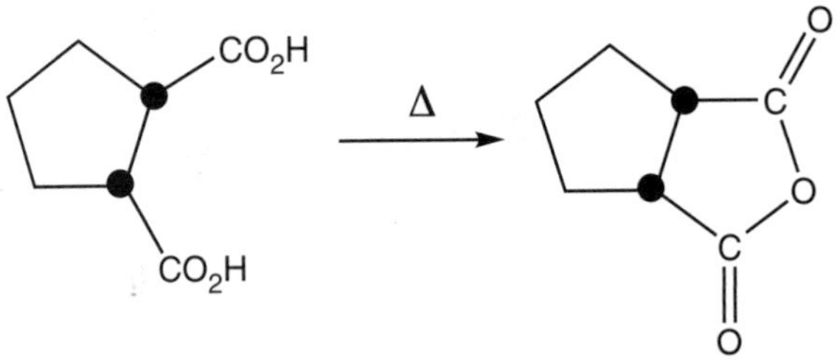

Figure 12-19: The thermal dehydration of a dicarboxylic acid to form a cyclic anhydride.

Uniting acids and alcohols to make esters

An ester consists of an alcohol portion and a carboxylic acid portion. In the synthesis of an ester, these two portions need to be brought together. The simplest method is to react an acid with an alcohol in the presence of another alcohol, but as you see in the following sections, other methods are useful as well.

Acid plus alcohol

This method is called the Fischer esterification. It's a condensation reaction where the loss of a water molecule accompanies the joining of the alcohol portion to the acid portion. The acid gives up the OH and the alcohol gives up the H to make the water molecule. All steps in the mechanism are reversible (that is, it establishes an equilibrium), so removing the ester as soon as it forms is helpful. Removal of the ester is normally easy since esters typically have lower boiling points than alcohols and carboxylic acids. Figure 12-20 illustrates the mechanism for the acid-catalyzed formation of an ester by the reaction of an alcohol with a carboxylic acid.

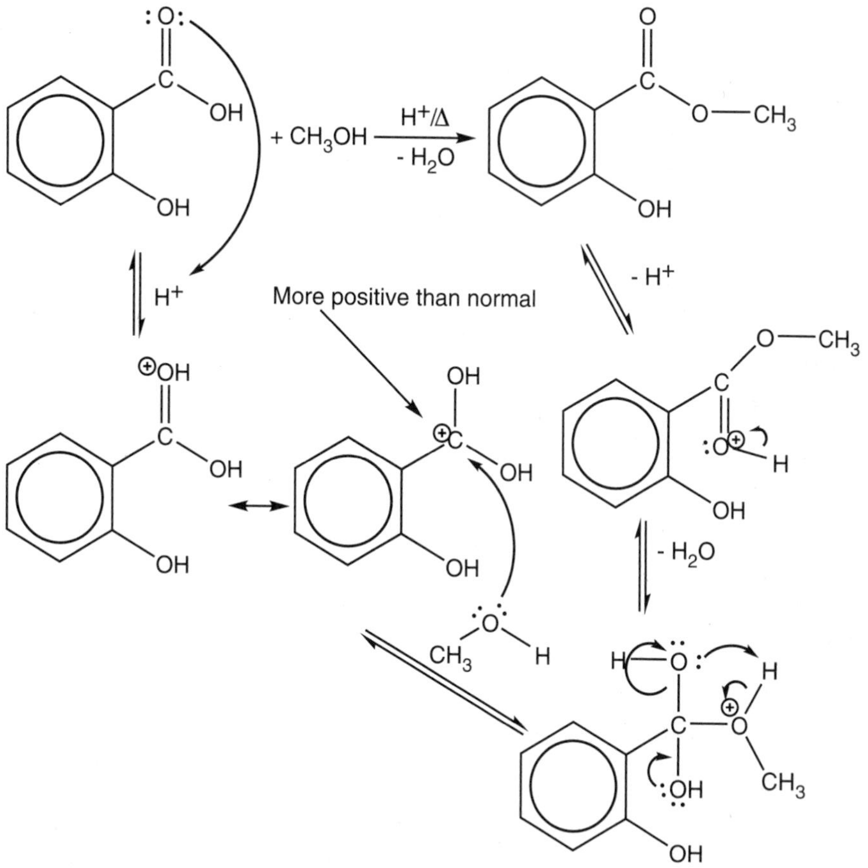

Figure 12-20: Acid-catalyzed formation of an ester from an alcohol and a carboxylic acid.

Acid chloride plus alcohol

This method is easier than the reaction of an alcohol with a carboxylic acid (described in the preceding section) because acid chlorides are more reactive than acids. The reaction forms HCl with either a hydroxide ion or pyridine, aiding in the removal of the HCl. The general reaction is in Figure 12-21 and the reaction using pyridine to scavenge the HCl is in Figure 12-22.

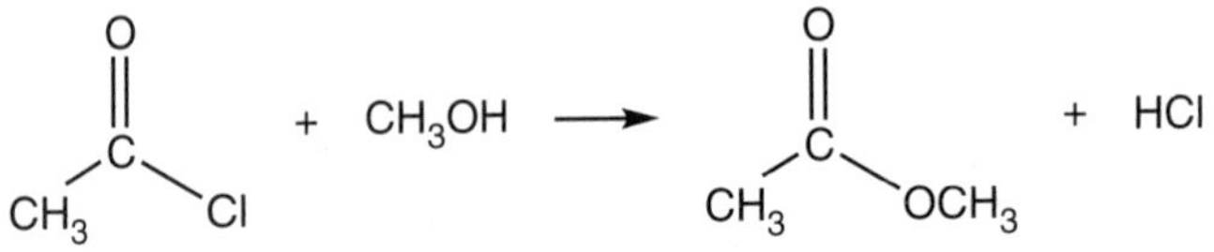

$CH_3C(=O)Cl + CH_3OH \longrightarrow CH_3C(=O)OCH_3 + HCl$

Figure 12-21: The general method for ester synthesis from an alcohol and an acid chloride.

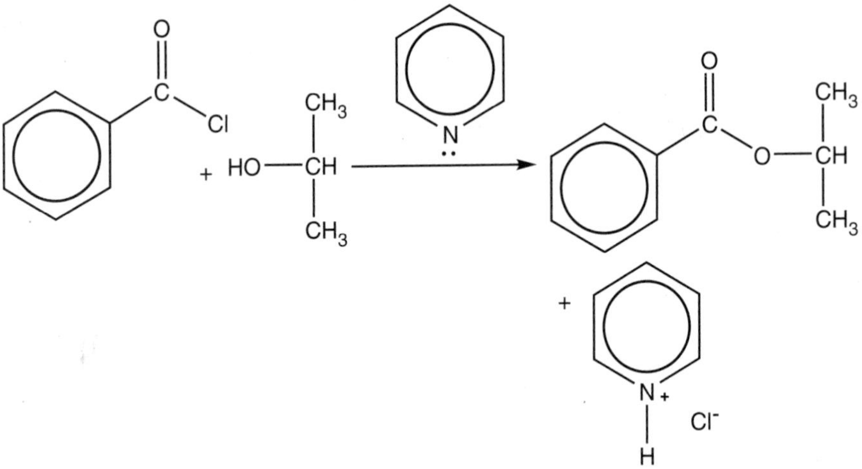

O, C, Cl, CH₃, HO, CH, N, O, CH₃, Cl⁻, N+, H

Figure 12-22: The reaction of an acid chloride with an alcohol, using pyridine to trap the HCl formed.

Acid anhydride plus an alcohol

Acid anhydrides are between acid chlorides and carboxylic acids in reactivity, so this reaction is more effective than the reaction with a carboxylic acid but less efficient than the reaction with an acid chloride. Half of the acid anhydride goes into forming the ester, while the other half becomes a carboxylic acid. Figure 12-23 illustrates this reaction, using salicylic acid as the alcohol and acetic anhydride as the acid anhydride to form aspirin, an ester.

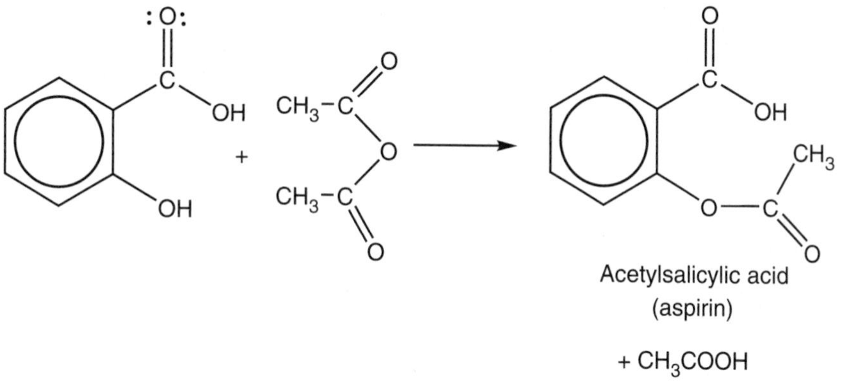

Figure 12-23: Forming an ester (aspirin) by the reaction of an alcohol (salicylic acid) with an acid anhydride (acetic anhydride).

Transesterification

Tranesterification involves the conversion of one ester into another. In this process, a less volatile alcohol replaces a more volatile alcohol. For example, heating an excess of ethanol with a methyl ester while rapidly removing the more volatile methanol as it forms results in transesterification. An acid catalyst facilitates the reaction, which is illustrated in Figure 12-24. To produce a propyl ester, the action of propanol on either a methyl ester or an ethyl ester would work.

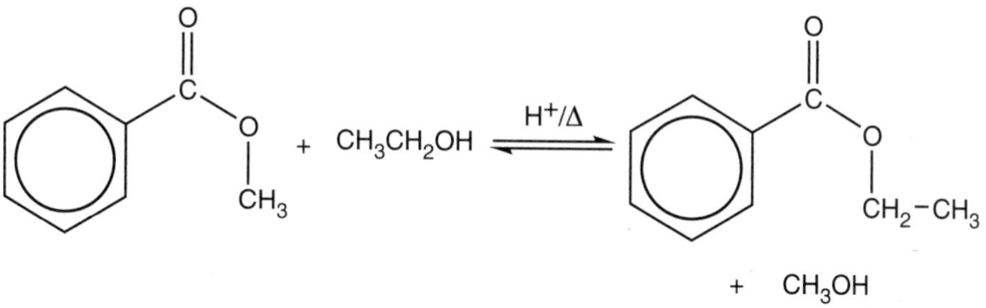

Figure 12-24: The transesterification of a methyl ester to an ethyl ester.

Methyl esters from diazomethane

The reaction of diazomethane with a carboxylic acid is an efficient way to produce a methyl ester; however, the procedure is dangerous. Figure 12-25 illustrates the formation of a methyl ester from benzoic acid and diazomethane.

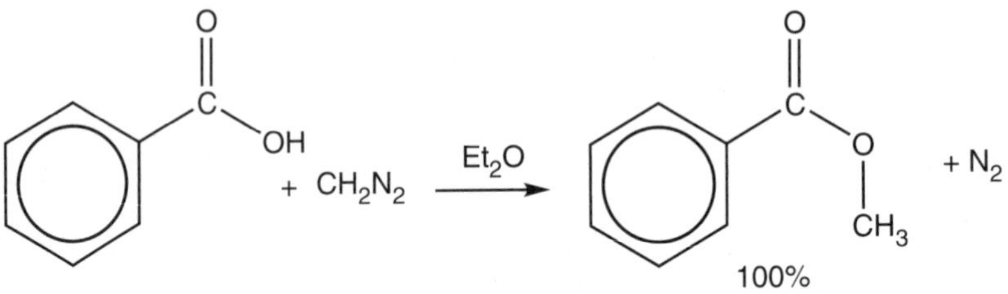

Figure 12-25: The formation of a methyl ester utilizing diazomethane.

Bringing acids and bases together to create amides

Amides contain an acid portion and an amine portion. However, unlike the formation of an ester, the reaction of a carboxylic acid with an amine is not an efficient method for preparing an amide, because, as you see in this section, the simple reaction of an acid (carboxylic acid) with a base (amine) causes interference. Fortunately, methods similar to many of the other ester synthesis methods are useful in the synthesis of amides.

From acid chlorides

Acid chlorides are very reactive, and they readily react with ammonia, primary amines, or secondary amines to form an amide. Figure 12-26 illustrates the reaction of an acid chloride with ammonia. Replacing one or two of the hydrogen atoms of ammonia with an organic group will result in an N-substituted amide. Tertiary amines react with acid chlorides to form a carboxylic acid and an ammonium salt.

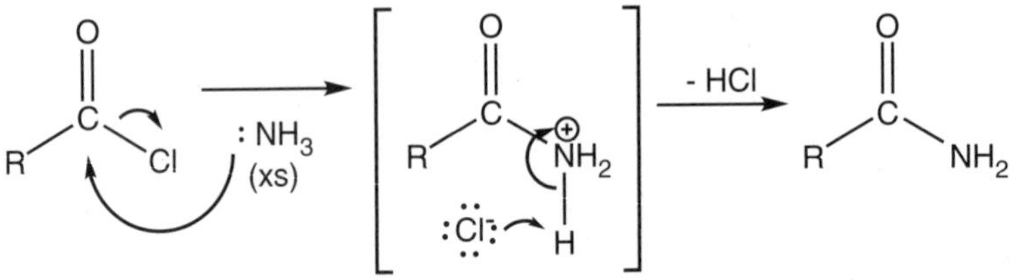

Figure 12-26: The formation of an amide by the reaction of an acid chloride with ammonia.

From acid anhydrides

This process is similar to the formation of an ester by the action of an acid anhydride on an alcohol (described in the earlier section "Acid anhydride plus an alcohol"). Half the acid anhydride forms the amide; the other half is a leaving group. Ammonia, primary amines, and secondary amines react to produce amides. Figure 12-27 shows the industrial preparation of phenacetin by the reaction of an amine with an acid anhydride. The mechanism for this reaction is similar to the mechanism for the reaction of an acid chloride with an amine (refer to Figure 12-26).

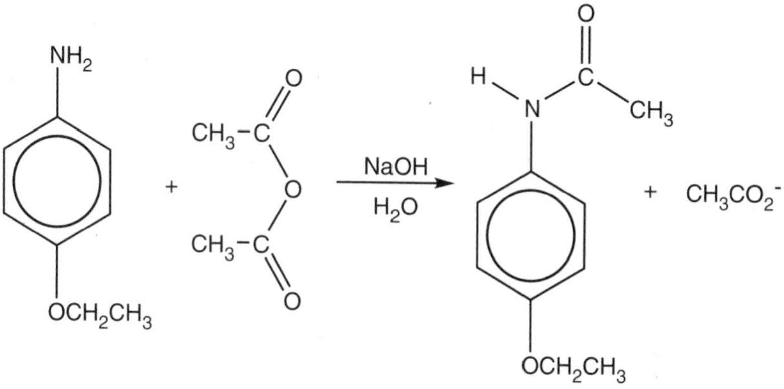

NH_2 CH_3–C(=O)–O–C(=O)–CH_3 OCH_2CH_3 + NaOH / H_2O → H–N–C(=O)–CH_3 OCH_2CH_3 + $CH_3CO_2^-$

Figure 12-27: The preparation of phenacetin by the reaction of an amine with an acid anhydride.

From esters

Amines also react with esters by a method similar to the reaction of an acid chloride with an amine (which was described in the previous section, "From acid chlorides"). Figure 12-28 illustrates the formation of benzamide by this type of reaction, using ammonia and methyl benzoate. Again, the mechanism is similar to the reaction of an acid chloride with an amine (Figure 12-26).

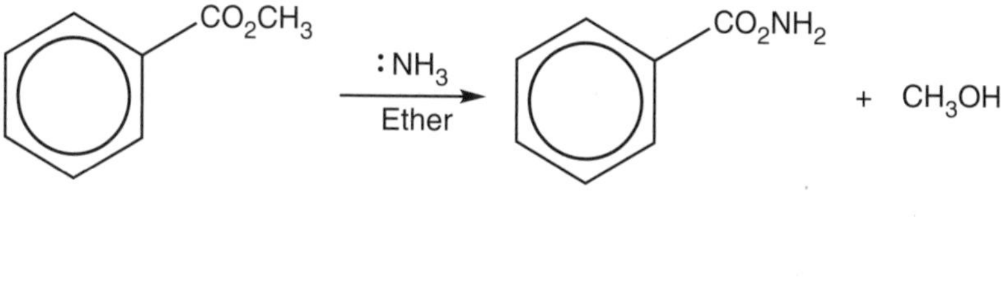

CO_2CH_3 —:NH_3 / Ether→ CO_2NH_2 + CH_3OH

Figure 12-28: The formation of benzamide by the reaction of ammonia with methyl benzoate.

From carboxylic acids and ammonium salts

The reaction of an amine or ammonia with a carboxylic acid first produces an ammonium salt, which upon heating loses water and produces an amide. This is a low yield process. Figure 12-29 shows an example of this type of reaction.

$$RC(=O)OH + NH_3 \rightarrow RC(=O)O^- + NH_4^+ \xrightarrow{\Delta} RC(=O)NH_2 + H_2O$$

Figure 12-29: The reaction of ammonia with a carboxylic acid to eventually form an amide.

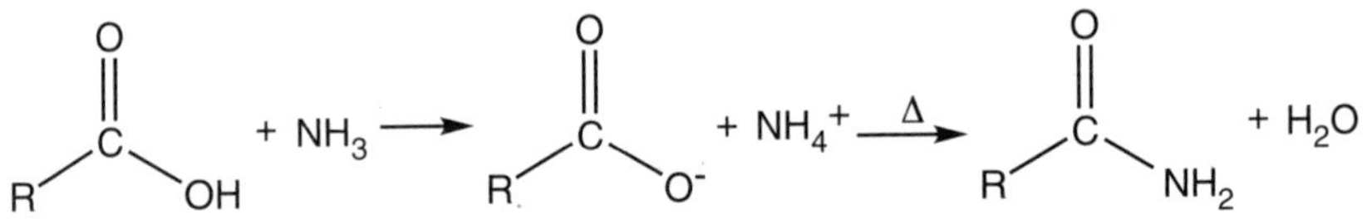

Exploring Reactions

Many of the reactions of carboxylic acids and derivatives involve nucleophilic substitution at the acyl carbon atom. This is a bimolecular process with the acyl group having a leaving group (L). The general mechanism is reminiscent of an S_N2 mechanism. Figure 12-30 illustrates the general process and Figure 12-31 gives more details on the mechanism.

$$Nu: + R{-}C(=O){-}L \rightarrow [Nu{\cdots}C(=O)(R){\cdots}L] \rightarrow Nu{-}C(=O){-}R + L$$

Figure 12-30: Nucleophilic attack at the acyl carbon atom showing its relationship to an S_N2 process.

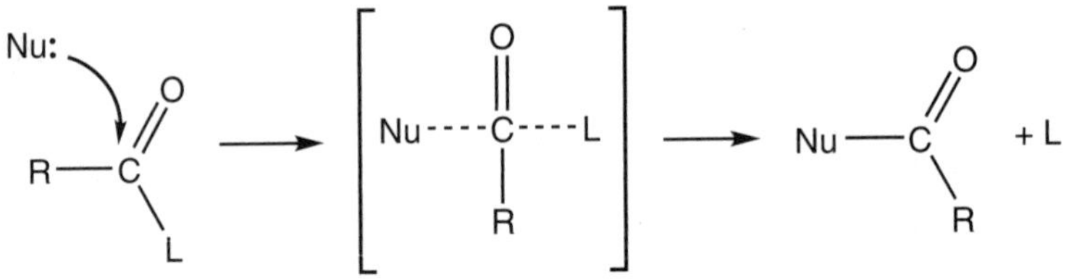

The various carboxylic acid derivatives vary in their reactivity (stability of the leaving group). Acid chlorides, for example, are more reactive than anhydrides (don't leave as easily). A summary of the relative reactivities appears in Figure 12-32.

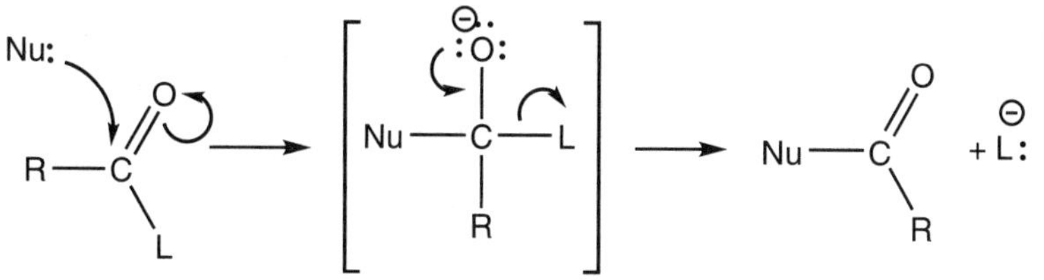

Figure 12-31: The general mechanism for the nucleophilic attack on an acyl carbon atom.

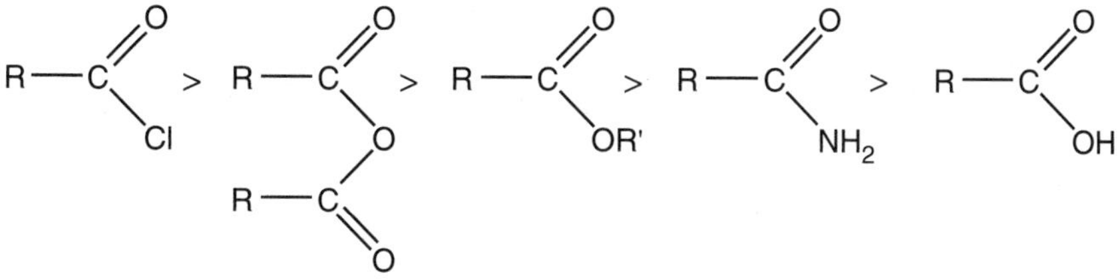

Figure 12-32: The relative reactivities of the carboxylic acid derivatives.

The sequence in Figure 12-32 not only represents the general reactivity of carboxylic acid derivatives but also gives information on the ease of synthesis. The more reactive a species is, the more difficult it is to prepare it (and vice versa). From this series, you can see that synthesizing a less reactive acyl compound from a more reactive acyl compound is always possible.

Generous carboxylic acids

We start this section by giving your brain a rest with some simple material: Carboxylic acids are acids. Acids donate a hydrogen ion, H^+, to other species. Therefore, that's the fundamental reaction of carboxylic acids. Rested? Good. As seen previously, these are weak acids (although they're stronger than most other organic acids). In this chapter we have seen a variety of other reactions, such as the formation of an ester, that utilize carboxylic acids as one of the reactants. Other reactions follow.

The Hell-Volhard Zelinsky reaction is a method for forming α-halo acid. This is a synthetically useful procedure because the α-halo acids are useful starting materials for other reactions. For example, the addition of hydroxide ion leads to the replacement of the halogen with an -OH group. The reaction with ammonia replaces the halogen with $-NH_2$. The reaction with cyanide ion, CN^-, converts the halide to a nitrile. Figure 12-33 illustrates this reaction.

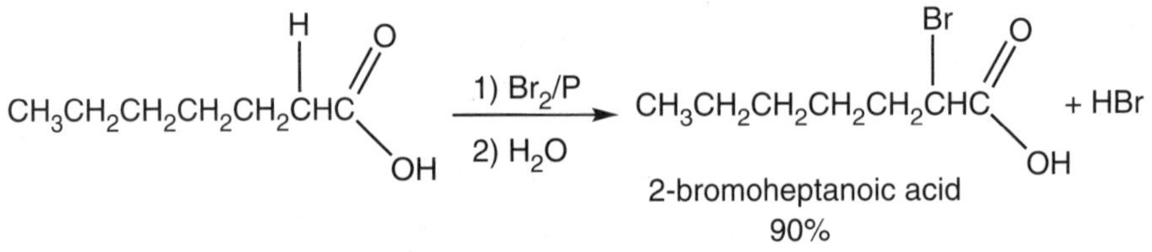

$$CH_3CH_2CH_2CH_2CH_2CH(H)C(=O)OH \xrightarrow[2)\ H_2O]{1)\ Br_2/P} CH_3CH_2CH_2CH_2CH_2CH(Br)C(=O)OH + HBr$$

2-bromoheptanoic acid
90%

Figure 12-33: An example of the Hell-Volhard Zelinsky reaction.

Simple acyl halide and anhydride reactions

Both acyl halides and anhydrides react with water (hydrolysis). Acyl halides react to form one mole of the carboxylic acid and one mole of the hydrohalic acid, HX. Anhydrides react to form two moles of carboxylic acid.

Acyl halides and anhydrides are important reactants for the formation of other carbonyl compounds, but you don't need to take up valuable brain space with information about any other acyl halide or anhydride reactions at this time.

Hydrolysis of esters

Esters can undergo hydrolysis using either an acid or a base as a catalyst. Hydrolysis always produces an alcohol from the alkyl portion of the ester. During acid hydrolysis, the acid portion of the ester yields a carboxylic acid. During base hydrolysis of an ester, which is called *saponification,* the acid portion of the ester yields the carboxylate ion.

Acid hydrolysis

Acid hydrolysis is the reverse of the Fischer esterification, seen earlier in the section "Acid plus alcohol." Figure 12-34 illustrates the mechanism.

Base hydrolysis (saponification)

Saponification (base hydrolysis) follows a simpler mechanism (see Figure 12-35).

In the reaction, one mole of hydroxide generates one mole of alcohol and one mole of carboxylate ion from one mole of ester. Based on this stoichiometry (the mole relationship as defined by the balanced chemical equation), if the number of moles of base is known, then the amount of ester is known. This stoichiometry is the saponification equivalent, used to determine the equivalents of ester.

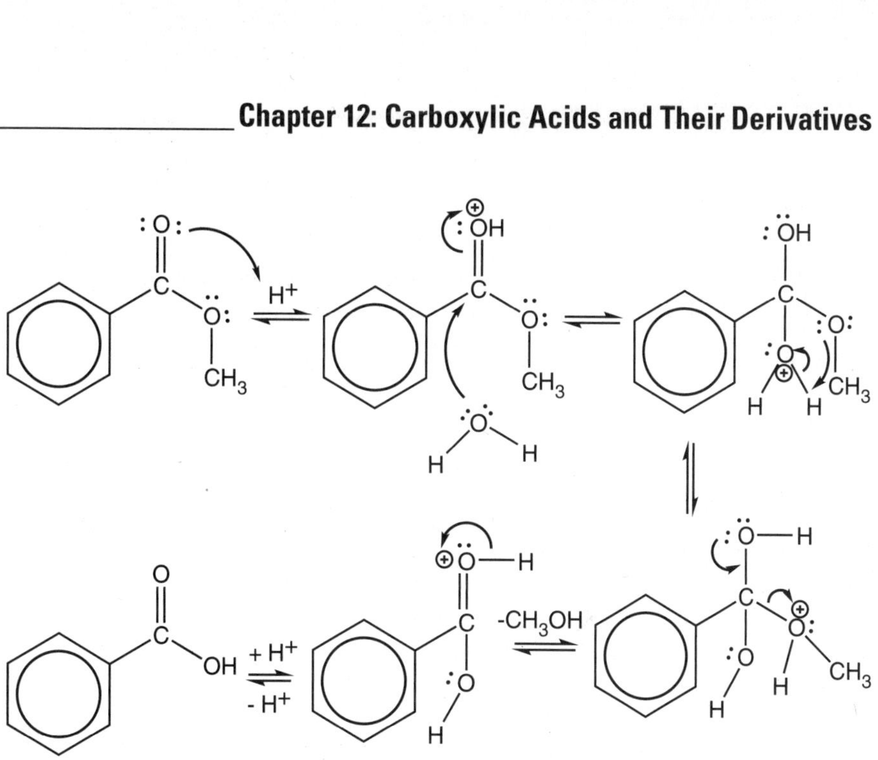

Figure 12-34: The mechanism for acid hydrolysis of an ester.

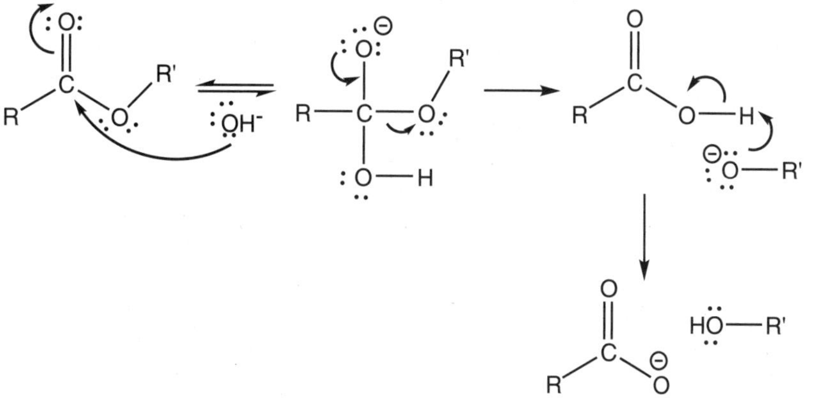

Figure 12-35: The mechanism for base hydrolysis of an ester.

The moles of base = moles of ester.

Amide reactions, ester's cousins

The reactions of amides have similarities to those of esters. Specifically, the reactions covered in this section involve the loss or gain of water (dehydration or hydrolysis).

Acid- or base-catalyzed hydrolysis

Acid hydrolysis of an amide yields a carboxylic acid and an ammonium ion. The mechanism for acid hydrolysis is shown in Figure 12-36. Base hydrolysis of an amide, on the other hand, yields ammonia and a carboxylate ion. You can see this mechanism in Figure 12-37. To identify similarities, compare these mechanisms to the mechanisms for the hydrolysis of esters (refer to Figures 12-34 and 12-35).

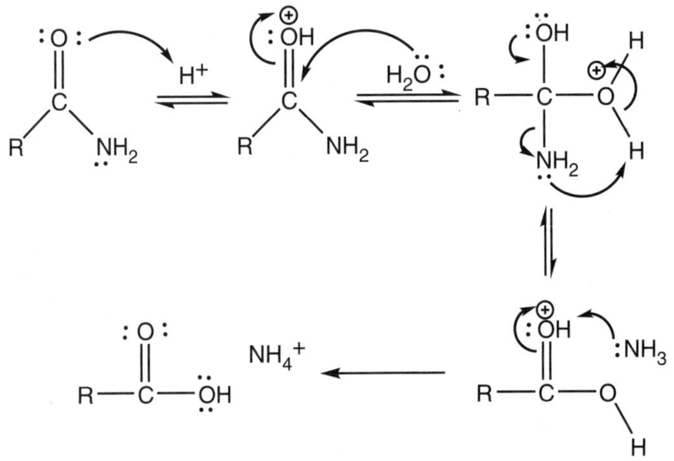

Figure 12-36: The mechanism for the acid hydrolysis of an amide.

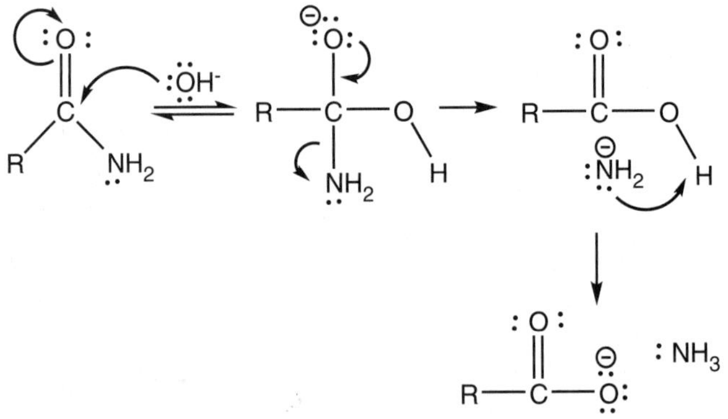

Figure 12-37: The mechanism for the base hydrolysis of an amide.

Dehydration

Amides undergo dehydration. Useful dehydrating agents include $SOCl_2$, P_4O_{10} (P_2O_5), $(AcO)_2O$, and $POCl_3/\Delta$. The product is a nitrile, and in fact, dehydration of an amide is one method to produce aryl nitriles. Figure 12-38 shows the synthesis of 2-ethylhexanenitrile from 2-ethylhexanamide with a 94-percent yield.

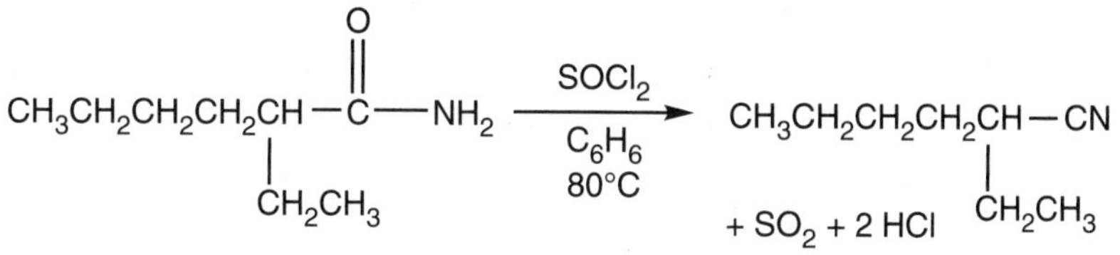

$$CH_3CH_2CH_2CH_2CH(CH_2CH_3)-C(=O)-NH_2 \xrightarrow[C_6H_6,\ 80°C]{SOCl_2} CH_3CH_2CH_2CH_2CH(CH_2CH_3)-CN + SO_2 + 2\ HCl$$

Figure 12-38: The synthesis of 2-ethylhexanenitrile from 2-ethylhexanamide.

Other reactions of carboxylic acids and derivatives

In this section we give you a quick look at a few additional and potentially useful reactions of carboxylic acids and compounds derived from carboxylic acids (derivatives).

Carbonic acid derivatives

Carbonic acid, H_2CO_3, is a diprotic acid. It's unstable and decomposes to carbon dioxide and water (see Figure 12-39).

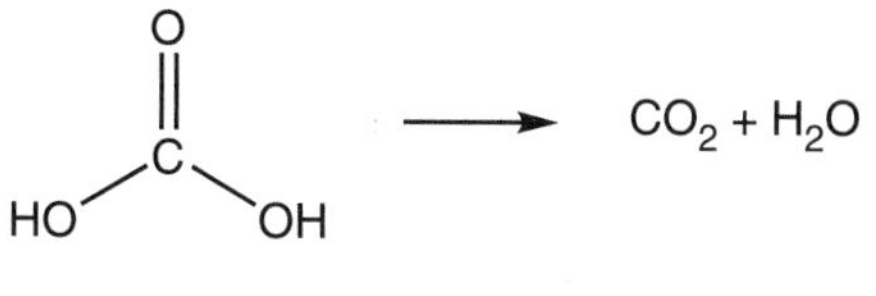

$$HO-C(=O)-OH \longrightarrow CO_2 + H_2O$$

Figure 12-39: The decomposition of carbonic acid.

The replacement of both -OH groups with chlorine produces carbonyl dichloride, also known as *phosgene,* a useful reactant. For example, phosgene reacts with two moles of alcohol to form a dialkyl carbonate. The reaction of phosgene with one mole of alcohol produces an alkyl chloroformate, which is a useful intermediate in organic syntheses. The reaction of phosgene with four moles of ammonia yields urea and two moles of ammonium chloride, NH_4Cl. Figure 12-40 shows the structures of some of these compounds.

One useful reaction utilizing alkyl chloroformate is the reaction with an amine in base to form a carbamate (urethane). Figure 12-41 illustrates this reaction.

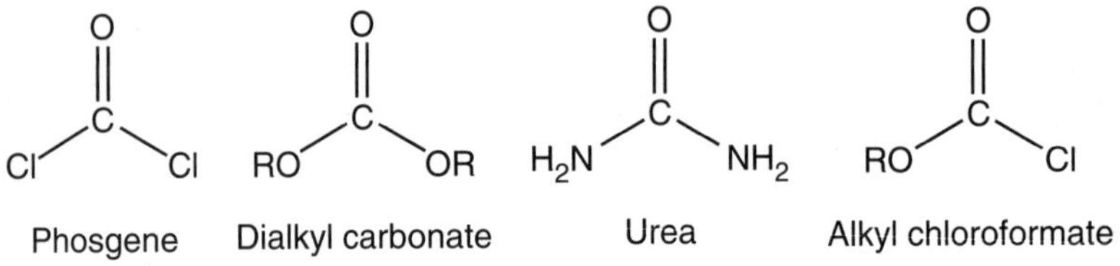

Figure 12-40: Some important carbonic acid derivatives.

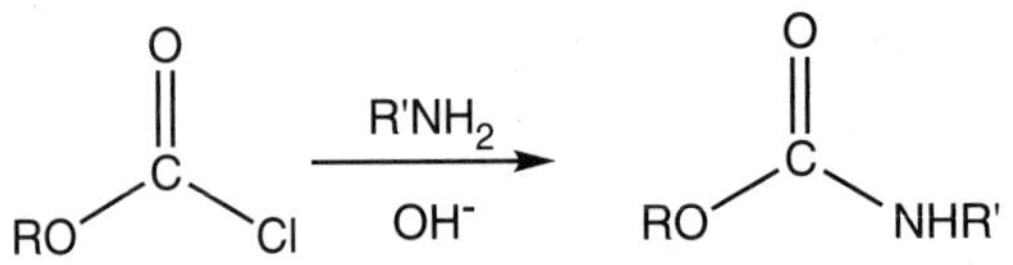

Phosgene — Dialkyl carbonate — Urea — Alkyl chloroformate

Figure 12-41: The reaction of an alkyl chloroformate with an amine in the presence of a base.

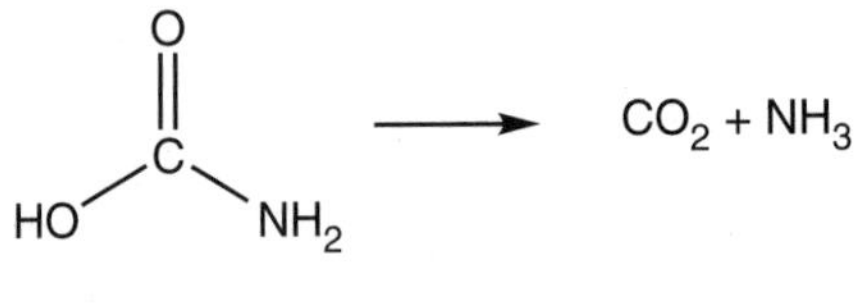

Figure 12-42: The decomposition of carbamic acid.

Another useful carbonic acid derivative is carbamic acid. Like carbonic acid, carbamic acid is unstable (see Figure 12-42).

Decarboxylation

Decarboxylation is the loss of carbon dioxide, which happens easily because of the stability of CO_2. Heating β-keto acids to between 100 and 150 degrees Celsius is one example of a decarboxylation reaction. The mechanism for the decarboxylation of a β-keto acid is in Figure 12-43.

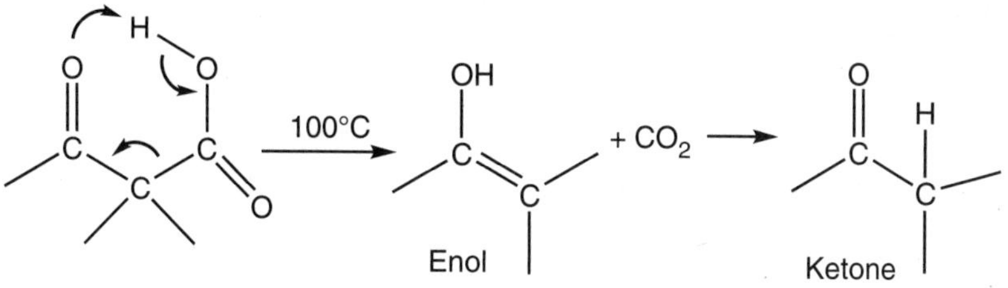

Figure 12-43: The mechanism for the decarboxylation of a β-keto acid.

Hunsdiecker reaction

The Hunsdiecker reaction is a free-radical reaction for the synthesis of an alkyl halide. The starting material comes from the reaction of a silver carboxylate with a solution of a halogen in a solvent such as carbon tetrachloride (see Figure 12-44). The overall free-radical mechanism is shown in Figure 12-45.

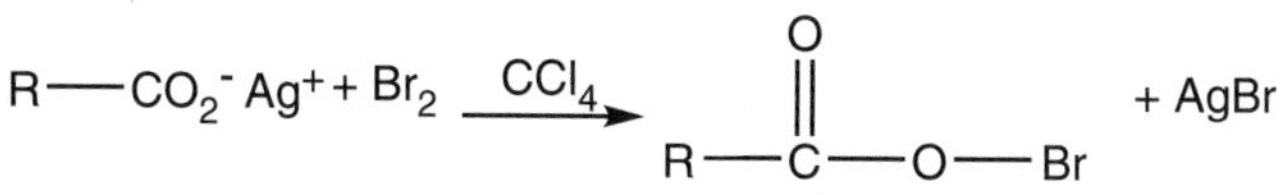

Figure 12-44: The formation of the starting material for the Hunsdiecker reaction.

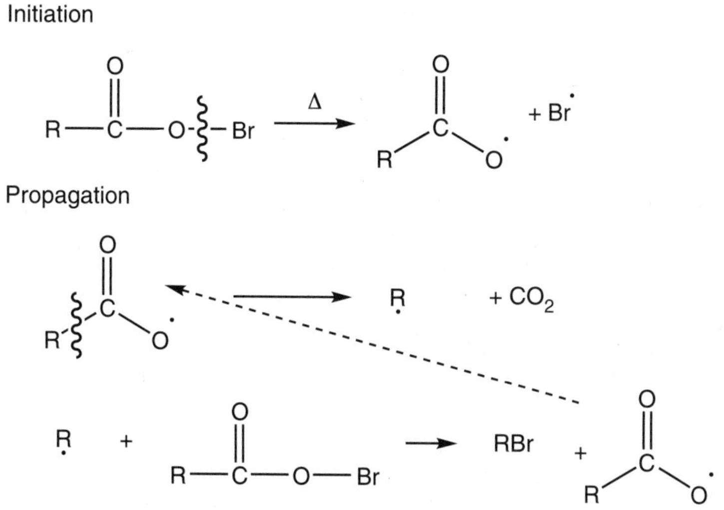

Figure 12-45: The free-radical mechanism of the Hunsdiecker reaction.

Other reagents can be used in the Hunsdiecker reaction, as shown in Figure 12-46.

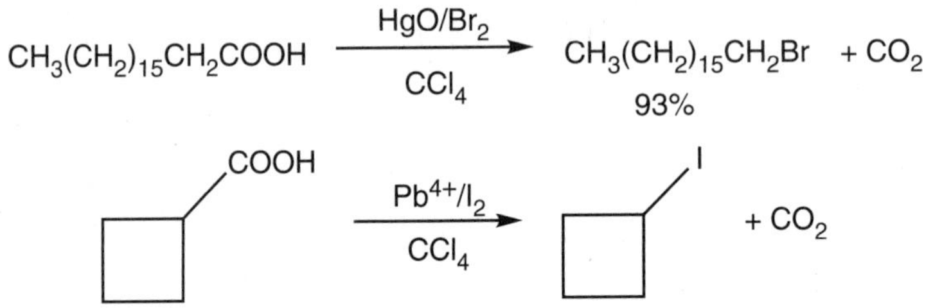
HgO/Br2
CH3(CH2)15CH2COOH
CCl4
CH3(CH2)15CH2Br + CO2
93%
COOH
Pb4+/I2
CCl4
I
+ CO2

Figure 12-46: Additional examples of the Hunsdiecker reaction.

The Reformatsky reaction

The Reformatsky reaction uses an organozinc intermediate to form β-hydoxy esters (see Figure 12-47). The general Reformatsky reaction is in Figure 12-48 and the mechanism is in Figure 12-49.

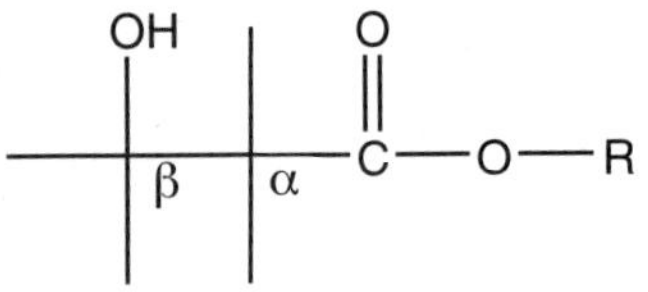
OH
O
β
α
C
O
R

Figure 12-47: A β-hydoxy ester.

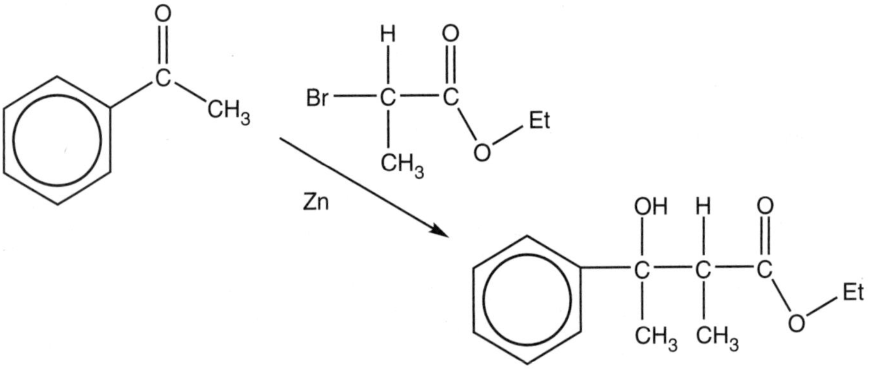
O
C
CH3
H
O
Br
C
C
Et
CH3
O
Zn
OH
H
O
C
C
C
Et
CH3
CH3
O

Figure 12-48: The Reformatsky reaction.

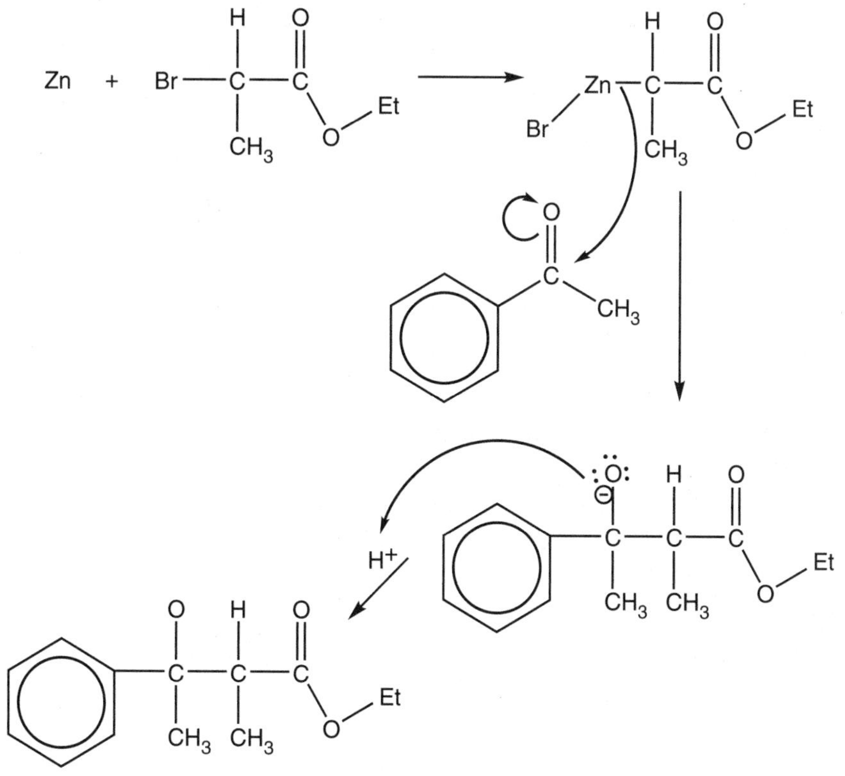

Figure 12-49: The mechanism for the Reformatsky reaction.

Taking a Look at Spectroscopy and Chemical Tests

The carbonyl stretch in the 1,700 cm^{-1} region of the infrared spectra of carbonyl compounds is a very obvious feature of the spectrum for these compounds. In this section we look at some other spectral features of carboxylic acids and their derivatives, and also at some chemical tests that can help you determine what you're dealing with.

Identifying compounds with spectral data

You can use the unique spectroscopy of carboxylic acids and derivatives, described in the following list, to help you identify those compounds.

- **Carboxylic acids:** In addition to the carbonyl stretch, the infrared spectra of carboxylic acids also have a broad OH stretch, which is often shifted to the 3,300–2,500 cm^{-1} region.

  The acid hydrogen is in the $\delta = 10–12$ region of the proton NMR spectrum.
- **Esters:** In addition to the carbonyl stretch, the infrared spectra contains two C-O stretches in the 1,300–1,050 cm^{-1} region. This is the result of having two different R groups attached to the singly bonded oxygen atom.
- **Acyl chlorides:** In acyl chlorides, the carbonyl stretch appears in the 1,850–1,780 cm^{-1} region.
- **Amides:** The carbonyl stretch of amides is in the 1,690–1,630 cm^{-1} region. If one hydrogen atom is attached to the nitrogen atom, the amide has an N-H stretch. If two hydrogen atoms are attached to the nitrogen, it has two N-H stretches. The N-H stretches are in the 3,500–3,300 cm^{-1} region.

Other derivatives have similar spectral properties. For example, acid anhydrides are similar to esters.

Using chemical tests

Carboxylic acids are soluble in either aqueous NaOH or $NaHCO_3$. The other common group of organic acids, phenols, are weaker than the carboxylic acids. Phenols are only soluble in aqueous NaOH. Di- and trinitrophenols are stronger acids than most other phenols, so they are also soluble in aqueous $NaHCO_3$.

The neutralization equivalent is a useful means of determining the molecular weight of a carboxylic acid. The process begins with a simple neutralization reaction of acid with standard base (usually sodium hydroxide). The reaction is

$$\text{Acid} + \text{NaOH} \rightarrow \text{Na}^+ \text{ carboxylate} + H_2O$$

Written in this form, you see that the equivalents (or milliequivalents) of acid are equal to the equivalents of base. The equivalent weight of the acid is the grams of acid divided by the equivalents of base. The equivalent weight of monoprotic acid is equal to the molecular weight. The equivalent weight of diprotic acid is equal to half the molecular weight.

Part IV

Advanced Topics (Every Student's Nightmare)

The 5th Wave
By Rich Tennant
Aaron was envied for being the first student in his chemistry class with an iSpectrometer.
©RICHTENNANT
CHEMISTRY

In this part . . .

Now, don't be scared — the topics in this part aren't all that advanced. First you look at nitrogen compounds, especially the amines. You study synthesis reactions and then the reactions of these nitrogen compounds. We even spend a little time on the synthesis of sulfa drugs.

Next you shift your focus to organometallics, those organic compounds that incorporate metals into their structure. Here you study one of the most useful organic reactions, the Grignard reaction.

Then you get a chapter on additional reactions of carbonyl compounds. In this chapter we take a look at more condensation reactions as well as some ether synthesis reactions.

Finally, this part finishes up with a chapter on biomolecules. If your Organic class doesn't cover this fun and exciting topic, feel free to skip this chapter.

Chapter 13

Amines and Friends

In This Chapter

- Looking at the structure, nomenclature, and properties
- Exploring how basicity affects nitrogen compounds
- Figuring out their synthesis and reactions
- Assessing multistep synthesis
- Analyzing and using spectroscopy to determine nitrogen compounds

Amines and amides are the most important nitrogen-containing organic compounds. Amides are carboxylic acid derivatives, which we cover in Chapter 12. In this chapter we focus on amines. Amines are nitrogen compounds where the nitrogen atom is attached to one to four organic groups.

Breaking Down the Structure and Nomenclature of Nitrogen Compounds

Amines can be split into two general types — primary/secondary/tertiary/quaternary (aliphatic/aryl) and heterocyclic. In aliphatic/aryl amines, a nitrogen atom is attached to one or more alkyl/aryl groups. In heterocyclic amines, the nitrogen atom is part of a ring system.

Amines may be one of the following:

- Primary (1°): One carbon atom connected directly to the nitrogen atom
- Secondary (2°): Two carbon atoms connected directly to the nitrogen atom
- Tertiary (3°): Three carbon atoms connected directly to the nitrogen atom
- Quaternary (4°): Four carbon atoms connected to the nitrogen atom

In quaternary amines (which are technically quaternary ammonium salts), the nitrogen atom has a positive charge.

Primary amines

The common names of primary amines consist of the name of the alkyl branch followed by the name *amine.* The systematic (IUPAC) name of primary amines consists of the name of the alkane with the *-e* replaced by the suffix *-amine.* Some examples of primary amines appear in Figure 13-1. If more than one amine group is present, you need to use the appropriate prefix. For example, $H_2NCH_2CH_2CH_2CH_2NH_2$ is 1,4-butandiamine, because the amino groups are attached to the first and fourth carbons.

$CH_3CH_2NH_2$
Ethyl amine
Ethanamine

NH_2
Cyclopentyl amine
Cyclopentanamine

CH_2-NH_2
Benzyl amine
Phenylmethanamine

Figure 13-1: Some examples of primary amines.

Aniline is a simple aromatic compound composed of an amino group attached to a benzene ring. Other aromatic amines are aniline derivatives. Some examples of aromatic amines are shown in Figure 13-2.

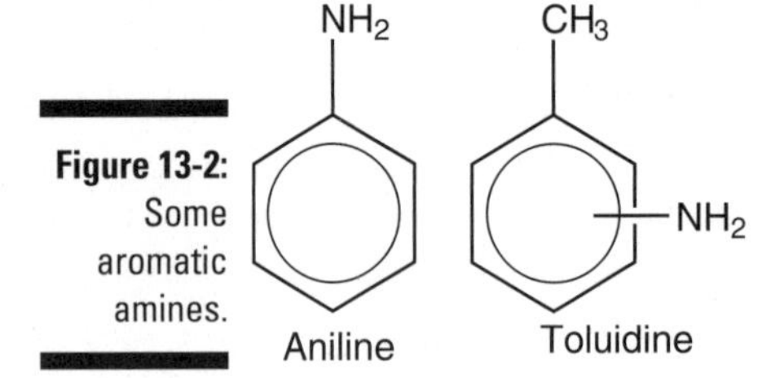

Figure 13-2: Some aromatic amines.

The amine group has a lower priority in numbering the ring positions around the ring than -OH and other oxygen-containing groups. Some examples showing the lower priority of the amine group are in Figure 13-3.

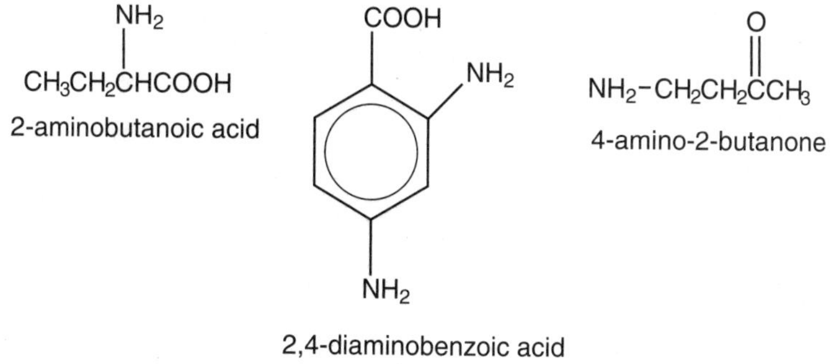

$$CH_3CH_2CH(NH_2)COOH$$

2-aminobutanoic acid

2,4-diaminobenzoic acid

$$NH_2-CH_2CH_2C(=O)CH_3$$

4-amino-2-butanone

Figure 13-3: Some primary amines containing other functional groups.

Secondary and tertiary amines

The common names of secondary and tertiary amines are an extension of the common names of primary amines, where the organic groups are named as branches followed by the word *amine.* For example, $(CH_3CH_2)_3N$ is triethyl amine. The systematic (IUPAC) names utilize the procedure for primary amines (the name of the alkane with the *-e* replaced by the suffix *-amine)* plus the names of the remaining organic groups preceded by an *N* to indicate attachment to the nitrogen atom. Some names of secondary and tertiary amines appear in Figure 13-4.

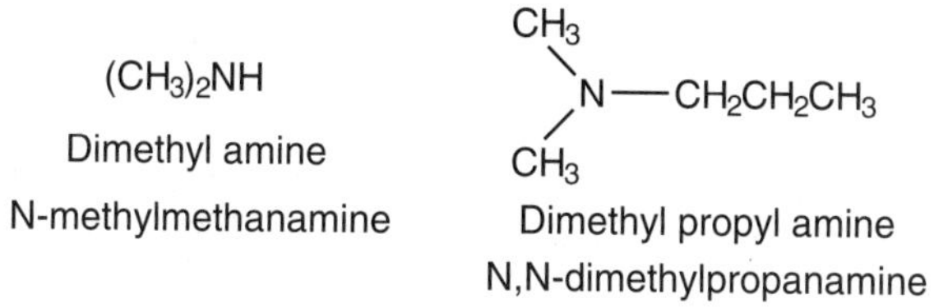

$(CH_3)_2NH$

Dimethyl amine

N-methylmethanamine

$(CH_3)_2N-CH_2CH_2CH_3$

Dimethyl propyl amine

N,N-dimethylpropanamine

Figure 13-4: Some secondary and tertiary amines.

Quaternary amines (quaternary ammonium salts)

Ammonium salts contain a nitrogen atom with four bonds that has a positive charge. Four-bonded nitrogen atoms derived from amines are ammonium ions (if they're derived from aniline, they're anilinium ions). If the four bonds are all to carbon atoms, the nitrogen atom is quaternary. Salts contain a cation (named first) and an anion (named last). Typical anions include Cl^- (chloride), Br^- (bromide), HSO_4^- (hydrogen sulfate or bisulfate), and NO_3^- (nitrate). Figure 13-5 shows two examples of ammonium ions.

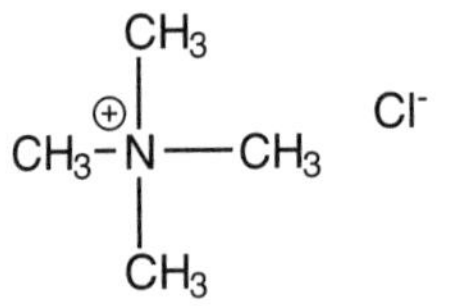

Tetramethylammonium chloride

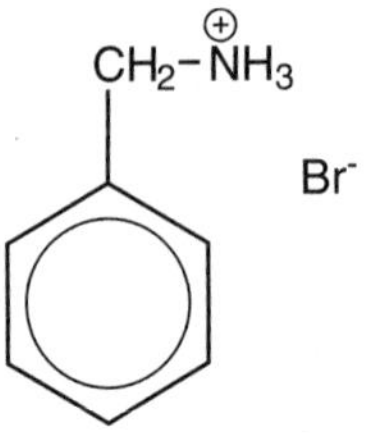

Benzylammonium bromide

Figure 13-5: Two examples containing ammonium ions.

Two additional examples of amine nomenclature are shown in Figure 13-6. Note that in the *p*-aminobenzoic acid the oxygen-containing group takes precedence in the naming, so that the compound is then named as a substituted benzoic acid and not a substituted aniline, as is done in the N,N-dimethylaniline beside it.

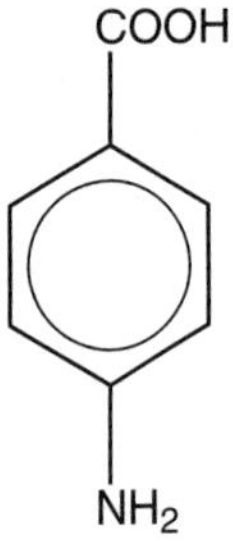

p-aminobenzoic acid 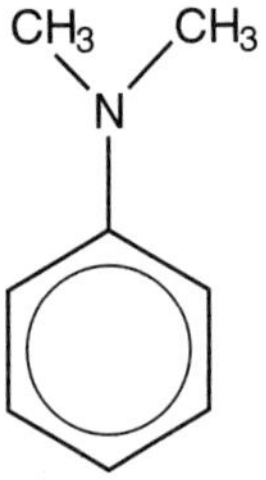 N,N-dimethylaniline

Figure 13-6: Two additional examples of amine nomenclature.

Heterocyclics

Heterocyclics are ring systems containing something other than carbon in the ring. The names of some nitrogen-containing heterocyclic compounds (with a single nitrogen) are listed in Figure 13-7. Heterocyclic systems may contain more than one nitrogen atom, though, and some examples are shown in Figure 13-8. These heterocyclics are important in biological and biochemical systems. For a detailed discussion of these topics, see *Biochemistry For Dummies.*

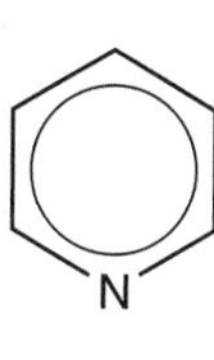

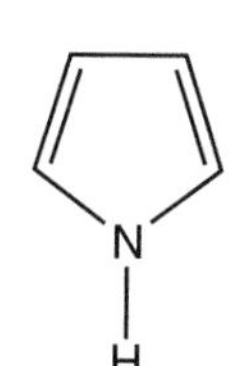

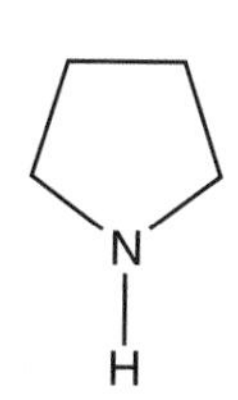

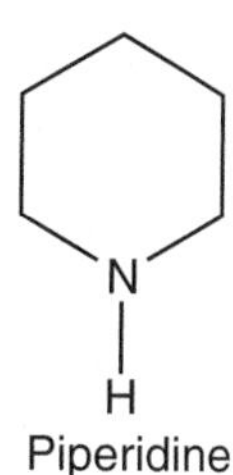

Figure 13-7: Some examples of nitrogen-containing heterocyclic compounds.

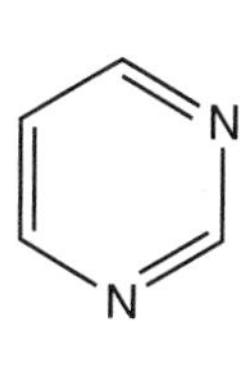

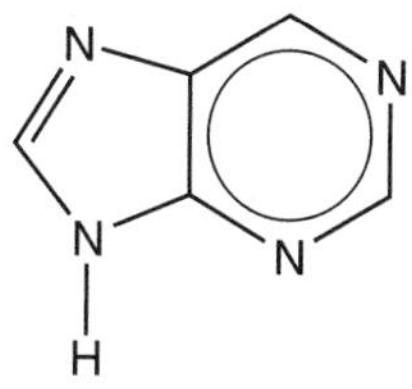

Figure 13-8: Some examples of nitrogen-containing heterocyclic compounds containing more than one nitrogen atom.

Sizing Up the Physical Properties

The high electronegativity of the nitrogen atom means that the carbon-nitrogen bond of amines is polar. This results in an attraction between two polar molecules (dipole-dipole intermolecular forces), which increases the melting and boiling points above those of hydrocarbons. However, in primary and secondary amines the ability to hydrogen bond overshadows the simple dipole-dipole forces present in tertiary amines. For this reason, primary and secondary amines have much higher melting and boiling points than tertiary amines. Hydrogen bonding in amines is not as strong as hydrogen bonding in alcohols; therefore, the melting and boiling points of amines are lower than those of comparable alcohols. (See Chapter 3 for a discussion of hydrogen-bonding in alcohols.)

In general, amines with up to about six carbon atoms are soluble in water. Primary, secondary, and tertiary amines react with acids to produce ammonium ions. These ammonium ions and quaternary ammonium ions are also soluble in water.

Amines can be separated from other organic compounds by treating the mixture with an aqueous acid, converting the amines to ammonium ions. These ammonium ions dissolve in the aqueous acid, leaving the other organic materials behind. Separating the aqueous layer results in an acid extract containing the ammonium ions. Adding base to the acid extract until the solution is basic converts the ammonium ions back to the free amines, which then separate from the aqueous layer.

Many amines have a very distinctive odor, like dead fish or worse.

Understanding the Basicity of Nitrogen Compounds

Primary, secondary, and tertiary amines react with acids to form amine salts. This is because of the basic nature of the amines.

Amines are both Brønsted-Lowry bases (they accept hydrogen ions from acids) and Lewis bases (they furnish an electron pair to Lewis acids). As Brønsted-Lowry bases they have K_b values. Aliphatic amines have K_b values of approximately 10^{-4}, and aromatic amines have values near 10^{-10}. (These values compare to a value of $\approx 10^{-5}$ for ammonia.)

The increase in the K_b values of aliphatic amines (thus making them stronger bases as compared to ammonia) is due to the electron-releasing nature of alkyl groups. This release of electrons "pumps" electron density back to the nitrogen atom, which stabilizes the positive charge.

Compared to the aliphatic amines, the aromatic amines have lower K_b values. This lower value indicates that the product of the protonation of aromatic amines is less stable. The decrease in stability is due to a loss in resonance stabilization of the protonated form.

An amine group attached to an aromatic ring is an activating *o-p*-director, as seen in Chapter 8.

The lone pair of electrons on the nitrogen atom makes the amines Lewis bases. As Lewis bases, they may behave as nucleophiles. Because aromatic amines are resonance stabilized, they're weaker nucleophiles than alkyl amines.

Synthesizing Nitrogen Compounds

Amines can be prepared a number of ways. These methods include nucleophilic substitution reactions, reduction reactions, and oxidation reactions.

Nucleophilic substitution reactions

Synthesizing amines with nucleophilic substitution reactions is normally an S_N2 process. This means that methyl amines react more readily than primary amines, and secondary and tertiary amines show very little reactivity. Figure 13-9 illustrates the basic reaction. The resultant amine may react further to give a mixed group of products as shown in the following reaction. Using a large excess of ammonia minimizes the chances for multiple alkylations.

$$H_3N: \curvearrowright R\text{—}X \xrightarrow{S_N2} \overset{\oplus}{N}H_3R \ X^- \xrightarrow[\text{or } NH_3]{OH^-} NH_2R$$

Figure 13-9: The basic reaction for a nucleophilic substitution reaction to produce an amine.

$$CH_3(CH_2)_6CH_2Br + NH_3 \rightarrow CH_3(CH_2)_6CH_2NH_2 \text{ octylamine (45\%)}$$
$$+ [CH_3(CH_2)_6CH_2]_2NH \text{ (dioctylamine [43\%])}$$
$$+ [CH_3(CH_2)_6CH_2]_3N \text{ (trioctylamine [trace])}$$
$$+ [CH_3(CH_2)_6CH_2]_4N^+ \text{ (trace)}$$

Aromatic halides don't react unless an electron-withdrawing group is attached to the ring. For example, bromobenzene doesn't react. An example showing the reaction when an electron-withdrawing group is present is illustrated in Figure 13-10, the nucleophilic substitution attack on *p*-bromonitrobenzene.

The azide ion is a better nucleophile than amines, but it has to be reduced to the amine after nucleophilic substitution. Lithium aluminum hydride ($LiAlH_4$) in ether followed by treatment with water reduces the azide ion to the amine. Figures 13-11 and 13-12 illustrate two examples of this reaction.

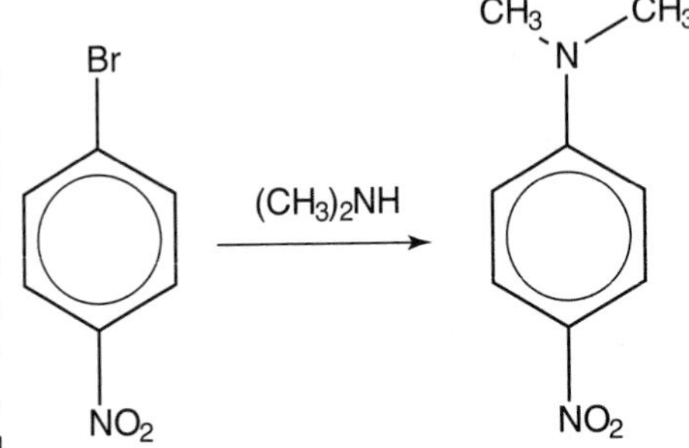

Figure 13-10: The nucleophilic substitution attack on *p*-bromonitrobenzene.

Azides are explosive, releasing nitrogen gas. That's why they're used to inflate car air bags.

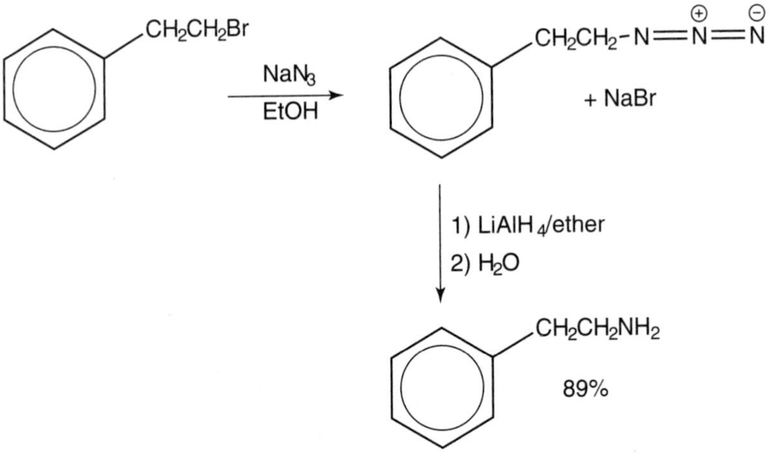

$C_6H_5CH_2CH_2Br \xrightarrow[EtOH]{NaN_3} C_6H_5CH_2CH_2{-}N{=}\overset{\oplus}{N}{=}\overset{\ominus}{N} + NaBr$

1) $LiAlH_4$/ether
2) H_2O

$C_6H_5CH_2CH_2NH_2$ 89%

Figure 13-11: The preparation of an amine from an azide.

$C_6H_{11}Br \xrightarrow[EtOH]{NaN_3} C_6H_{11}N_3 + NaBr$

1) $LiAlH_4$/ether
2) H_2O

$C_6H_{11}NH_2$ 54%

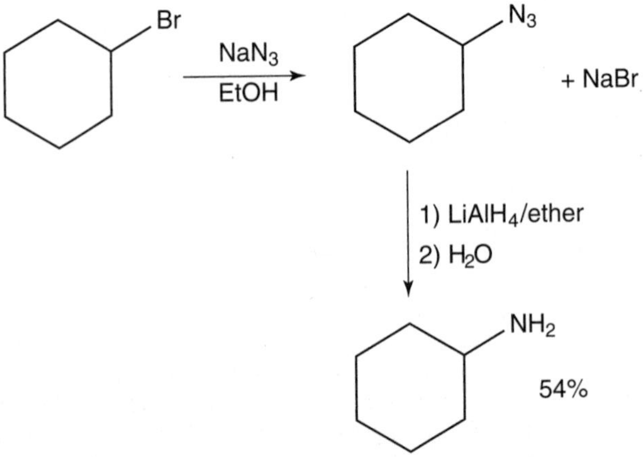

Figure 13-12: Another preparation of an amine from an azide.

The *Gabriel synthesis* of amines uses potassium phthalimide (prepared from the reaction of phthalimide with potassium hydroxide). The structure and preparation of potassium phthalimide is shown in Figure 13-13. The extensive conjugation (resonance) makes the ion very stable. An example of the Gabriel synthesis is in Figure 13-14. (The N_2H_4 reactant is hydrazine.) The Gabriel synthesis employs an S_N2 mechanism, so it works best on primary alkyl halides and less well on secondary alkyl halides. It doesn't work on tertiary alkyl halides or aryl halides.

Figure 13-13: The preparation of potassium phthalimide from phthalimide and potassium hydroxide.

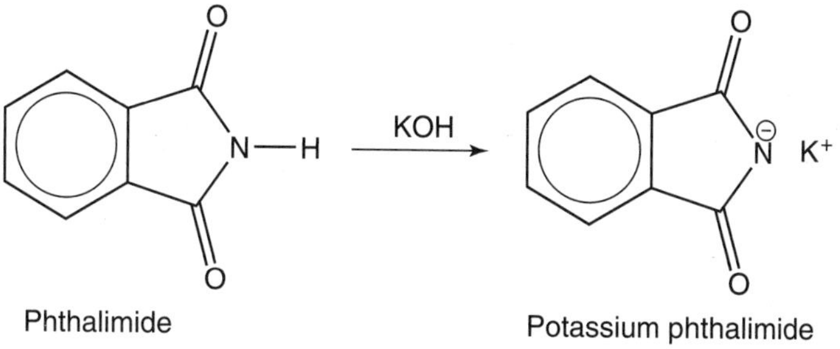

Phthalimide —KOH→ Potassium phthalimide

Figure 13-14: Using the Gabriel synthesis to produce an amine.

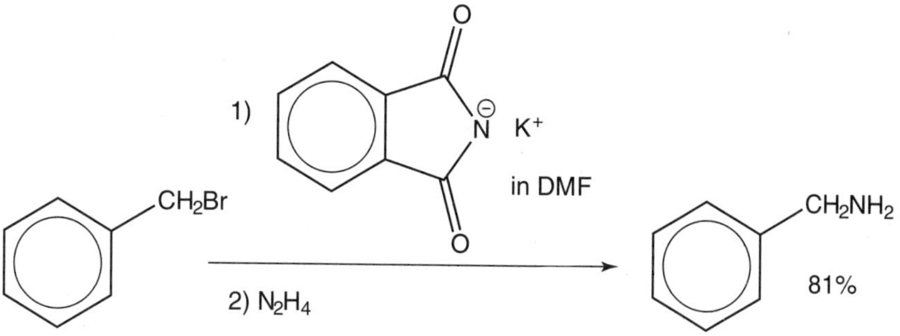

$C_6H_5CH_2Br$ —1) potassium phthalimide in DMF; 2) N_2H_4→ $C_6H_5CH_2NH_2$ 81%

Reduction preparations

We can reduce a number of nitrogen species to an amine. In the following sections we take a look at some of the methods that can be used.

Nitro reductions

Organic nitro compounds, RNO_2, can be reduced to amines. The R may be either alkyl or aryl. Aromatic nitro compounds are easy to prepare and reduce. Their preparation utilizes a mixture of nitric acid and sulfuric acid to nitrate the aromatic ring. (However, multiple nitrations may occur, potentially causing problems.) The nitro group can be reduced with a number of simple methods, including catalytic hydrogenation of the nitro compound or the reaction of the nitro compound with a metal (Fe, Zn, or Sn) in the presence of hydrochloric acid followed by the addition of excess base. The generic symbol for all of these reductions is [H]. An example of the formation of a nitro compound followed by reduction is shown in Figure 13-15.

HNO_3/H_2SO_4 → NO_2 [H] → NH_2

Figure 13-15: The preparation of a nitro compound followed by reduction to an amine.

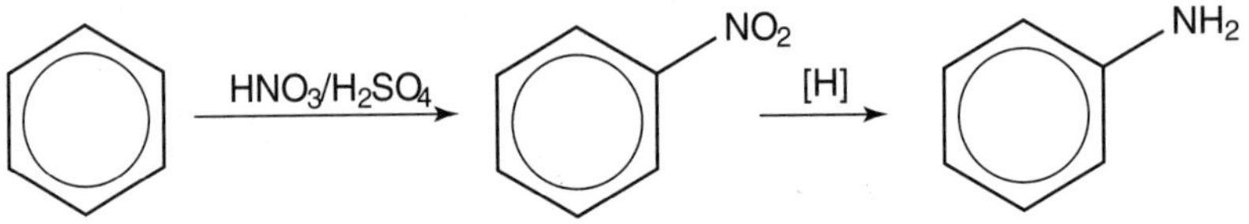

Reductive amination

A number of organic species, including amides, oximes, and nitriles, undergo *reductive amination,* a variety of reduction reactions that produce amines. In general, these processes involve imines, R=N-R, or related species. Reduction processes include hydrogenation using Raney nickel as the catalyst (for nitriles), the reaction with sodium/EtOH (for oximes), and the use of lithium aluminum hydride, $LiAlH_4$ (for amides or nitriles). Figure 13-16 illustrates the preparation of amphetamine by reductive amination.

CH_2 C O CH_3 $:NH_3$ $-H_2O$ CH_2 CH_3 C N H

CH_2 CH_3 CH NH_2 H_2/Raney nickel

Amphetamine

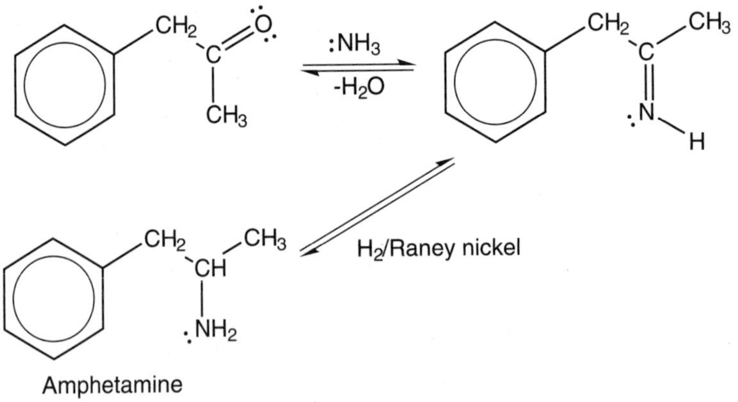

Figure 13-16: The preparation of amphetamine by reductive amination.

In some cases, you may run into problems with reductive amination. The upper pathway in Figure 13-17 illustrates one of the problems (a secondary alkyl halide and a weak nucleophile leads to elimination instead of substitution), which necessitates the use of the lower pathway.

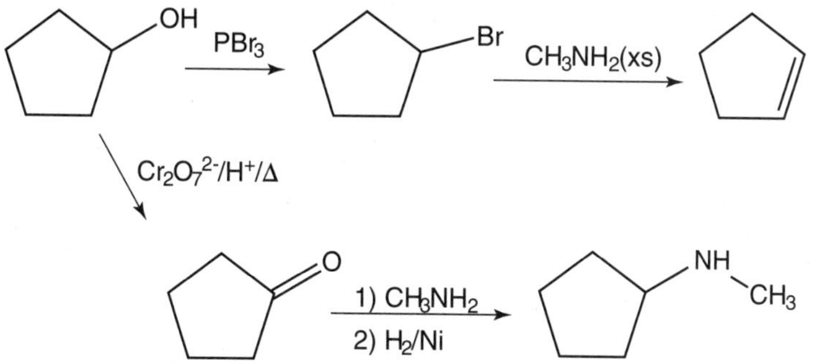

Figure 13-17: In the upper pathway, reductive amination fails. The lower pathway works.

Another type of reductive amination is shown in Figure 13-18. This reaction illustrates the formation of an amine from a ketone through the formation of an intermediate oxime. Figure 13-19 shows the conversion of a nitrile to an amine. (The nitrile can be formed by the action of cyanide ion, CN^-, on a halide via an S_N2 mechanism.)

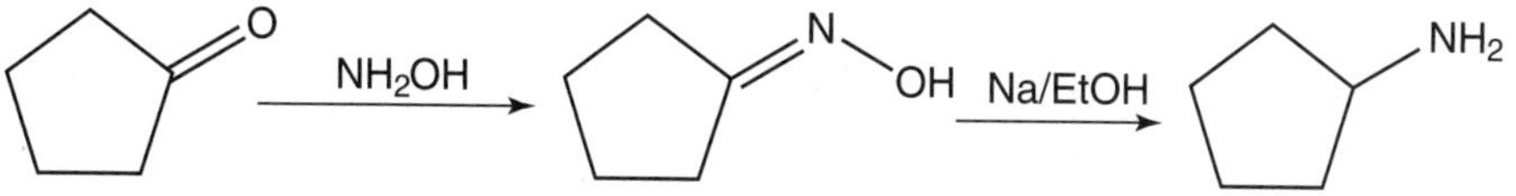

Figure 13-18: The formation of an amine from a ketone via an oxime.

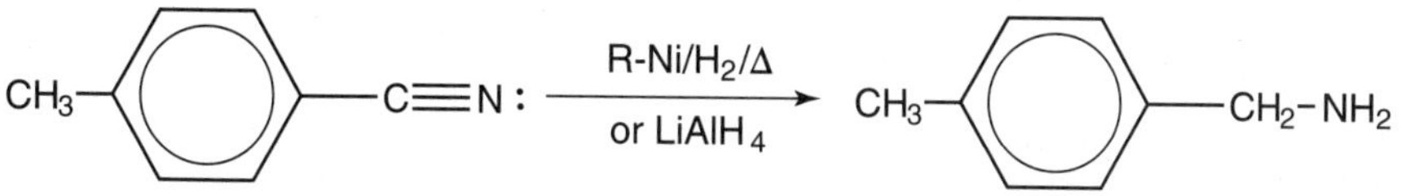

Figure 13-19: The formation of an amine from a nitrile.

Hofmann rearrangement (degradation)

The *Hofmann rearrangement* is a useful means of converting an amide to an amine. The nitrogen of the amide must be primary. The reaction results in the loss of one carbon atom. Figure 13-20 illustrates the generic Hofmann rearrangement reaction, and the generic mechanism of the Hofmann rearrangement reaction is shown in Figure 13-21. In the reaction, an intermediate isocyanate forms. The R in Figure 13-21 may be alkyl or aryl. The first intermediate in the mechanism is resonance stabilized, which promotes the reaction (it's similar in structure to an enolate). In addition, the third intermediate is also resonance stabilized.

$$R{-}C(=O)NH_2 \xrightarrow[Br_2]{OH^-} RNH_2 + CO_3^{2-} + H_2O + Br^-$$

Figure 13-20: The generic Hofmann rearrangement reaction.

$$R{-}C(=O)NH_2 \xrightarrow{:\ddot{O}H^-} R{-}C(=O)\ddot{N}H^- + H_2O \xrightarrow{Br{-}Br} R{-}C(=O)N(H)Br + Br^- \xrightarrow{:\ddot{O}H^-} R{-}C(=O)\ddot{N}^-Br \longrightarrow R{-}C(=O)\ddot{N}: \longrightarrow R{-}N{=}C{=}O$$

An isocyanate

$$R{-}N{=}C{=}O \xrightarrow{:\ddot{O}H^- / H_2O} CO_3^{2-} + H_2NR$$

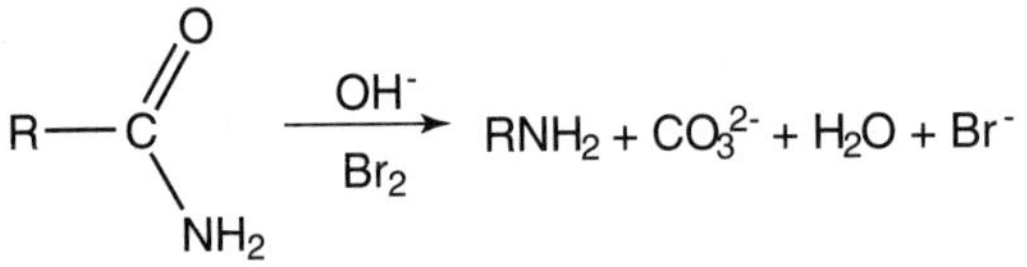

Figure 13-21: The mechanism for the Hofmann rearrangement reaction.

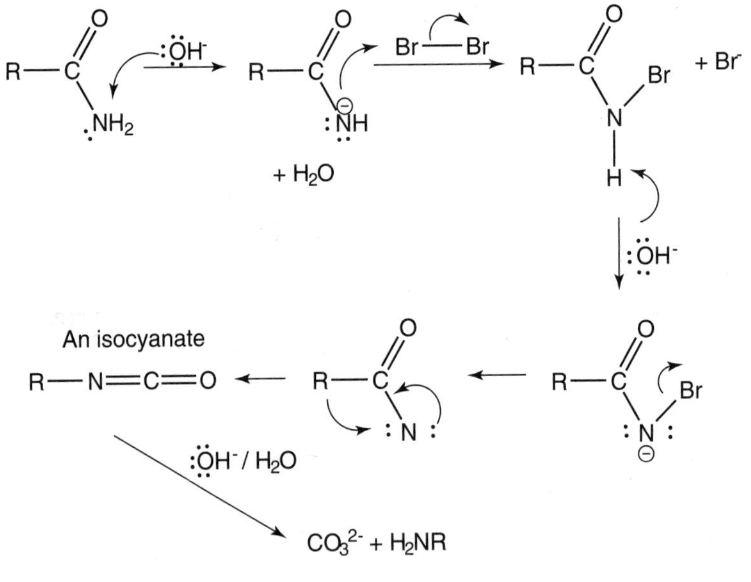

A related reaction is the *Curtius rearrangement,* which replaces the amide with an azide, RCO-N_3. The azide can be formed by the reaction of an acyl chloride with sodium azide.

Seeing How Nitrogen Compounds React

Primary, secondary, and tertiary amines behave as Brønsted-Lowry bases. These amines react like ammonia, adding H^+ to produce an ammonium ion.

Amines may also behave as nucleophiles (Lewis bases). Primary amines are stronger nucleophiles than secondary amines, which, in turn, are stronger nucleophiles than tertiary amines. As nucleophiles, amines attack acid chlorides to form amides. Later in this chapter you see that they're important in the formation of sulfonamides.

Reactions with nitrous acid

Amines react with nitrous acid (formed by the reaction $NaNO_2 + H^+ \rightarrow HNO_2$) to give a variety of products. The nitrous acid isn't very stable, so generating it in place from sodium nitrite is necessary. (Sodium nitrite is a meat preservative and a color enhancer.) Under acidic conditions, nitrous acid dehydrates to produce the nitrosonium ion, NO^+. The NO^+ ion is a weak electrophile that's resonance stabilized. (See Chapter 7.) Figure 13-22 illustrates the dehydration of nitrous acid.

Figure 13-22: The dehydration of nitrous acid.

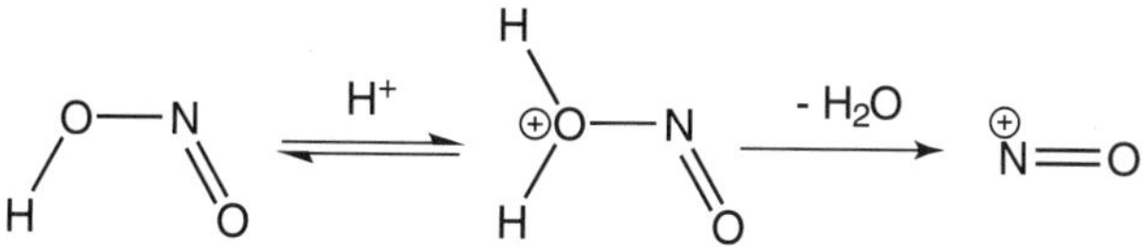

Tertiary amines don't react directly with acidic sodium nitrite. However, as seen in Figure 13-23, even though the tertiary amine doesn't react, its presence activates an aromatic system leading to attack by NO^+.

Figure 13-23: The attack of an activated aromatic system by NO^+.

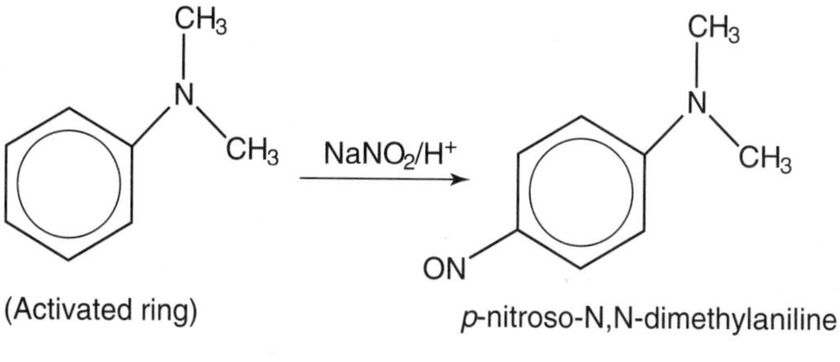

(Activated ring)

p-nitroso-N,N-dimethylaniline

As seen in Figure 13-24, secondary amines react directly with acidic sodium nitrite to form a nitrosamine. (These compounds are very, very toxic.) Primary amines react under similar conditions to form unstable diazonium salts (see Figure 13-25). Diazonium salts readily lose the very stable N_2 to form reactive carbocations that are useful in a number of synthetic pathways. Figure 13-26 shows the resonance stabilization of a diazonium ion.

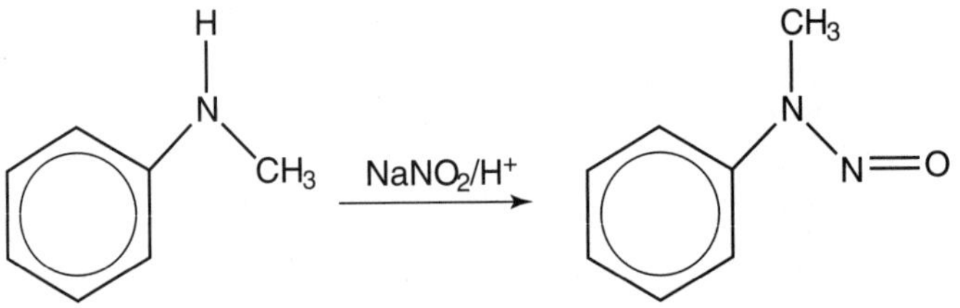

Figure 13-24: The formation of a nitrosamine by the reaction of a secondary amine with acidic sodium nitrite.

H
N
CH_3
$NaNO_2/H^+$
CH_3
N
N=O

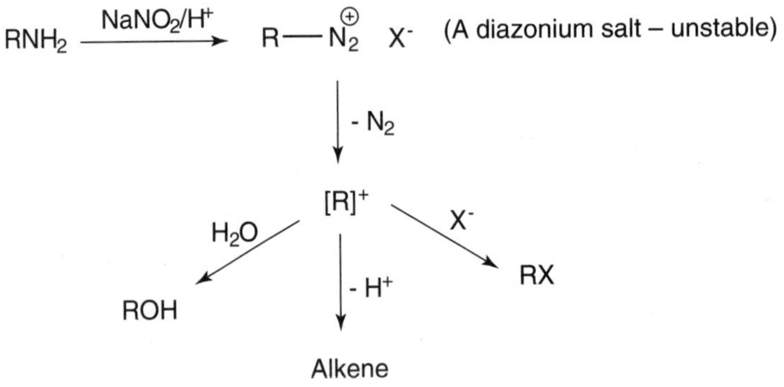

Figure 13-25: The formation of a diazonium salt, its decomposition, and several possible outcomes of the carbocation formed by decomposition.

RNH_2
$NaNO_2/H^+$
$R—N_2^{\oplus}$ X^-
(A diazonium salt – unstable)
$- N_2$
$[R]^+$
H_2O
ROH
$- H^+$
Alkene
X^-
RX

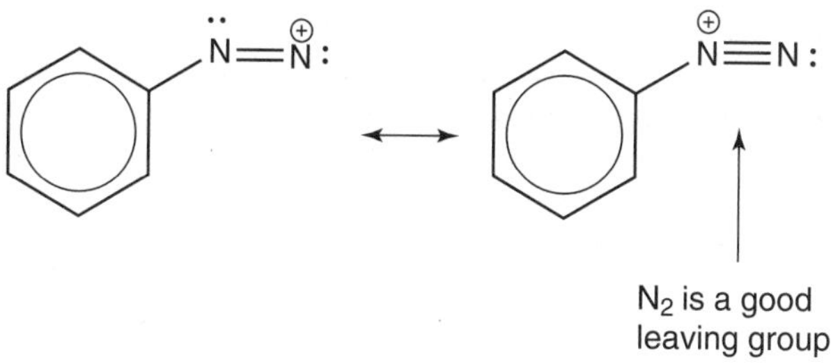

Figure 13-26: Resonance stabilization of a diazonium ion.

N_2 is a good leaving group

Replacement reactions

Many kinds of replacement reactions involve nitrogen compounds. A good many of these processes, described in the following sections, utilize diazonium salts.

Sandmeyer reaction

The *Sandmeyer reaction* utilizes a diazonium salt to produce an aryl halide. The process begins by converting an amine to a diazonium salt. Decomposition of the diazonium salt in the presence of a copper(I) halide places the halide ion into the position originally occupied by the amine. The most useful copper(I) halides are CuCl and CuBr; in addition, the copper(I) pseudohalides, such as CuCN, also works by placing a nitrile in the position originally occupied by the amine. Figure 13-27 shows an example of the Sandmeyer reaction.

NH_2 CH_3 HNO_2 H_2SO_4 $N_2^{\oplus}$ HSO_4^- CH_3 CuBr Br CH_3 73%

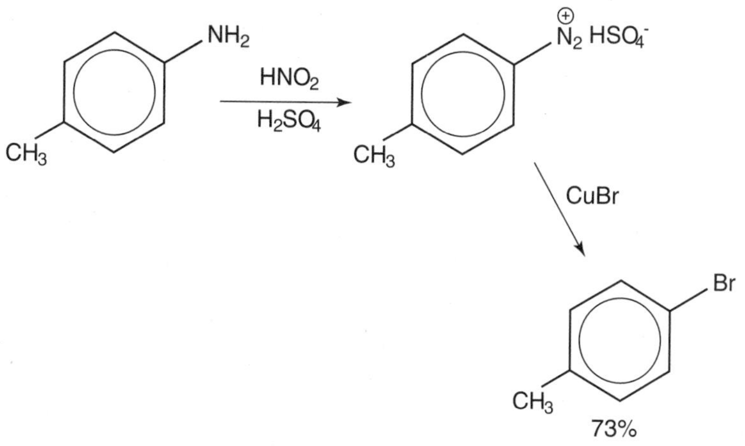

Figure 13-27: An example of the Sandmeyer reaction.

Replacement by iodide ion

This reaction is similar to the Sandmeyer reaction, but the halide source is potassium iodide (KI). Figure 13-28 illustrates this reaction.

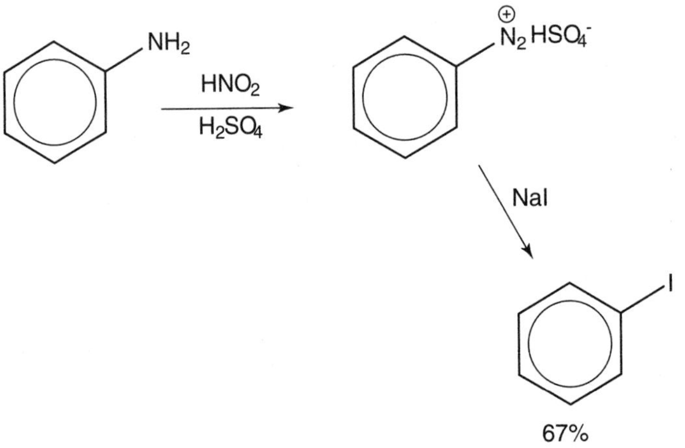

$$C_6H_5NH_2 \xrightarrow[H_2SO_4]{HNO_2} C_6H_5N_2^{+}\,HSO_4^{-} \xrightarrow{NaI} C_6H_5I \text{ (67\%)}$$

Figure 13-28: The preparation of an aryl iodide.

Schiemann reaction

The *Schiemann reaction* is a means of preparing aryl fluorides. The process is similar to the Sandmeyer reaction. The source of the fluoride is fluoroboric acid, HBF_4. Figure 13-29 illustrates the Schiemann reaction.

$$C_6H_5NH_2 \xrightarrow[HCl,\ Cold]{HNO_2} C_6H_5N_2^{+}\,Cl^{-} \xrightarrow{HBF_4} C_6H_5N_2^{+}\,BF_4^{-} \xrightarrow{\Delta} C_6H_5F + N_2 + BF_3$$

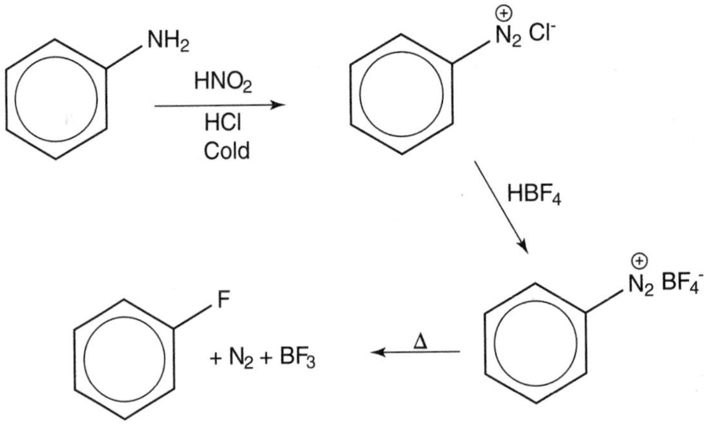

Figure 13-29: The Schiemann reaction.

Formation of ethers and phenols

Diazonium salts can also be used to form ethers and phenols. Reaction of diazonium salt with an alcohol generates an ether, while thermal hydrolysis of the diazonium salt yields a phenol. Figure 13-30 illustrates both formations. As seen in Figure 13-31, this process also works on substituted aromatic systems.

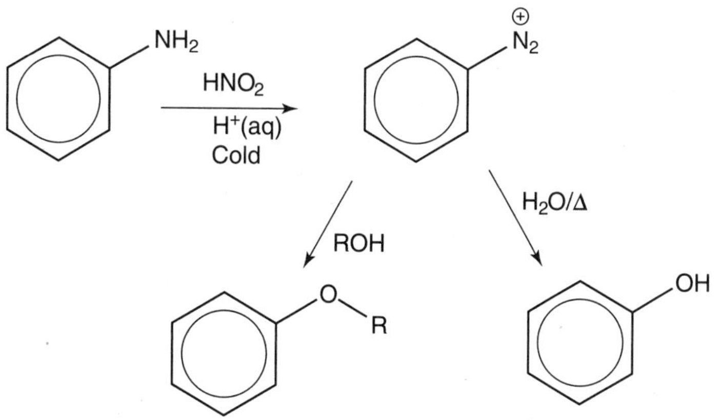

Figure 13-30: Using a diazonium salt to produce an ether and a phenol.

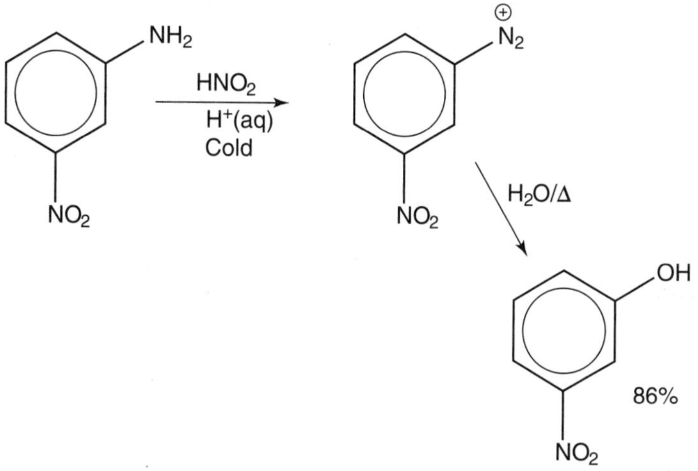

Figure 13-31: The conversion of a substituted aromatic amine to a substituted phenol.

Deamination

Deamination replaces the amine group with a hydrogen atom. This process normally uses hypophosphorous acid, H_3PO_2. The general process for deamination is in Figure 13-32. This is a synthetically useful technique that leads to different products than other replacement methods. Figure 13-33 illustrates the formation of two different dibromotoluenes.

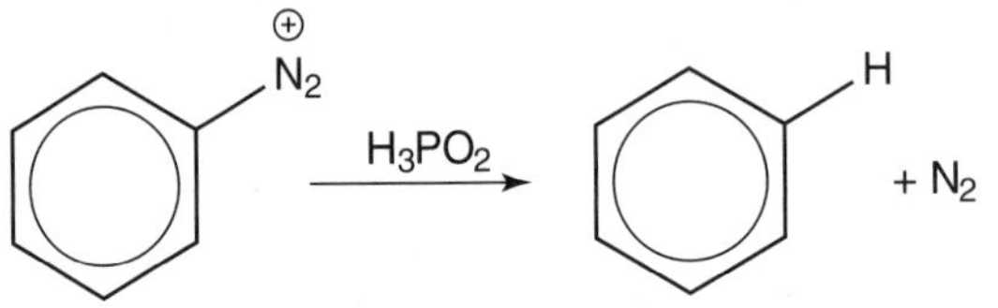

Figure 13-32: The general process for deamination.

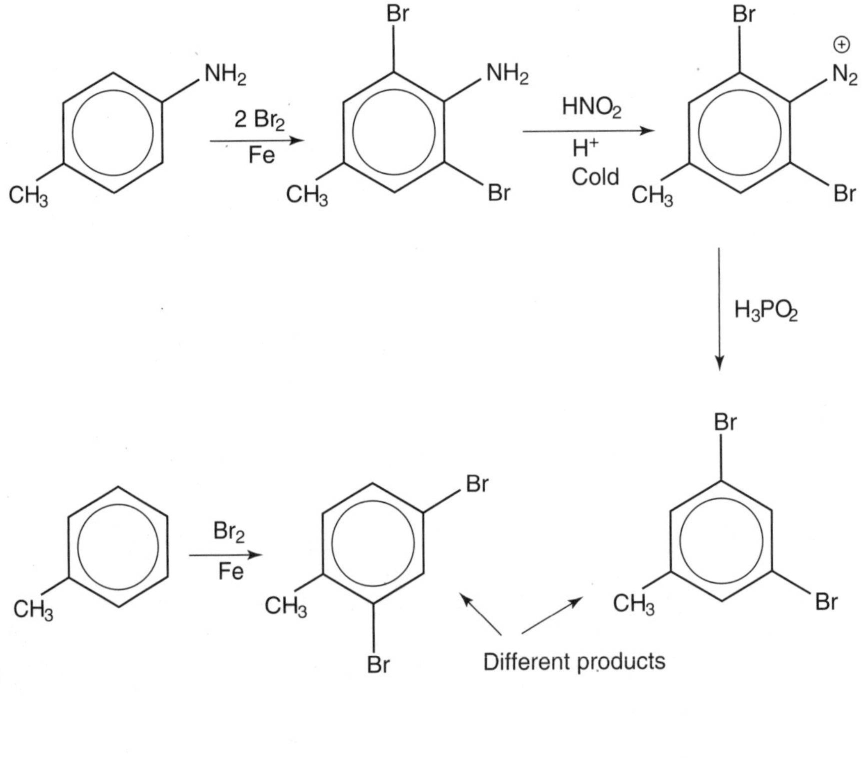

Figure 13-33: The formation of two different dibromotoluenes.

Coupling reactions of diazonium salts

Diazonium salts can attack an aromatic ring that's been activated by certain groups, including -OH and -NR_2. The product is an azo compound. Figure 13-34 illustrates the basic reaction, with Y representing the activator. Normally para attack occurs because the ortho position is crowded; however, if the para position is blocked, then ortho attack may occur. In many cases, extended conjugation is present, leading to absorption of light in the visible portion of the spectrum. Azo dyes are examples of these compounds.

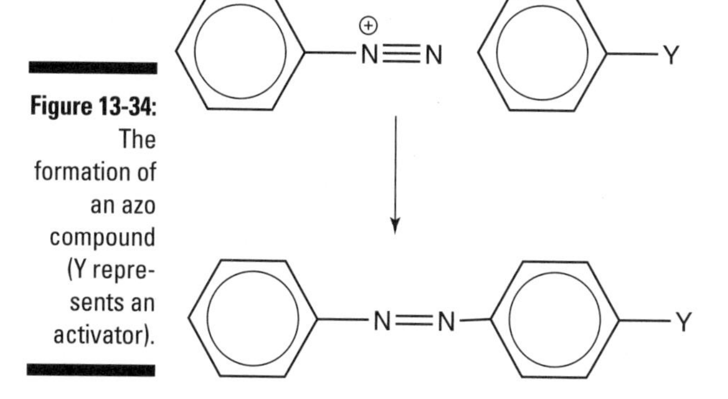

Figure 13-34: The formation of an azo compound (Y represents an activator).

Figure 13-35 illustrates the mechanism for the formation of *p*-hydroxyazobenzene beginning with the reaction of benzenediazonium chloride with an aromatic system activated by an -OH. The mechanism for the reaction with an amine is similar. Figure 13-36 illustrates the reaction of a diazonium salt with an amine. The product of the reaction in Figure 13-36 is *p*-(dimethylamino) azobenzene.

Reactions with sulfonyl chlorides

Amines can react with sulfonyl chlorides, but the product of the reaction depends upon the type of amine.

- Primary amines react to form an N-substituted sulfonamide (as shown in Figure 13-37).
- Secondary amines react to form an N,N-disubstituted sulfonamide (Figure 13-38).
- Tertiary amines react to form salts (Figure 13-39).

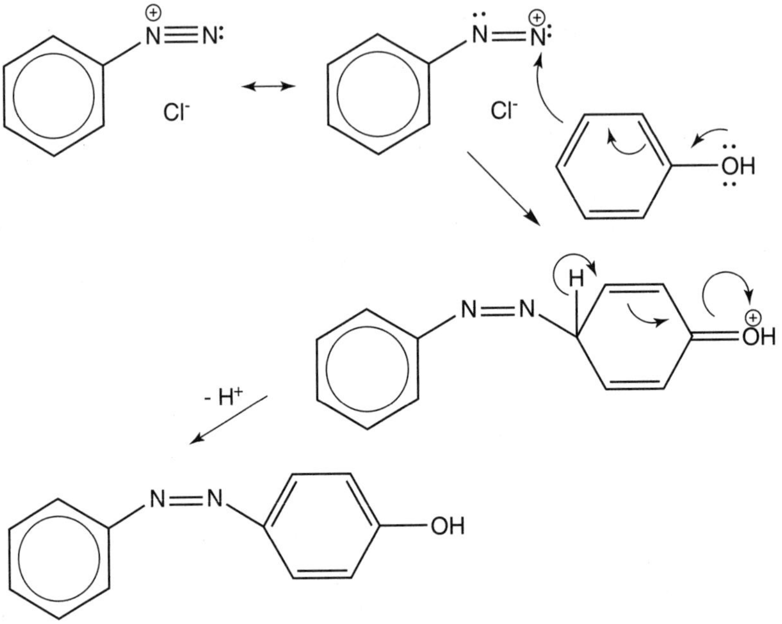

Figure 13-35: The mechanism for converting benzene-diazonium chloride to *p*-hydroxy-azobenzene.

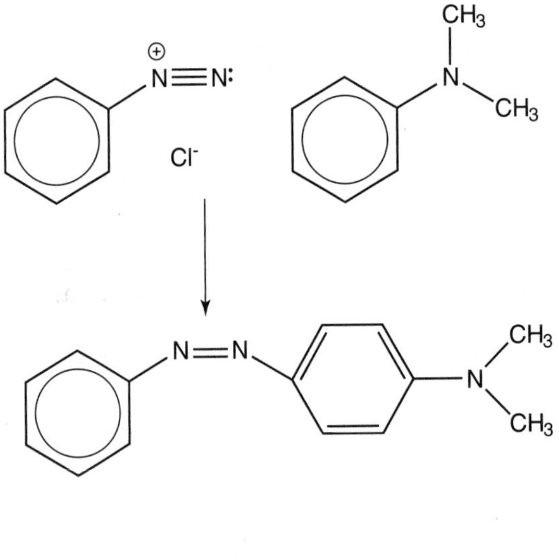

Figure 13-36: The reaction of a diazonium salt with an amine.

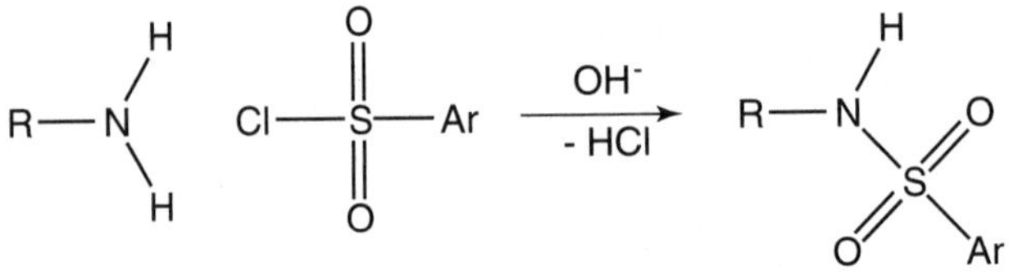

Figure 13-37: The reaction of a primary amine with a sulfonyl chloride.

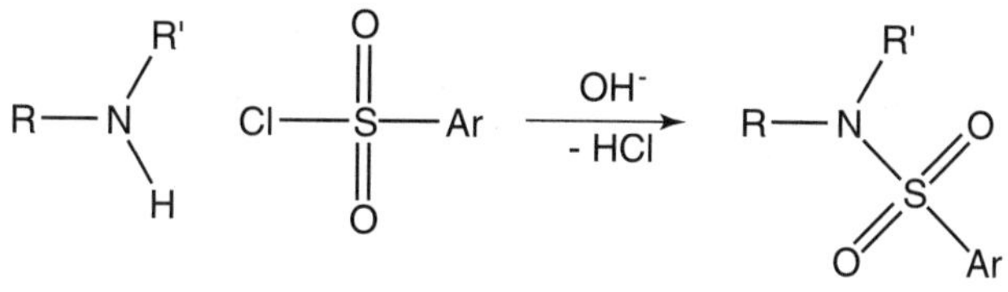

Figure 13-38: The reaction of a secondary amine with a sulfonyl chloride.

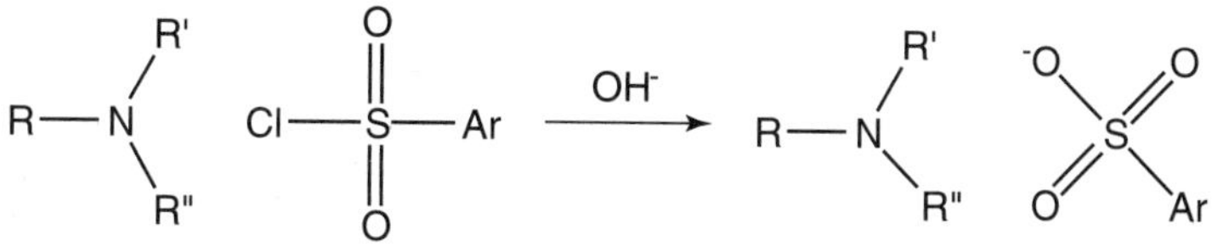

Figure 13-39: The reaction of a tertiary amine with a sulfonyl chloride.

Exploring elimination reactions

Elimination reactions involving amines are important synthetic methods. They can be used to make a variety of useful organic compounds, including alkenes. We examine a few of them in this section.

Hofmann elimination

The *Hofmann elimination* converts an amine into an alkene. The process begins by converting an amine to a quaternary ammonium salt (that is, it has a nitrogen atom with four bonds). The general mechanism for the elimination step is in Figure 13-40. Figure 13-41 illustrates a sample reaction scheme for the Hofmann elimination.

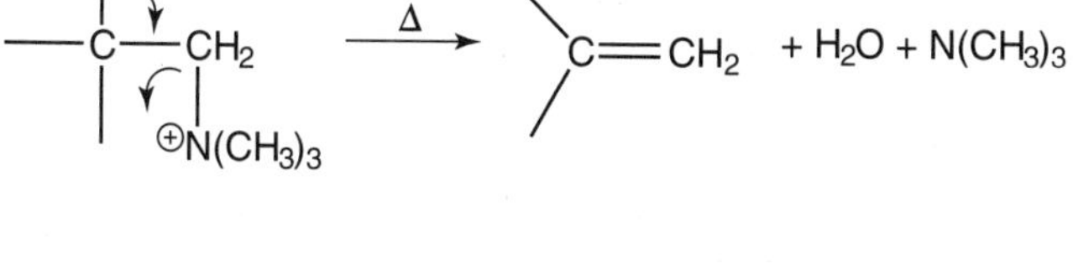

Figure 13-40: The general mechanism for the elimination in the Hofmann elimination.

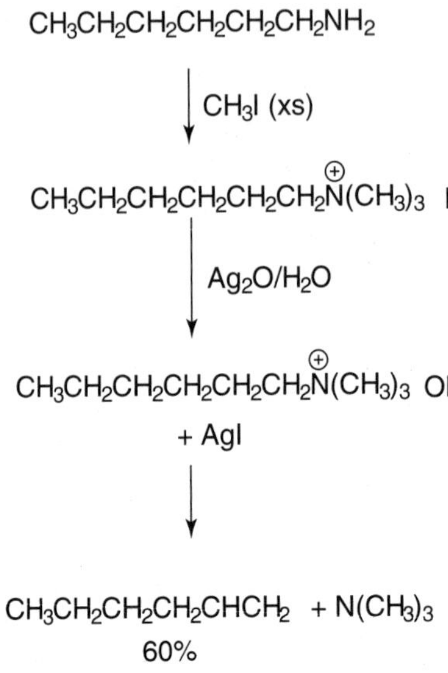

$$CH_3CH_2CH_2CH_2CH_2CH_2NH_2 \xrightarrow{CH_3I\ (xs)} CH_3CH_2CH_2CH_2CH_2CH_2\overset{\oplus}{N}(CH_3)_3\ I^- \xrightarrow{Ag_2O/H_2O} CH_3CH_2CH_2CH_2CH_2CH_2\overset{\oplus}{N}(CH_3)_3\ OH^- + AgI \longrightarrow \underset{60\%}{CH_3CH_2CH_2CH_2CHCH_2} + N(CH_3)_3$$

Figure 13-41: A sample reaction scheme for the Hofmann elimination.

As you no doubt learned in Organic I, a more highly substituted double bond is more stable (Zaitsev's rule). As Figure 13-42 shows, the reaction does the opposite of Zaitsev's rule. That is, the least-substituted product predominates. This happens because the transition state has carbanion character (shown in Figure 13-43). (A 1° carbanion is more stable than a 2°, which is more stable than a 3°.) The process involves anti-elimination (opposite side elimination as shown in Figure 13-44). The geometry requires the bulky groups to have the greatest separation.

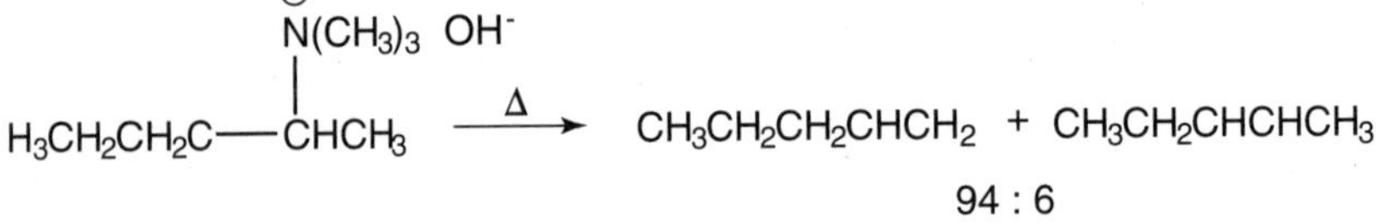

$$H_3CH_2CH_2C\!-\!\underset{}{\overset{\overset{\oplus}{N}(CH_3)_3\ OH^-}{|}}{CHCH_3} \xrightarrow{\Delta} CH_3CH_2CH_2CHCH_2 + CH_3CH_2CHCHCH_3 \quad 94:6$$

Figure 13-42: Product distribution resulting from the Hofmann elimination.

$$\left[CH_3\!-\!CH_2\!-\!CH\!-\!\overset{\overset{H-OH\ \delta-}{\vdots\,\delta-}}{CH_2} \text{ with } CH\!-\!\overset{\oplus}{N}(CH_3)_3 \right]^{*}$$

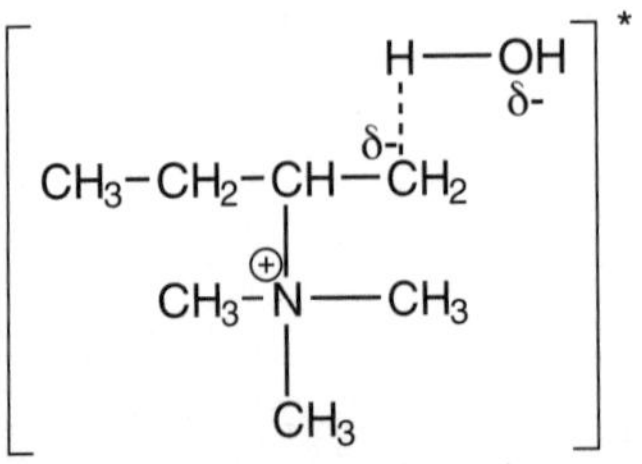

Figure 13-43: The intermediate in the Hofmann elimination.

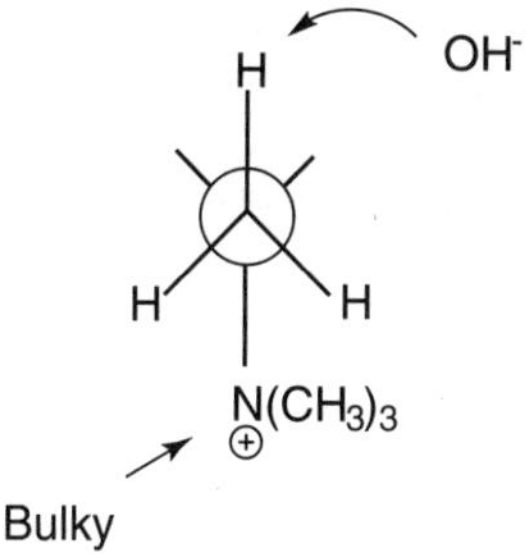

H OH⁻
H H
$\overset{\oplus}{N}(CH_3)_3$
Bulky

Figure 13-44: Anti-elimination in the Hofmann elimination.

Cope elimination

In the *Cope elimination,* thirty percent hydrogen peroxide, H_2O_2, is used to produce an amine oxide, which upon heating undergoes elimination. This is a *syn*-elimination process. Figure 13-45 illustrates the general reaction, while Figure 13-46 shows the mechanism of the *syn*-elimination step.

α β $CH_2-\ddot{N}-CH_3$ H CH_3
30% H_2O_2
Cold
O^- $CH_2-\overset{\oplus}{N}-CH_3$ CH_3
An amine oxide
Δ
CH_2 HO N CH_3 CH_3

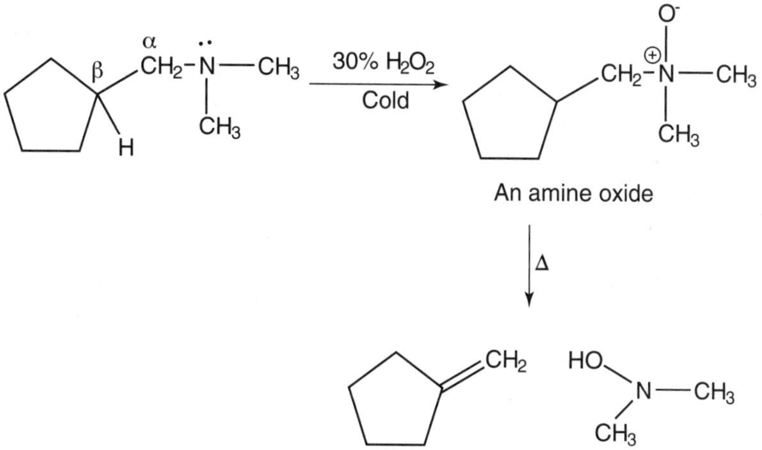

Figure 13-45: An example of the Cope elimination.

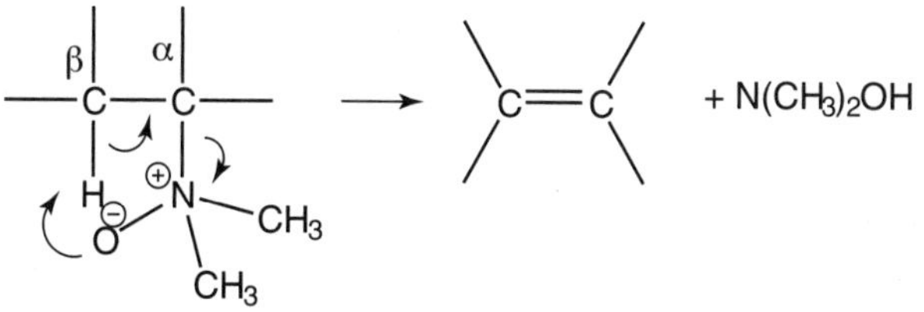

β α C C H $\overset{\ominus}{O}$ $\overset{\oplus}{N}$ CH_3 CH_3
C=C
+ $N(CH_3)_2OH$

Figure 13-46: The mechanism of the *syn*-elimination step in the Cope elimination.

Mastering Multistep Synthesis

In many cases, a desired compound cannot be synthesized directly from readily available materials. In these cases, a multistep synthesis must be performed. Figure 13-47 illustrates a multistep synthesis. (A similar type of problem appears on many Organic Chemistry II exams; they're retrosynthetic analysis problems.)

For aromatic amines, $-NR_2$ is activating *o-p*-directors, and $-N^+R_2H$ is deactivating (that is, it may interfere with the reactions).

Br

1) Mg/ether
2) CO_2
3) H^+

O
C
OH

$SOCl_2$

O
C
Cl

$HN(CH_3)_2$(xs)

O
C
N
CH_3
CH_3

1) $LiAlH_4$
2) OH^-/H_2O

CH_2
N
CH_3
CH_3

Figure 13-47: An example of a multistep synthesis.

When attacking a retrosynthetic analysis problem, you often know only the formula of the starting material and the desired product (in addition, the instructor may impose a few other rules). The answer to the problem should resemble Figure 13-47.

You have many options for attacking multistep synthesis problems. In general, begin with the desired product and work backwards. For example, what reactant can produce the final tertiary amine in Figure 13-47? After you determine the identity of the reactant, you back up one step and determine which reactant can produce the amide given in Figure 13-47. After this, you back up another step and repeat the procedure until you reach the starting material. If you get lost, you may need to retrace your steps and redo one or more steps. Only try to work from the beginning as a last resort.

The formation of sulfa drugs is another example of a multistep synthesis. The sulfa drugs are bactericides, effective against a wide variety of bacteria because they mimic *p*-aminobenzoic acid (Figure 13-48). Many bacteria require *p*-aminobenzoic acid, which they are unable to synthesize, and need to synthesize folic acid. Many types of sulfa drugs exist, and most of them involve the substitution of one of the hydrogen atoms on the -SO_2-NH_2. Prontosil (Figure 13-49) was the first commercially available sulfa drug. The metabolism of prontosil produced sulfanilamide.

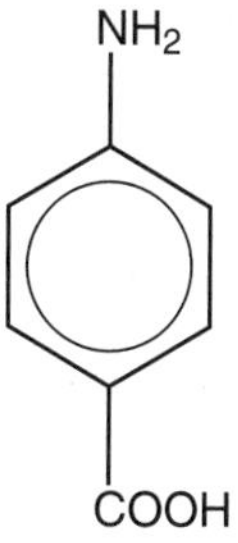

NH_2

COOH

Figure 13-48: The structure of *p*-amino-benzoic acid.

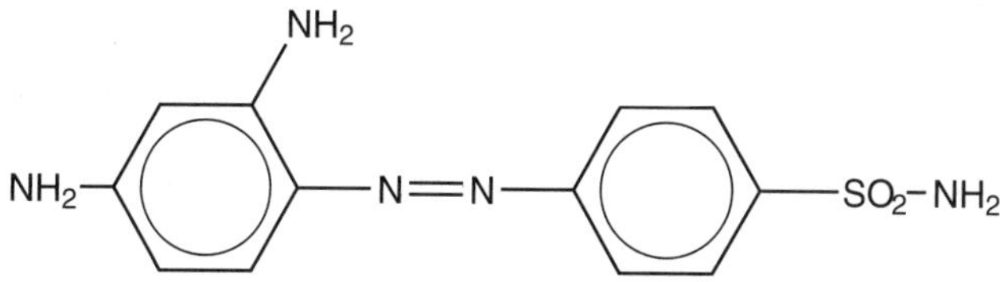

NH_2

NH_2 —N=N— SO_2–NH_2

Figure 13-49: The structure of prontosil.

The procedure for synthesizing sulfanilamide (a sulfa drug) is a multistep procedure as illustrated in Figure 13-50. The first step also works if you substitute an acyl chloride for the acid anhydride. The conversion of the amine to an amide converts the strong activator into a medium activator, limiting multiple attacks. The last step converts the amide back into an amine.

In the general reactivity scheme, acyl chlorides are more reactive than acid anhydrides, which are more reactive than carboxylic acids.

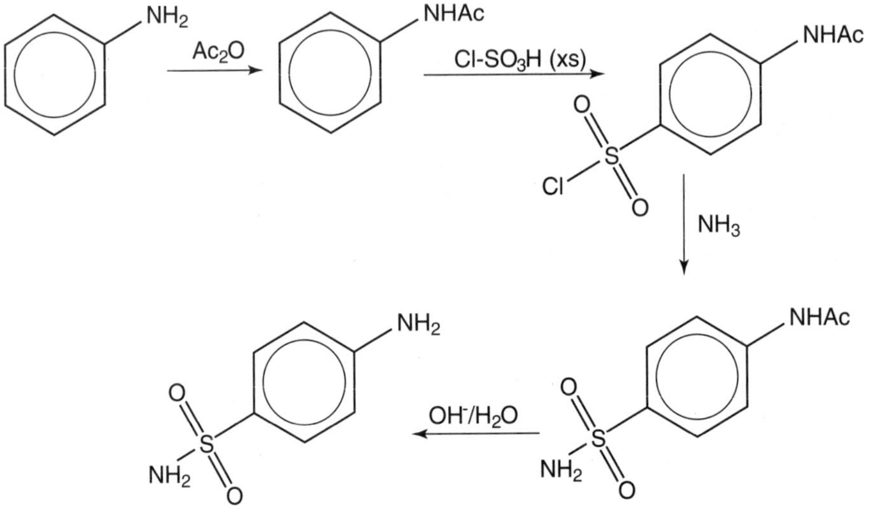

Figure 13-50: The multistep synthesis of sulfanilamide.

Identifying Nitrogen Compounds with Analysis and Spectroscopy

Amines are easily identified because they're readily soluble in dilute acid. Sodium fusion converts the amine to the cyanide ion, which is detectable by a variety of methods. The ready formation and decomposition of diazonium salts (discussed in the earlier section "Reactions with nitrous acid") leads to the identification of primary amines. The Hinsberg test (see the nearby sidebar) is useful in identifying amines.

The infrared spectra of amines show one or two N-H stretches in the 3500–3200 cm^{-1} region. Primary amines usually have two bands, while secondary amines usually have one band. Obviously, since there are no N-H bonds, tertiary amines have no N-H stretch. The bands are small and sharp in comparison the corresponding alcohol peaks.

The hydrogen attached to the nitrogen appears in the $\delta = 1–5$ region of the proton NMR.

Hinsberg test

Though it has largely been replaced by spectroscopic methods, at one time the Hinsberg test was useful in the characterization of amines. The first step in the test was the reaction of the amine with a benzenesulfonyl chloride in base. The second step was to acidify the reaction mixture. The results of the two steps indicated the type of amine present. Normally this was a good test for up to eight carbon atoms. This table lists the results for each type of amine.

Type of Amine	***Results of the First Step***	***Results of the Second Step***
Primary	Solution	Precipitation
Secondary	Precipitation	No visible change (No reaction)
Tertiary	No reaction	Solution

Chapter 14

Metals Muscling In: Organometallics

In This Chapter

- Coping with Grignard reagents
- Deciphering organolithium reagents
- Taking a quick look at other organometallic compounds
- Testing what you know

A metal or semimetal atom can form a covalent bond to carbon in many situations, and the resultant compound is an organometallic compound. The metal-to-carbon bond in these compounds is polar covalent: $X_3C{:}^{-}M^{+}$. (Metals such as sodium and potassium form ionic organometallic compounds in which the metal forms a cation and the organic portion is a carbanion.) We look at a couple of types of these organometallic compounds in this chapter, particularly those containing magnesium (Grignard reagents) and lithium (organolithium reagents).

Grignard Reagents: Grin and Bear It

Grignard reagents (organomagnesium compounds) are extremely useful in many organic reactions. These materials are relatively easy to prepare, but they're very sensitive to trace amounts of moisture and air and decompose if either is present.

Don't use a Grignard reagent when the starting material has an active hydrogen, such as carboxylic acids, alcohols, amines, and sulfonic acids, because Grignard reagents behave as bases towards them. In addition, the starting material can't contain other groups that may be attacked by the reagent, including other carbonyl groups, epoxide groups, and nitrile groups.

Preparation of Grignard reagents

The preparation of a Grignard reagent begins with magnesium metal and dry ether (in most cases, either diethyl ether or THF, tetrahydrofuran). The ether cleans the surface of the metal and takes the reagent into solution for reaction. (If either the ether or the reaction vessel contains moisture, the yield is poor.) The magnesium then reacts with either an alkyl halide or an aryl halide. The ease of reactivity decreases in the order RI > RBr > RCl. Iodides may react too rapidly, but chlorides may react too slowly. Thus bromides are usually the best. The general reaction is

$$RX \text{ or } ArX + Mg(ether) \rightarrow R^{\delta-\delta+}MgX \text{ or } Ar^{\delta-\delta+}MgX$$

For example, the preparation of ethyl magnesium bromide is

$$CH_3CH_2Br + Mg^0/Et_2O \rightarrow CH_3CH_2{}^{\delta-\delta+}MgBr$$

Pay particular attention to the bond polarity ($\delta^-\delta^+$). The carbon atom has a partially negative charge (and therefore is a nucleophile) and the magnesium atom has a partially positive charge (so is an electrophile).

Reactions of Grignard reagents

Grignard reagents behave both as bases and as nucleophiles. The basicity leads to the requirement that water be rigorously excluded from the reaction mixture both in the preparation and in the use of the Grignard reagent.

Basicity

Basicity refers to the ability of a Grignard reagent to react with proton donors, including weak donors like water. The carbanion is the conjugate base of a very weak acid ($K_a \approx 10^{-40} - 10^{-50}$). This process is

$$R\text{–}H \rightarrow H^+ + R^-$$

Weak acids (WA) have strong conjugate bases (SB), while strong acids (SA) have weak conjugate bases (WB). The weaker the acid or base, the stronger its conjugate.

A Grignard reagent behaves as a base towards water (an acid). An example of this reaction is

$$\underset{SB}{CH_3MgBr} + \underset{SA}{H_2O} \rightarrow \underset{WA}{CH_4} + \underset{WB}{MgBr(OH)}$$

As with other acid-base reactions, the strong acid and strong base react to produce a weaker acid and a weaker base. The strong acid, in this case H_2O, reacts to form the weaker base MgBr(OH), while the strong base, in this case the Grignard reagent CH_3MgBr, reacts to form the weak acid CH_4. Water (a conjugate acid, CA) and MgBr(OH) (a conjugate base, CB) constitute a conjugate acid-base pair, while the Grignard reagent (CB) and methane (CA) constitute another conjugate acid-base pair.

Another example illustrating the basicity of Grignard reagents is

$$\underset{SB}{CH_3MgBr} + \underset{SA}{CH_3C{\equiv}CH} \rightarrow \underset{WA}{CH_4} + \underset{WB}{CH_3C{\equiv}CMgBr}$$

This reaction has two conjugate acid-base pairs, and the strong acid and base react to produce the weak acid and base. The terminal alkyne ($K_a \approx 10^{-25}$, $pK_a \approx 25$) is a significantly stronger acid than an alkane.

The acidity of hydrocarbons increases in the order $sp^3 < sp^2 < sp$.

Nucleophilicity

The nucleophilic reactions of Grignard reagents include reactions that create carbon-carbon bonds and the formation of alcohols.

One way to create a carbon-carbon bond is to react a Grignard reagent with a carbonyl compound. The result of this reaction is an alcohol derived from an aldehyde. Formaldehyde gives a primary alcohol, but any other aldehyde gives a secondary alcohol. Ketones and esters both react to form tertiary alcohols.

The general reaction of a Grignard reagent with a carbonyl is shown in Figure 14-1. The product of the first step in the mechanism in Figure 14-1 is a salt, which undergoes hydrolysis in the next step. The hydrolysis (an acid-base reaction) is shown in Figure 14-2.

Figure 14-1: The general reaction of a Grignard reagent with a carbonyl.

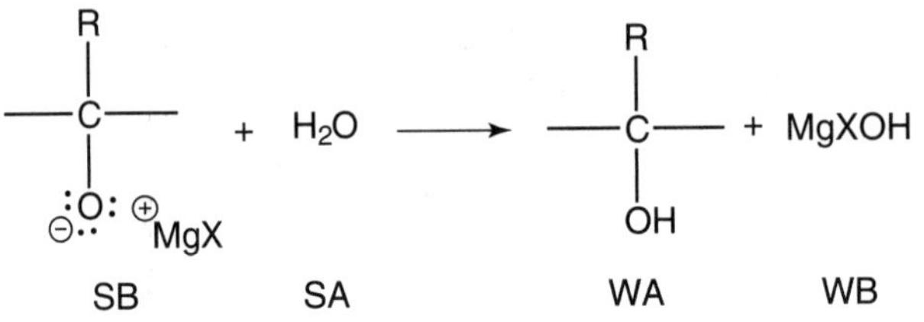

Figure 14-2: The hydrolysis of the salt formed in Figure 14-1.

Only formaldehyde yields a primary alcohol by reaction with a Grignard reagent. Figure 14-3 illustrates the reaction of ethylmagnesium bromide with formaldehyde to form 1-propanol. More-complicated alcohols, such as cyclopentylmethanol, can be synthesized by this means (as shown in Figure 14-4).

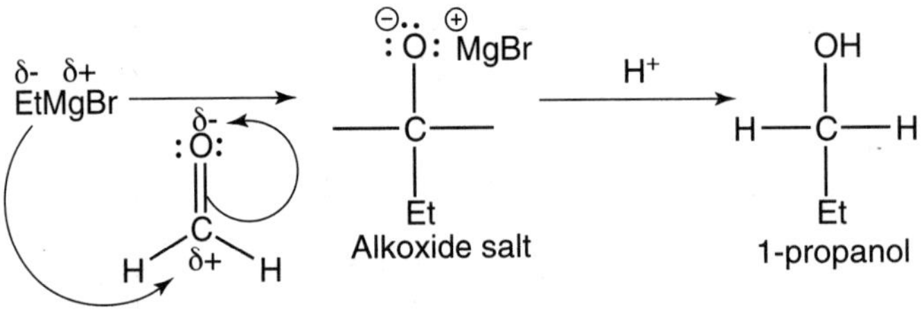

Figure 14-3: The formation of a primary alcohol.

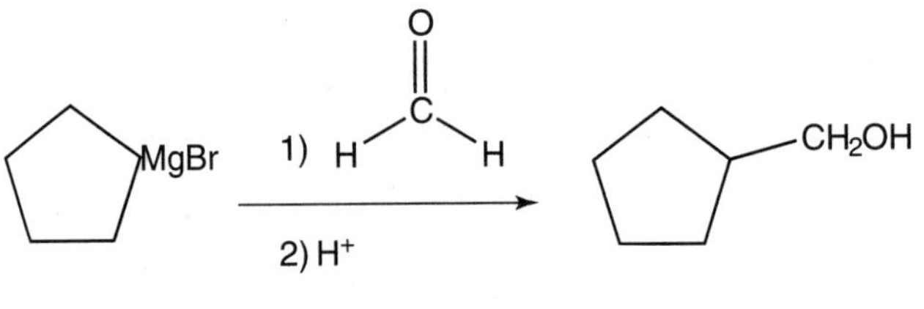

Figure 14-4: The synthesis of cyclopentylmethanol.

Any aldehyde other than formaldehyde reacts with a Grignard reagent to produce a secondary alcohol (see Figure 14-5). The general mechanism is the same for any aldehyde (see Figure 14-6). As shown in Figure 14-7, more than one path may lead to the same alcohol.

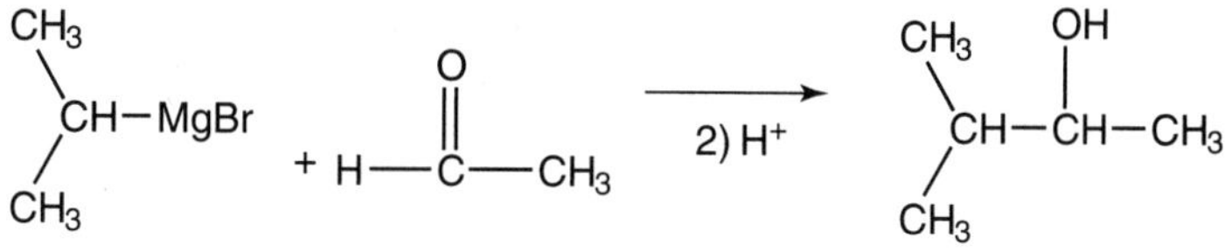

Figure 14-5: The synthesis of a secondary alcohol by the reaction of a Grignard reagent with an aldehyde.

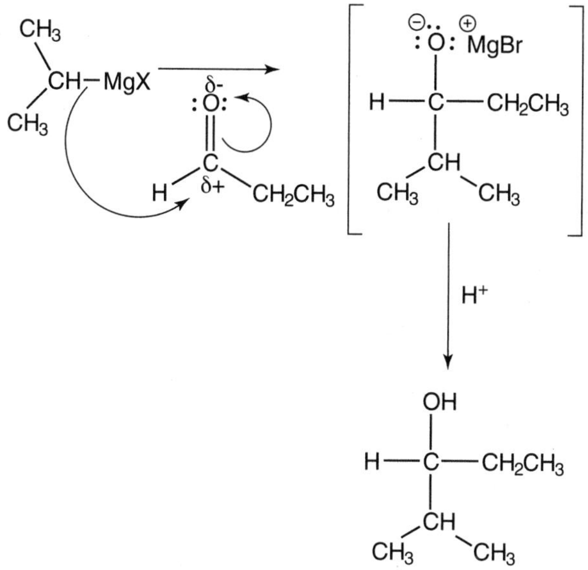

Figure 14-6: A partial mechanism for the synthesis of a secondary alcohol.

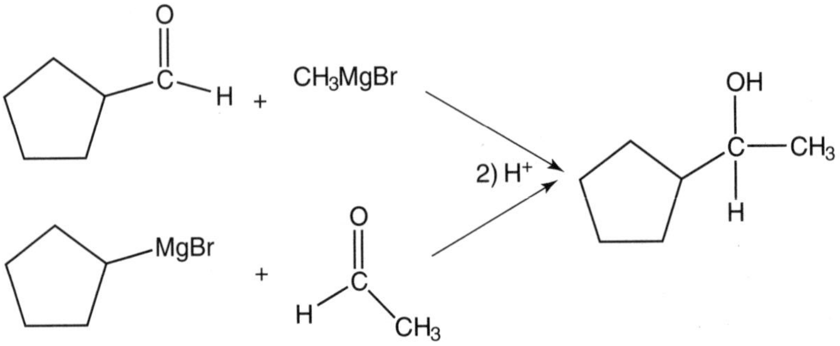

Figure 14-7: Two synthetic paths to the same alcohol.

A ketone reacts with a Grignard reagent to produce a tertiary alcohol. Examples of this process are shown in Figure 14-8 and Figure 14-9.

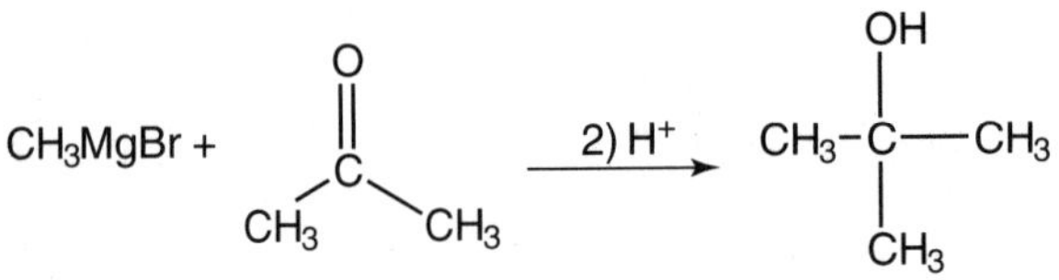

Figure 14-8: The synthesis of a tertiary alcohol through the reaction of a ketone with a Grignard reagent.

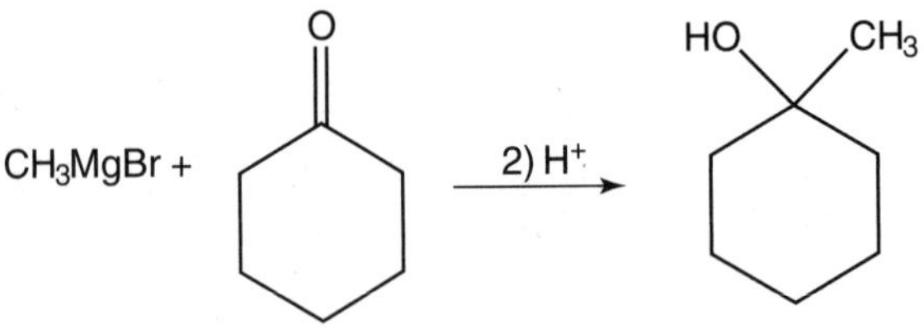

Figure 14-9: The synthesis of a tertiary alcohol containing a cyclohexane ring.

An alcohol can be formed through the reaction of ethylene oxide with a Grignard reagent. As usual, nucleophilic attack is carried out by the Grignard reagent. This procedure produces a primary alcohol with the addition of two carbon atoms. Figure 14-10 illustrates the reaction of ethylene oxide with a Grignard reagent.

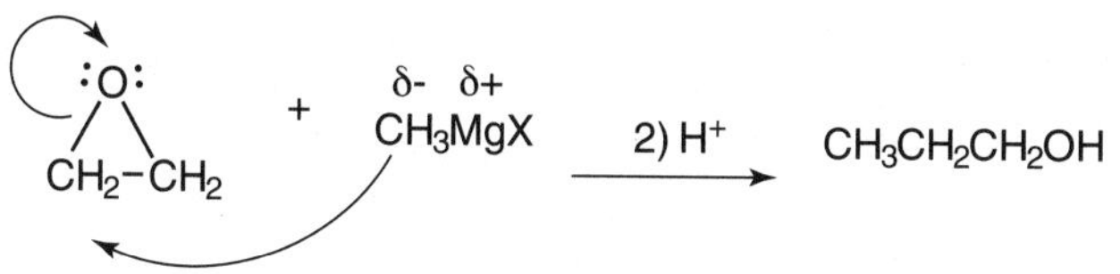

Figure 14-10: The synthesis of 1-propanol through the reaction of ethylene oxide with a Grignard reagent.

The reaction of a Grignard reagent with esters is a substitution reaction involving addition and then elimination. The process gives a tertiary alcohol. Figure 14-11 shows an example of the formation of an isotopically labeled tertiary alcohol. This reaction involves the attack of the ester by two molecules of the Grignard reagent, which adds two identical R groups to the carbonyl carbon.

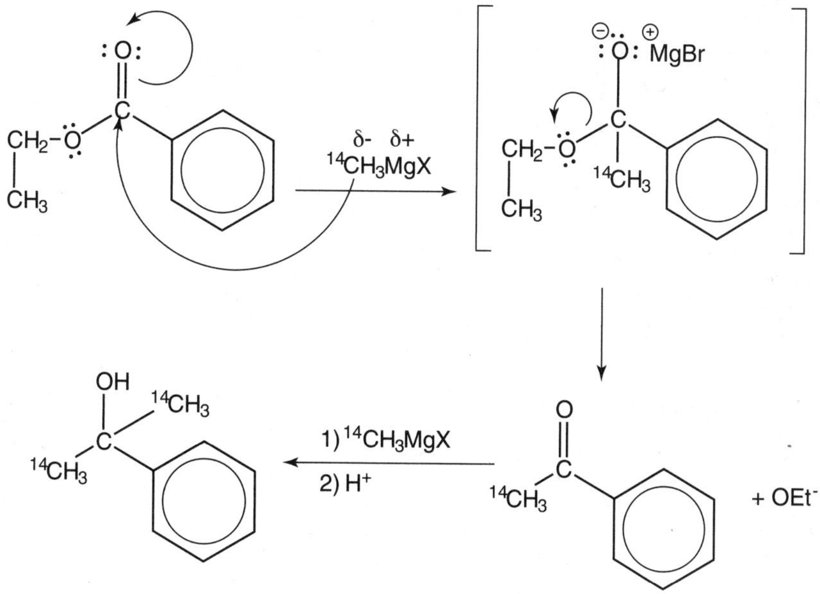

Figure 14-11: The reaction of an ester with a Grignard reagent.

Organolithium Reagents

In general, organolithium compounds behave like Grignard reagents. However, organolithium compounds tend to be more reactive, so they're useful in situations where the reaction with a Grignard reagent is slow or results in a low yield.

The preparation of an organolithium compound (shown in Figure 14-12) is similar to the preparation of a Grignard reagent except that lithium replaces the magnesium. Two moles of lithium per mole of halide are necessary: One lithium yields an organolithium, and the other yields a lithium halide.

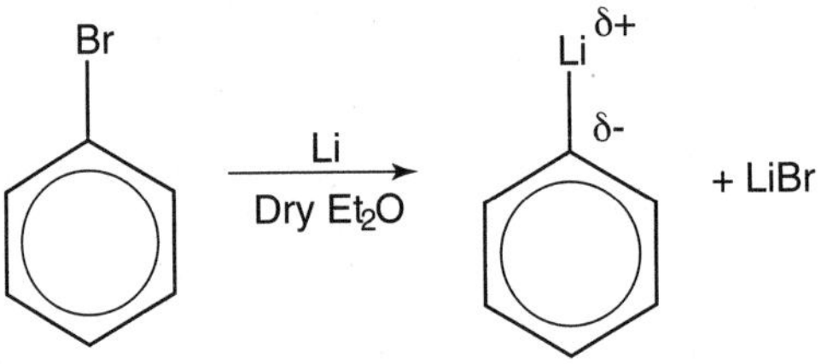

Br

Li

Dry Et_2O

Li $\delta+$

$\delta-$

+ LiBr

Figure 14-12: The formation of an organolithium reagent.

Organolithium reagents, like Grignard reagents, are bases that react with proton (or deuteron) donors. Figure 14-13 illustrates this reaction. In this reaction D_2O (heavy water) is the deuterated form of water in which the hydrogen atoms (H) are replaced with deuterium atoms (D).

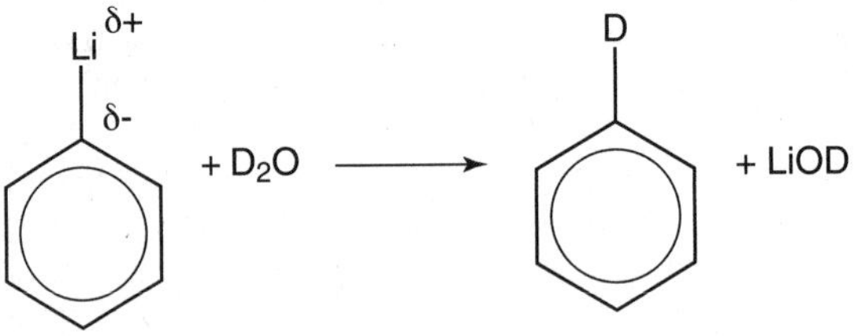

Li $\delta+$

$\delta-$

+ D_2O

D

+ LiOD

Figure 14-13: An organolithium reagent behaving as a base towards a proton (deuteron) donor.

Like Grignard reagents, organolithium compounds react with formaldehyde to produce a primary alcohol. Figure 14-14 illustrates the reaction of an organolithium reagent with formaldehyde.

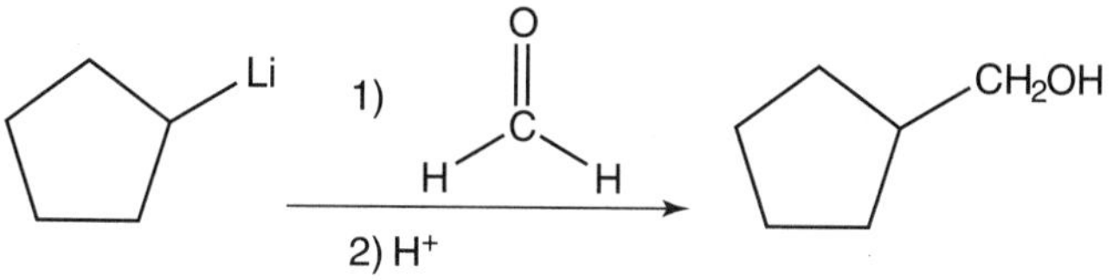

Figure 14-14: The preparation of a primary alcohol from an organolithium compound and formaldehyde.

Formation of Other Organometallics

Grignard reagents react with halides of less electropositive metals to give other organometallic compounds. The less electropositive (more electronegative) metals include Hg, Zn, Cd, Si, and the nonmetal P. An example of this type of reaction is

$$n\ RMgX + MX_n \rightarrow MR_n + n\ MgX_2$$

Less electronegative metals, such as sodium, tend to form ionic organometallic compounds. These compounds have limited use in synthesis. One example of using a different organometallic in a synthesis is

$$CH_3{-}C{\equiv}C{:}^{-}Na^{+} + CH_3{-}CHO \rightarrow + 2\ H^{+} \rightarrow CH_3{-}C{\equiv}C{-}CH(OH){-}CH_3$$

In general, the organometallic compounds presented in this chapter have a metal atom behaving as just another functional group. After you consider the polarity difference, the compounds behave simply as odd organic compounds. However, some organometallic compounds, such as ferrocene, are very different. Ferrocene forms from the reaction of iron(II) ions with cyclopentadienyl ions. The cyclopentadienyl ion comes from cyclopentadiene. This compound is acidic because the loss of a hydrogen ion, shown in Figure 14-15, leads to a stable aromatic system. Once formed, cyclopentadienyl ions readily react with iron(II) ions to produce the orange, stable compound ferrocene, $Fe(C_5H_5)_2$. When first synthesized, the formation of ferrocene was interesting because of its unusual structure.

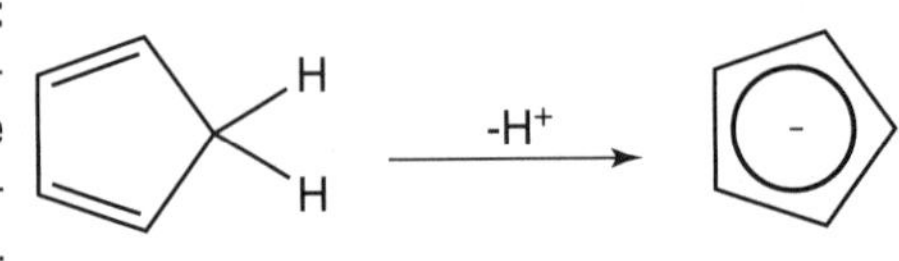

Figure 14-15: The formation of the cyclopentadienyl ion.

The structure of ferrocene is shown in Figure 14-16. The iron ion is sandwiched between two cyclopentadienyl rings, which are freely rotating in solution. The cyclopentadienyl rings react like aromatic systems because, well, they're aromatic. The bond is between the iron ion and the π-electron cloud of the aromatic system. Another metal ion may replace the iron in ferrocene.

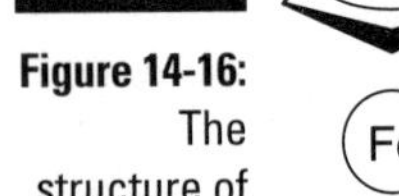

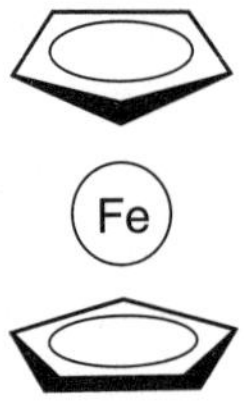

Figure 14-16: The structure of ferrocene.

Putting It Together

No matter how well you think you understand the material, the real test of understanding is your ability to answer questions involving the material. For example, what if someone asks you, "What are three methods of preparing the compound in Figure 14-17 via a Grignard reagent?"

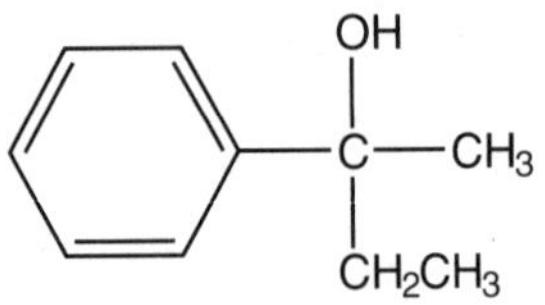

Figure 14-17: Synthesize this compound using a Grignard reagent.

In a question such as this one, you need to know not only the basic reactions of Grignard reagents but also the limitations. For example, if you wish to prepare a Grignard reagent containing an aromatic ring (Figure 14-18), the substituent, Z, may be –R, –Ar, –OR, or –Cl. However, Z may not be $-CO_2H$, –OH, $-NH_2$, $-SO_3H$, $-CO_2R$, –CN, or $-NO_2$, because acidic groups generally cause problems by undergoing acid-base reactions with the Grignard reagent.

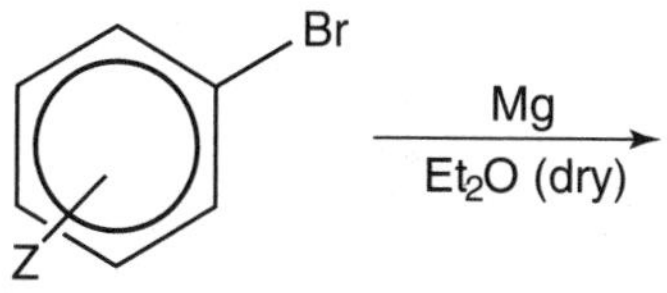

Br
Mg
Et_2O (dry)
Z

Figure 14-18: The preparation of a Grignard reagent is dependent upon the identity of Z.

Chapter 15

More Reactions of Carbonyl Compounds

In This Chapter

- Shedding light on the Claisen condensation
- Deciphering acetoacetic and malonic ester synthesis
- Understanding the Knoevenagel condensation
- Tackling the Mannich reaction
- Studying the Stork enamine synthesis
- Exploring barbiturates

In Chapter 11 you see that a hydrogen atom between two carbonyl groups is weakly acidic (K_a values of 10^{-10} to 10^{-14}). These compounds are β-dicarbonyl compounds. In the same chapter you see that Michael additions can add enolate ions to these compounds. In this chapter we examine additional reactions of β-dicarbonyl compounds.

As you examine the reactions in this chapter, keep in mind this basic info: The acidity of a hydrogen atom between two carbonyl groups is partly due to the inductive effect (weakening of the bond to the hydrogen atom due to the electron density of the bond being pulled towards the nearby electronegative oxygens) and the resonance stabilization of the anion formed (see Figure 15-1). The partial negative charge on the carbon atom makes it a nucleophile that is subject to attack by an electrophile.

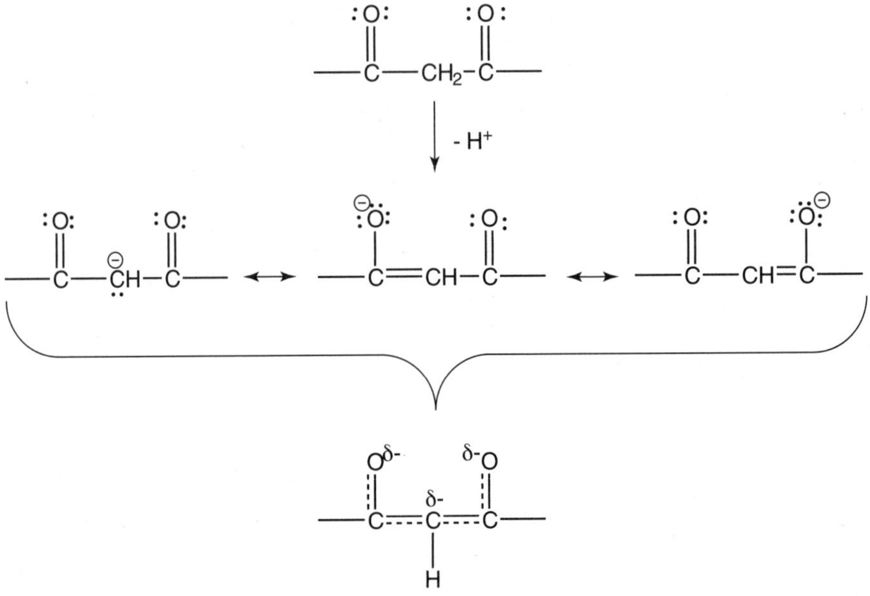

Figure 15-1: Resonance stabilization of the anion formed from a β-dicarbonyl compound.

Checking Out the Claisen Condensation and Its Variations

The term *condensation* refers to the joining of two molecules with the splitting out of a smaller molecule. The Claisen condensation is used extensively in the synthesis of dicarbonyl compounds. In biochemistry it is used to build fatty acids in the body. The Dieckmann condensation, the crossed Claisen condensation, and others (with other carbanions) are variations of the Claisen condensation. In this section we briefly look at these variations.

Doing the two-step: Claisen condensation

The Claisen condensation is one method of synthesizing β-dicarbonyl compounds, specifically a β-keto ester. This reaction begins with an ester and occurs in two steps. In the first step, a strong base, such as sodium ethoxide, removes a hydrogen ion from the carbon atom adjacent to the carbonyl group in the ester. (Resonance stabilizes the anion formed from the ester.) The anion can then attack a second molecule of the ester, which begins a series of mechanistic steps until the anion of the β-dicarbonyl compound forms, which, in the second reaction step (acidification), gives the product.

The general mechanism of the Claisen condensation, with ethoxide as the base, is shown in Figure 15-2. Sodium ethoxide is necessary because the starting material is an ethyl ester. If the starting material were a methyl ester, then the base would be sodium methoxide. Choosing a base that matches the type of ester minimizes the formation of other products.

Note that the final product in Figure 15-2 is stabilized as illustrated in Figure 15-3. This is an example of stabilization by keto-enol tautomerization (see Chapter 11 to review). The driving force is the result of the stability of the carbanion as shown in Figure 15-1. In this example of condensation, two ester molecules join, and alcohol, a smaller molecule, splits out.

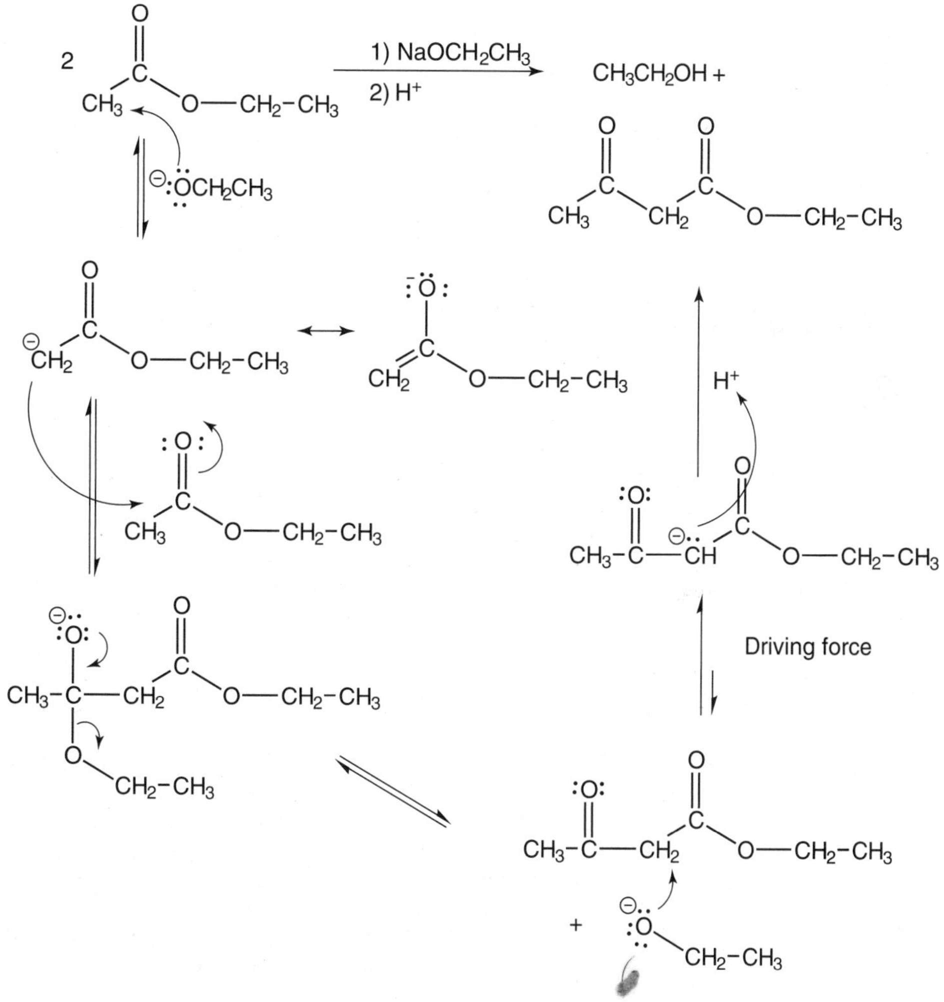

Figure 15-2: The general mechanism for the Claisen condensation.

Figure 15-3: Stabilization of the product from Figure 15-2.

$CH_3-CO-CH_2-CO-O-CH_2-CH_3 \leftrightarrow CH_3-C(OH)=CH-CO-O-CH_2-CH_3$

Enol

The Claisen condensation bears some resemblance to the Aldol condensation seen in Chapter 11. The initial step in the mechanisms are very similar in that in both cases a resonance-stabilized ion is formed.

Circling around: Dieckmann condensation

Dieckmann condensation is a cyclic Claisen condensation where a molecule attacks itself to form a ring structure. Figure 15-4 shows an example beginning with dimethyl hexanedioate, and Figure 15-5 shows what happens when the carbon chain increases by one carbon atom. This process favors the formation of five- or six-membered rings because they're the most stable rings.

Figure 15-4: The Dieckmann condensation beginning with dimethyl hexanedioate.

1) $Na^+ {}^-OCH_3/MeOH$
2) H^+

82%

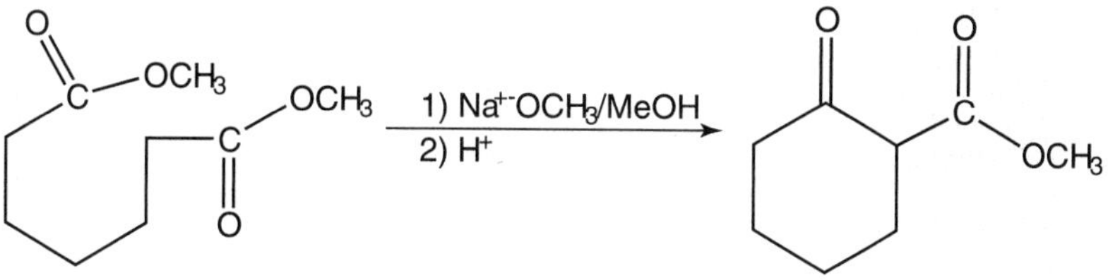

Figure 15-5: The Dieckmann condensation producing a six-membered ring.

Doubling Up: Crossed Claisen condensation

A crossed Claisen condensation employs two different esters. If the esters are A and B, the possible products are AA, AB, BA, and BB. To minimize the complicated mixture of products, one of the reactants must have no α-hydrogen atoms. If this is ester B, then the products would be AA and AB. If the concentration of A is very low, then only a small quantity of AA can form. Figure 15-6 illustrates an example of a crossed Claisen condensation, where the product is of the AB form.

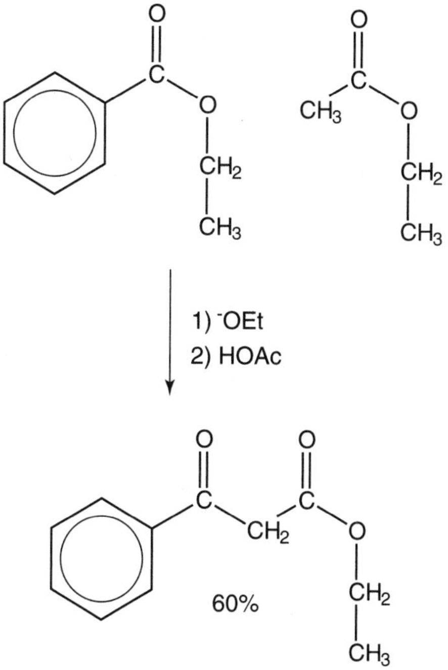

Figure 15-6: A crossed Claisen condensation.

Other carbanions

Very strong bases can be used in other reactions that are similar to the Claisen condensation. In these reactions, carbanions are formed instead of enolates. Figures 15-7, 15-8, and 15-9 show examples of various reactions employing three different strong bases: sodium ethoxide, sodium amide, and sodium triphenylmethanide. Sodium hydride, NaH, is a strong base that would also work. Any of these very strong bases could be used in each of the specific reactions.

Figure 15-7: The use of sodium ethoxide for a Claisen condensation–type reaction.

1) NaOEt/EtOH
2) H^+

91%

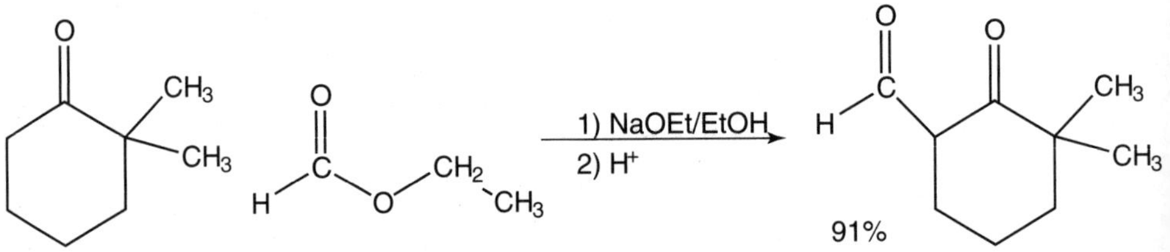

Figure 15-8: The use of sodium amide for a Claisen condensation–type reaction.

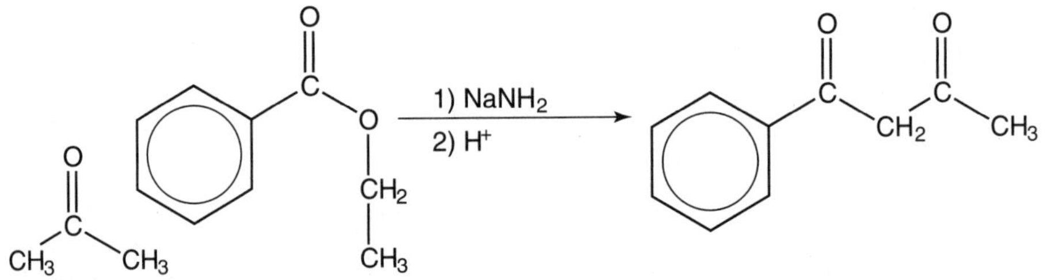

1) $NaNH_2$
2) H^+

Write the complete mechanism for each of the reactions in Figures 15-7, 15-8, and 15-9, and compare these to the Claisen condensation in order to identify similarities and differences.

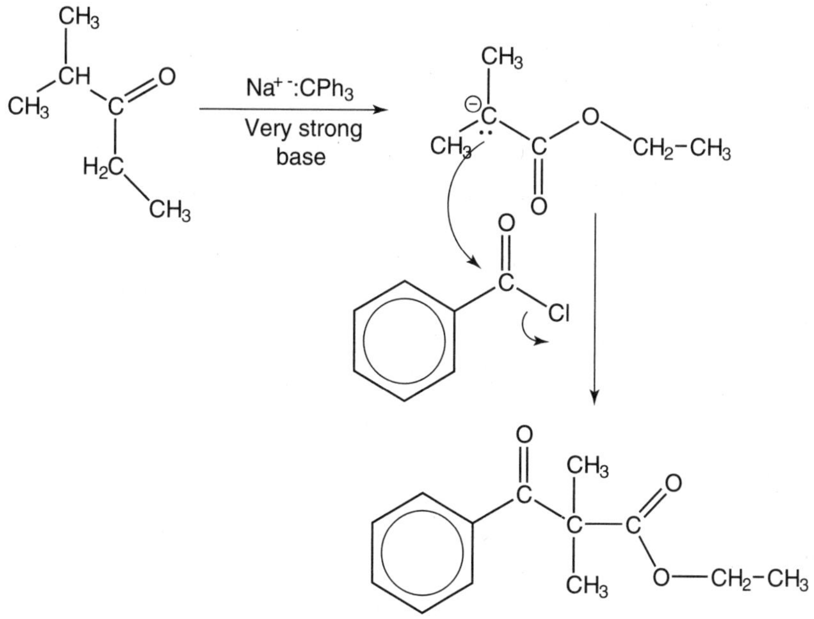

Figure 15-9: Using sodium triphenylmethanide for a Claisen condensation–type reaction.

Exploring Acetoacetic Ester Synthesis

Acetoacetic ester synthesis is the preparation of substituted acetones, and it's an important method for creating a variety of products. It begins with the reaction of acetoacetic ester (a dicarbonyl) or a similar compound with a strong base to produce a carbanion, which then reacts with alkyl halide, RX. The structure of acetoacetic ester is in Figure 15-10. Figure 15-11 illustrates an example of an acetoacetic ester synthesis and two possible outcomes. Figure 15-12 shows the preparation of 2-heptanone with a 65 percent yield via the acetoacetic ester synthesis. Figure 15-13 presents the preparation of 2-benzylcyclohexanone with a 77 percent yield.

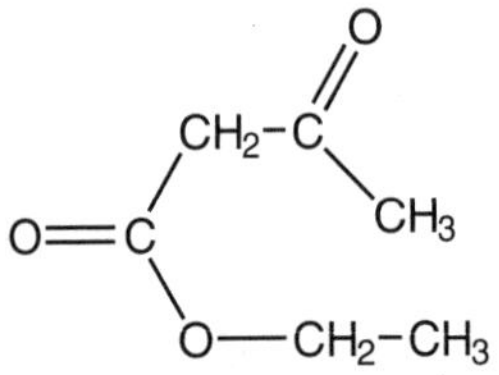

Figure 15-10: The structure of acetoacetic ester.

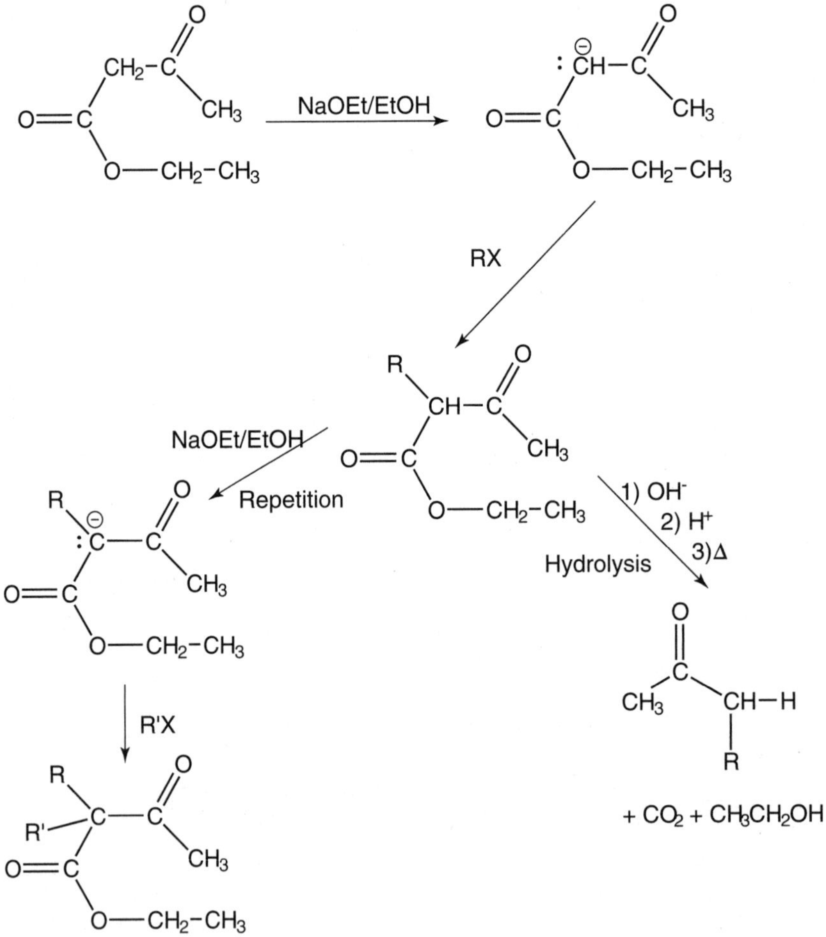

NaOEt/EtOH

RX

NaOEt/EtOH

Repetition

R'X

1) OH^-
2) H^+
3) Δ

Hydrolysis

$+ CO_2 + CH_3CH_2OH$

Figure 15-11: An example of an acetoacetic ester synthesis and two possible outcomes.

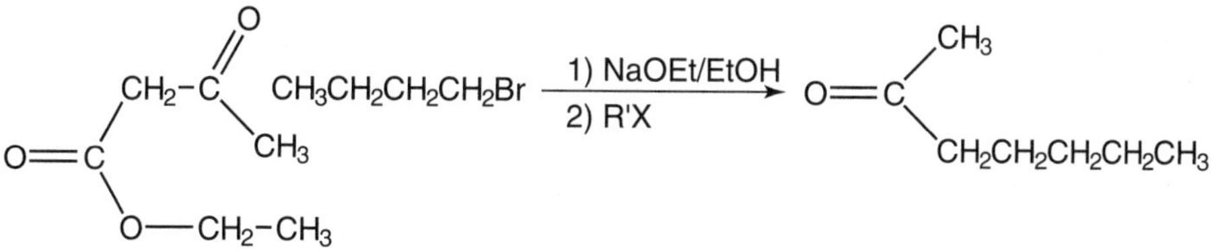

Figure 15-12: The preparation of 2-heptanone (65% yield).

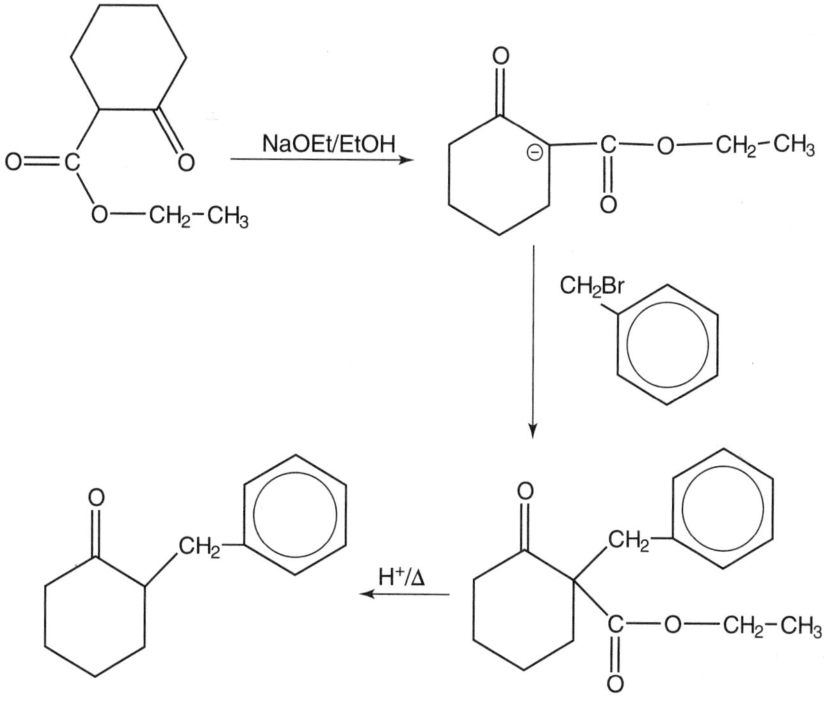

Figure 15-13: The preparation of 2-benzylcyclohexanone (77% yield).

Defining Malonic Ester Synthesis

Malonic esters have two ester groups, each of which may react as in the acetoacetic ester synthesis due to their similar structure (see the preceding section). The malonic ester synthesis provides a method for preparing a substituted acetic acid. Figure 15-14 shows the structure of one type of malonic ester. Figure 15-15 outlines the basic malonic ester synthesis. *May repeat* in that figure refers to the reaction with a second molecule of RX (or R'X).

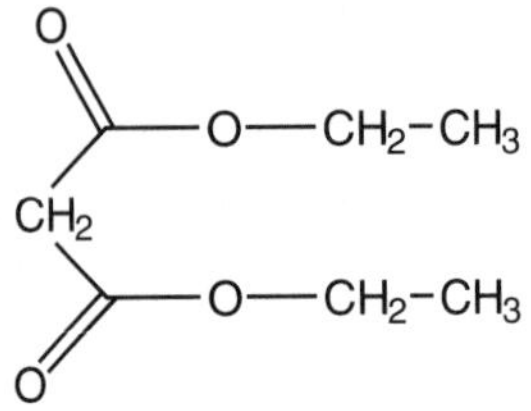

Figure 15-14: A typical malonic ester.

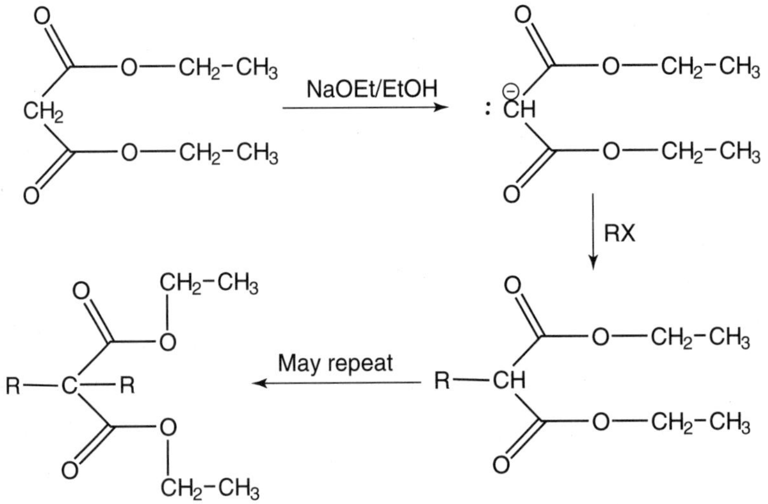

Figure 15-15: An outline of the basic malonic ester synthesis.

In most cases, the product of the malonic ester synthesis isn't the final product you're looking for. Commonly, the next step after the reaction in Figure 15-15 is hydrolysis and decarboxylation. Figure 15-16 shows this step.

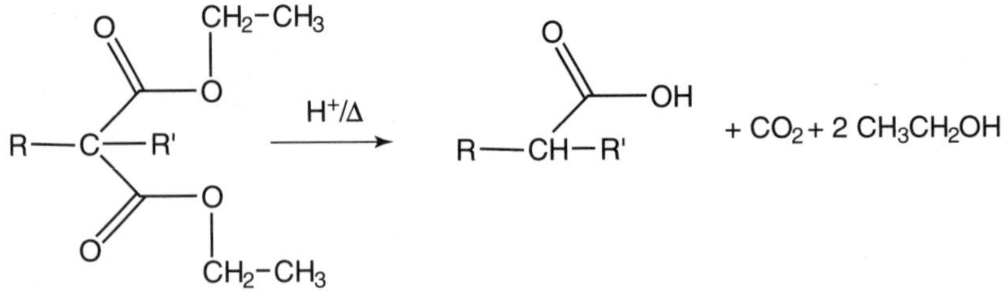

Figure 15-16: Hydrolysis and decarboxylation.

Synthesis of the anion shown in Figure 15-17 is very important since the anion produced is a nucleophile and can undergo further reaction. Other strong bases, such as sodium hydride, NaH, can replace the ethoxide ion used in Figure 15-17.

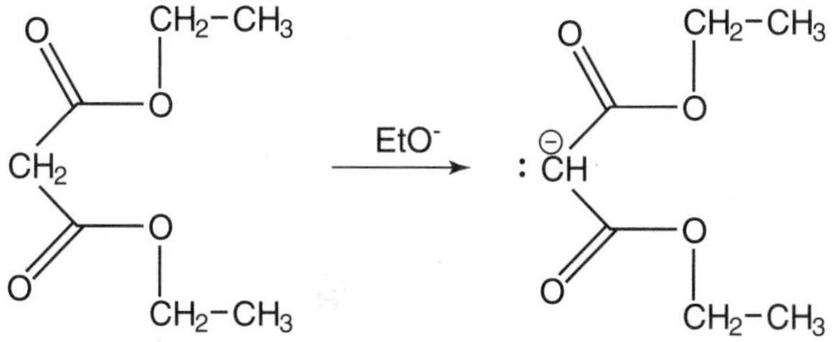

Figure 15-17: Formation of the anion.

Once formed, the anion has several uses, and the following figures illustrate how versatile malonic ester synthesis and its subsequent anion can be. Here we are using a typical molecule in a series of reactions. Any compound of similar structure could be used. Figure 15-18 shows the use of the anion to form compound A. Hydrolysis and decarboxylation of compound A yields a carboxylic acid. Figure 15-19 illustrates another synthesis involving compound A. A slight change in the procedure presented in Figure 15-18 is shown in Figure 15-20.

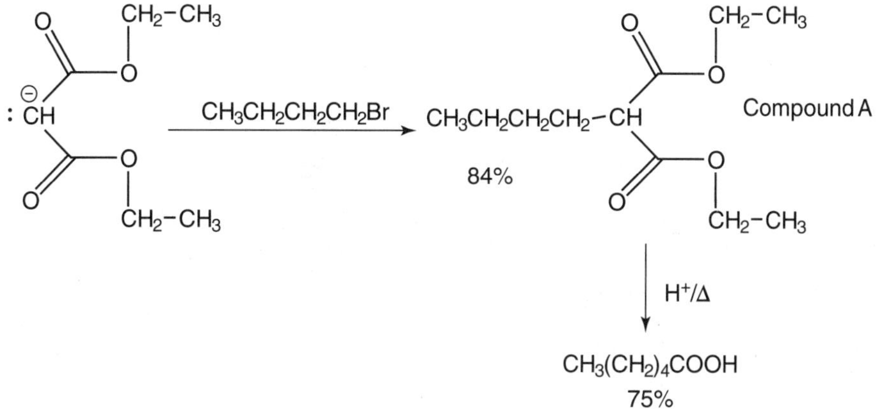

Figure 15-18: Formation of compound A from the anion in Figure 15-17.

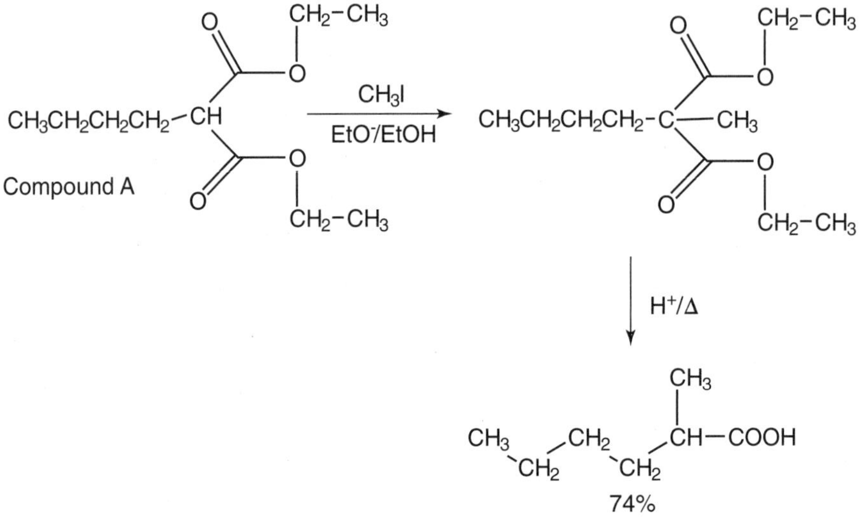

Figure 15-19: Another example of the use of compound A in synthesis.

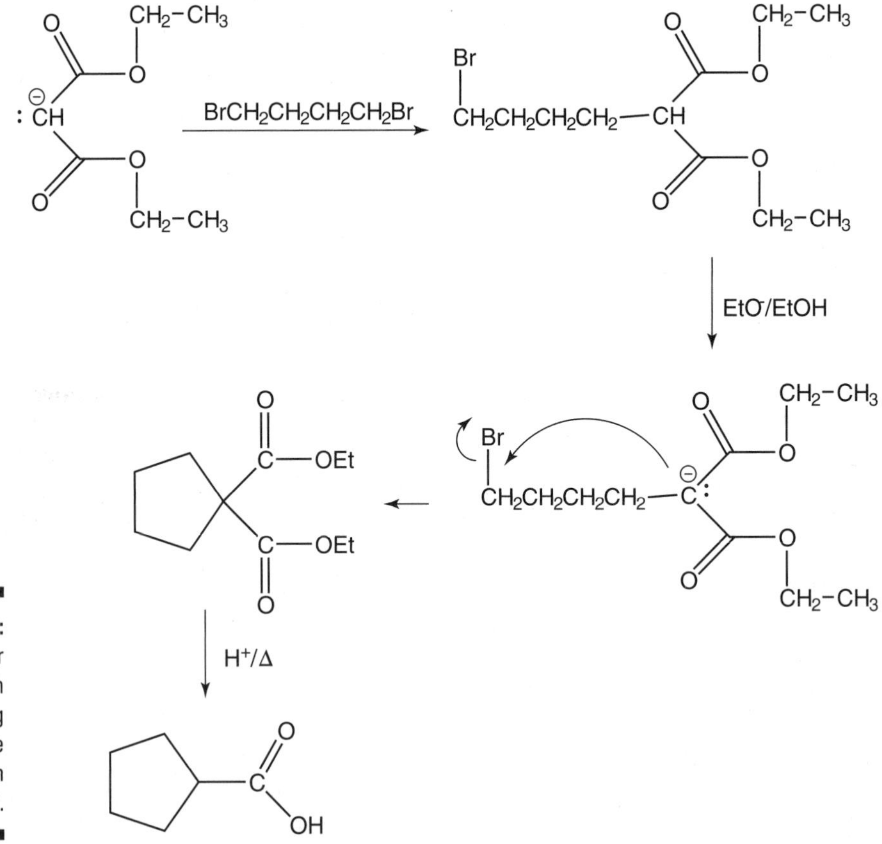

Figure 15-20: Another reaction beginning with the anion in Figure 15-17.

Working with Other Active Hydrogen Atoms

Compounds whose structures are similar to β-dicarbonyl compounds also have active hydrogens. These compounds have a CH_2 sandwiched between two electron-withdrawing groups, some examples of which are in Figure 15-21. The loss of a hydrogen ion from the sandwiched carbon leaves an anion, which can then behave as a nucleophile similar to other nucleophiles seen in this chapter.

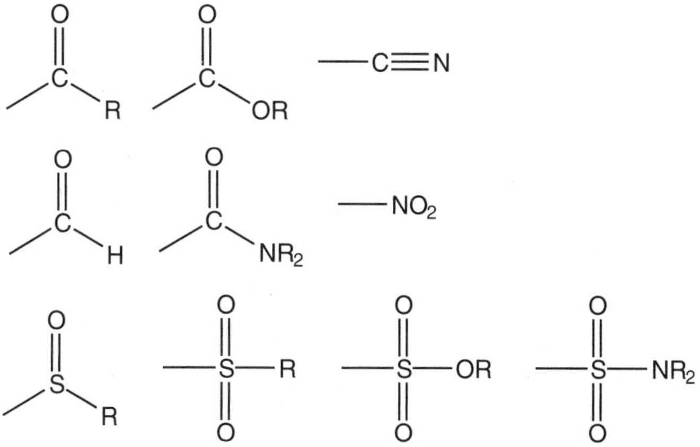

Figure 15-21: Some electron-withdrawing groups.

Reacting with Knoevenagel Condensation

The condensation of aldehydes and ketones with active hydrogen atoms is called Knoevenagel condensation. It is related to an aldol condensation and commonly is used to produce enones (a compound with a carbon-carbon double bond adjacent to a carbonyl). The process requires a weak base (an amine). A typical example and mechanism are presented in Figure 15-22.

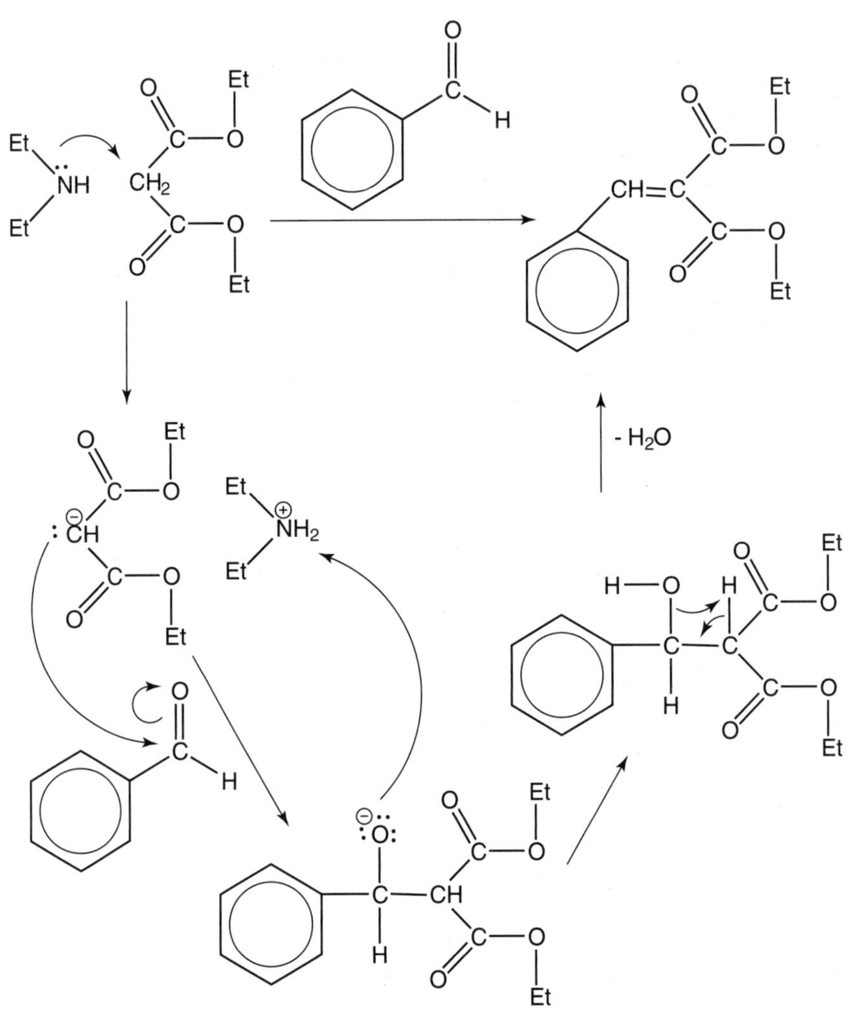

Figure 15-22: An example of a Knoevenagel condensation.

Looking at Mannich Reactions

A Mannich reaction is the reaction of formaldehyde with a primary or secondary amine and a compound with an active hydrogen atom. The product, an amine with a γ-carbonyl, is called a Mannich base, useful in a number of synthesis reactions. An example is in Figure 15-23, and the mechanism is in Figure 15-24.

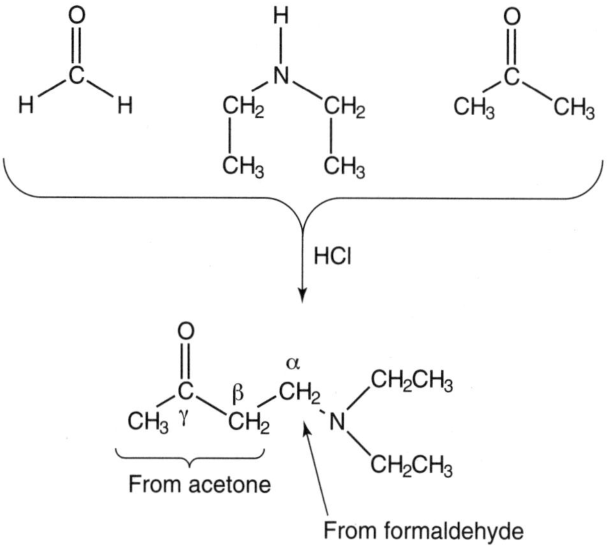

Figure 15-23: The formation of a Mannich base.

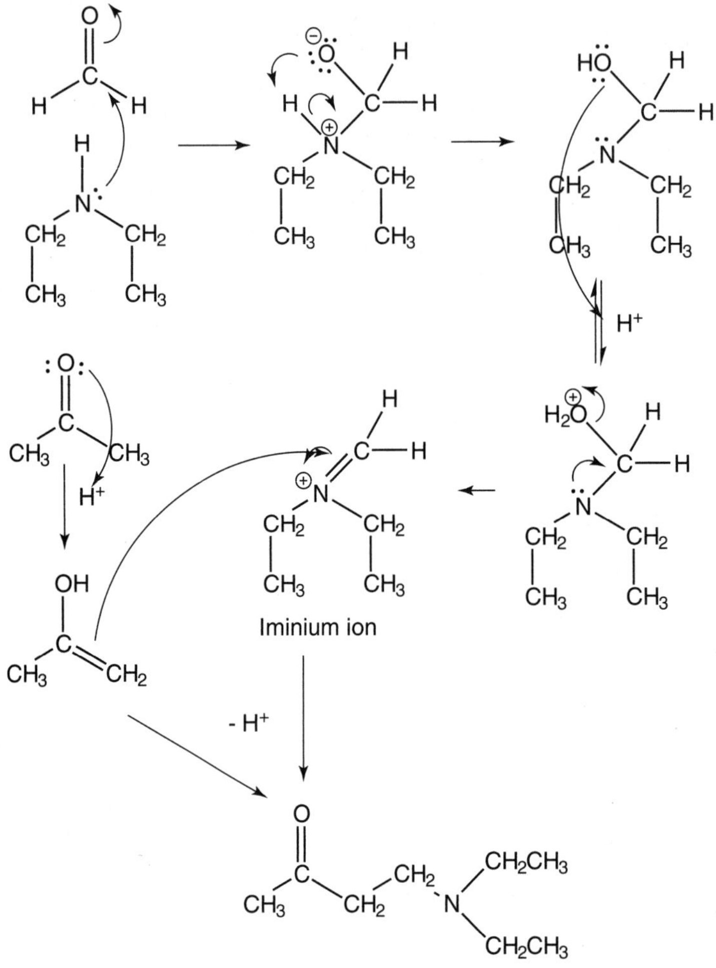

Figure 15-24: The mechanism for the formation of a Mannich base.

Creating Enamines: Stork Enamine Synthesis

Stork enamine synthesis takes advantage of the fact that an aldehyde or ketone reacts with a secondary amine to produce an enamine. Enamines are resonance stabilized (see Figure 15-25) and have multiple applications. In the first resonance structure, the nitrogen is the nucleophile, while in the second resonance structure, the carbanion is the nucleophile. Some commonly used secondary amines, pyrrolidine, piperidine, and morpholine, are shown in Figure 15-26.

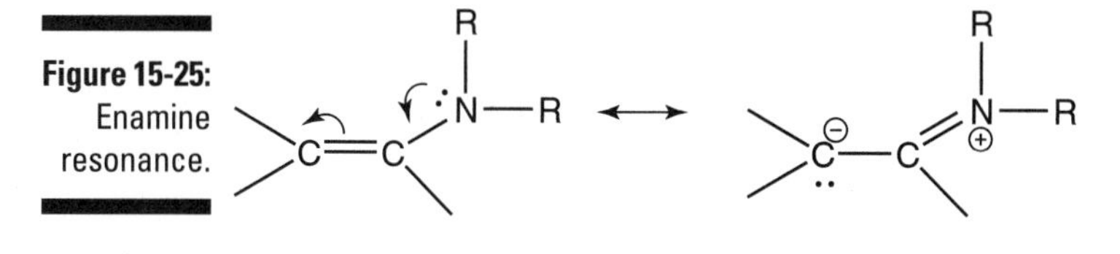

Figure 15-25: Enamine resonance.

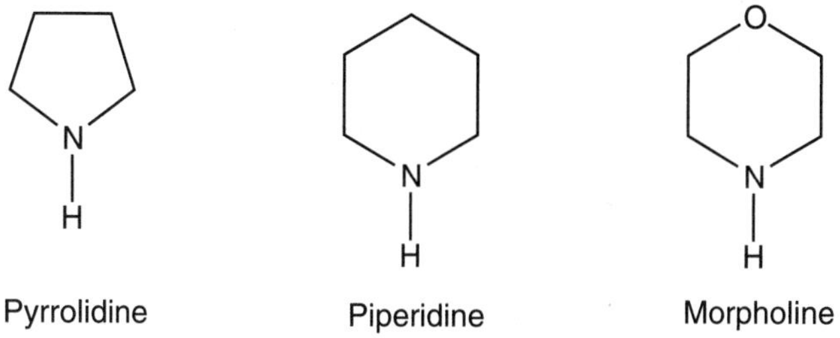

Pyrrolidine

Piperidine

Morpholine

Figure 15-26: Some secondary amines commonly used in the Stork enamine synthesis.

Figure 15-27 illustrates the formation of an enamine. The mechanism is shown in Figure 15-28. Figure 15-29 illustrates the formation of a 1,5-diketone (with a 71 percent yield).

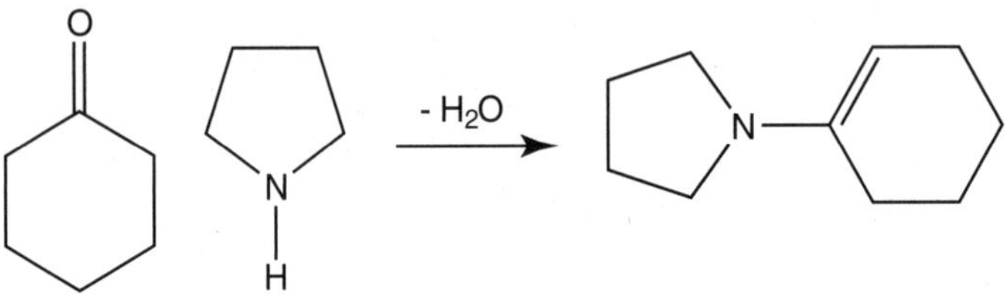

- H_2O

Figure 15-27: Preparation of an enamine.

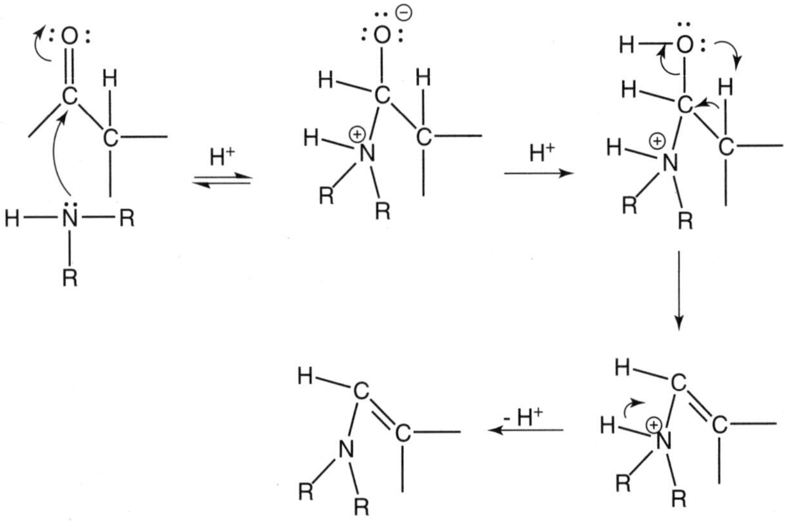

H^+

H^+

- H^+

Figure 15-28: The mechanism of enamine formation.

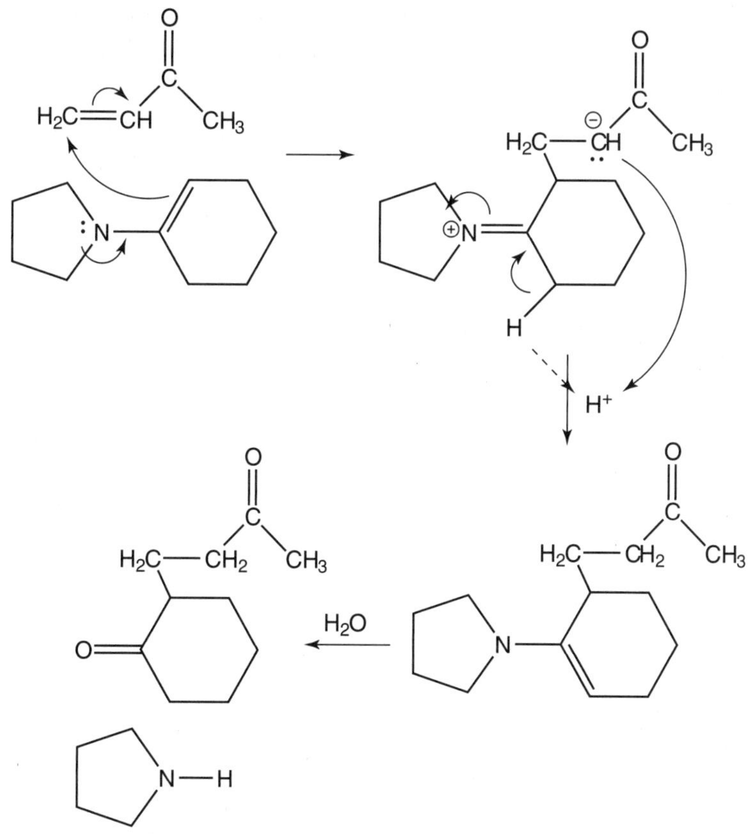

Figure 15-29: The formation of a 1,5-diketone (71% yield).

Putting It All Together with Barbiturates

The preparation of barbiturates illustrates many of the synthetic methods covered in this chapter. The preparation employs the reaction of urea ($CO(NH_2)_2$) with malonic ester to form barbituric acid. The general reaction is presented in Figure 15-30. The stable pyrimidine and other resonance forms help drive the reaction. By alternating the substituent at carbon number five (C5), various pharmacologically active substances can be formed. Barbital, a sedative, and phenobarbital, a sleeping aid, are shown in Figure 15-31.

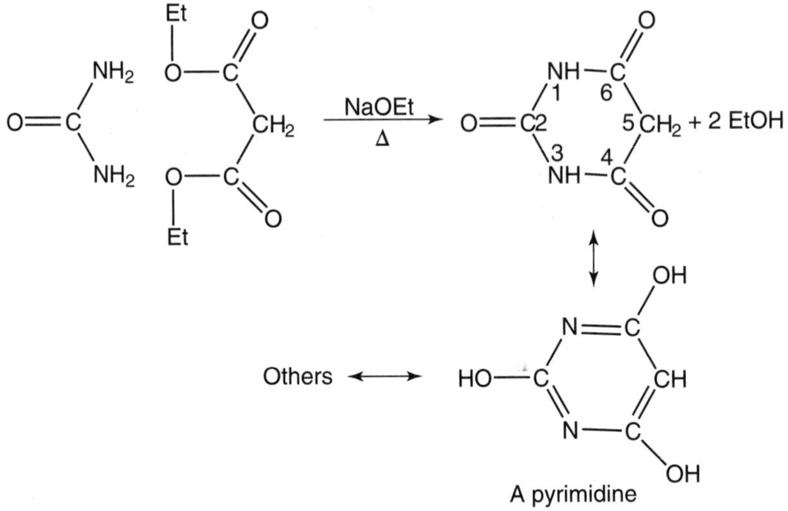

Figure 15-30: The preparation of barbiturates.

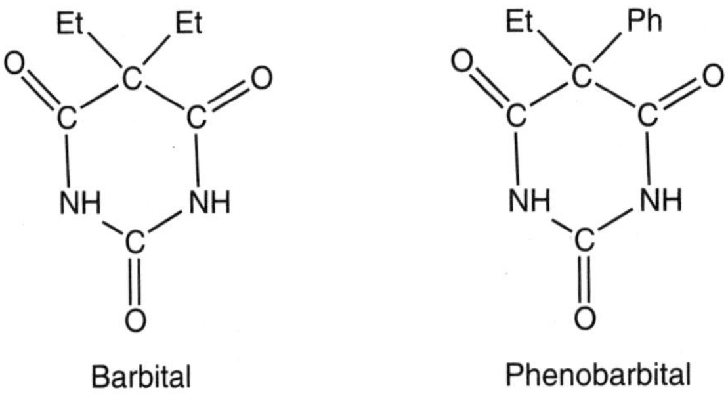

Figure 15-31: Two examples of barbiturates.

Chapter 16

Living Large: Biomolecules

In This Chapter

- Exploring carbohydrates, from the simple to the complex
- Observing the reactions, synthesis, and degradation of monosaccharides
- Checking out lipids, specifically fats
- Deciphering amino acid structure and how these protein building blocks form

A *biomolecule* is any molecule produced by living cells. Some biomolecules include the following (all of which we cover in this chapter):

- **Carbohydrates:** Ultimately, carbohydrates are the product of *photosynthesis,* the process in plants that combines carbon dioxide, water, and energy with chlorophyll and other biomolecules to produce carbohydrates and release oxygen gas. The major carbohydrate formed during photosynthesis is glucose. Plants and animals sometimes combine simple carbohydrates such as glucose into more complicated carbohydrates such as starch, glycogen, and cellulose.
- **Lipids:** Lipids are biomolecules that are insoluble in water but soluble in low-polarity organic solvents such as Et_2O and $CHCl_3$. Lipids include fats and oils, as well as many other biologically important molecules (think waxes and steroids).
- **Proteins:** Proteins are polymers of amino acids. Some of the most biologically important proteins are enzymes, which act as biological catalysts, allowing reactions to occur without the harsh conditions and reagents commonly used in organic chemistry. Virtually everything that happens in your body is associated with one or more enzymes.

Nucleic acids, which are polymers of nucleotides, are another class of biomolecules, but they're beyond the scope of this book; for more info about nucleic acids, refer to a more advanced text, such as *Biochemistry For Dummies* (written by us and published by Wiley).

Delving into Carbohydrate Complexities

Carbohydrates are either polyhydroxy aldehydes and ketones or substances that form these compounds after hydrolysis. The general formula is $[C_x(H_2O)_y]$. Normally carbohydrates occur as hemiacetals or acetals (hemiketals or ketals).

Because carbohydrates are so essential to the existence of living things (no organism would have the energy to perform even the basic functions of life without them), we spend a great deal of time in this chapter exploring the numerous complexities of carbohydrates. First, we introduce you to the kinds of carbohydrates that exist. Then we explore the reactions, synthesis, and degradation that affect the simplest carbohydrates — monosaccharides. Next, we get to know some particular monosaccharides (the D-aldose family) before looking at more complex sugars (including those containing nitrogen).

Introducing carbohydrates

Monosaccharides, which can't be broken down through hydrolysis, are the simplest carbohydrates of them all. Every other, more complex carbohydrate that exists can be broken down into two or more monosaccharides via hydrolysis, as you can see from the following:

- Disaccharides break down into two monosaccharides.
- Oligosaccharides break down into anywhere from two to ten monosaccharides.
- Polysaccharides break down into numerous monosaccharides.

In general, the names of carbohydrates end in *-ose.* If an aldehyde group is present, then the carbohydrate is an aldose. If a ketone group is present, then the carbohydrate is a ketose. However, you can also classify carbohydrates in terms of the number of carbon atoms present, which leads to names such as triose, tetrose, pentose, and hexose. Combinations of these two systems are possible; for example, glucose (pictured in Figure 16-1) is an aldohexaose. Common names for glucose include *blood sugar, grape sugar,* and *dextrose.*

The sections that follow describe the process of mutarotation in glucose and how glycoside formation can inhibit it.

Mutarotation

Mutarotation is when the "straight" chain forms of glucose (as shown by the Fischer projections) are in equilibrium with the cyclic forms of glucose (as shown by the Haworth projections). For more details on Fischer and Haworth projections, see *Organic Chemistry I For Dummies*. Figure 16-2 illustrates the general process occurring during mutarotation; notice that two different forms result.

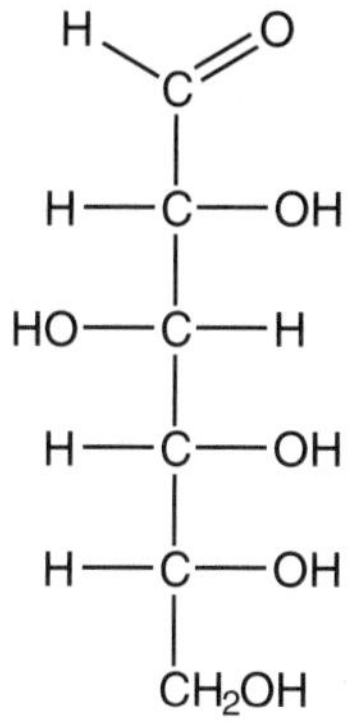

Figure 16-1: The structure of glucose.

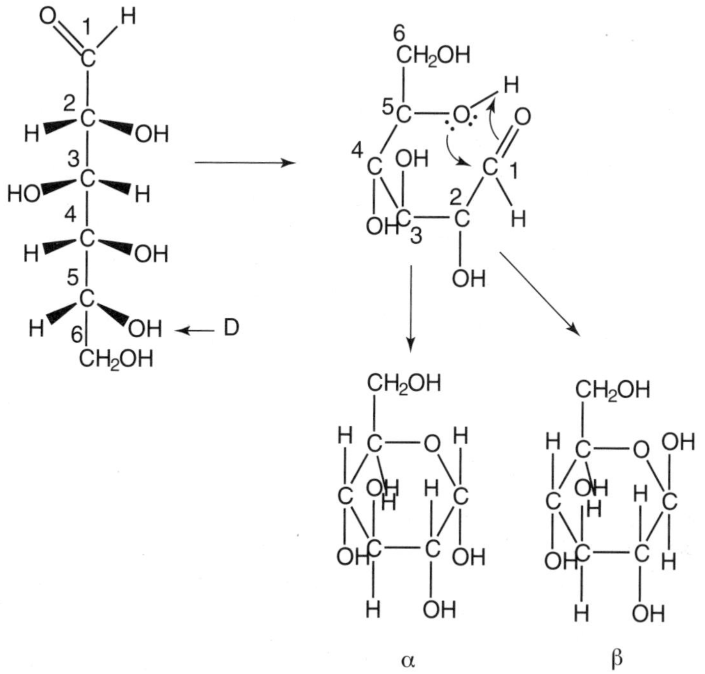

Figure 16-2: Mutarotation in glucose.

Mutarotation produces two types of cyclic forms called *anomers* (α and β), which differ in their arrangement about the anomeric carbon atom (originally the carbonyl carbon atom). If the -OH on the anomeric carbon atom is down, then the structure represents the α anomer; if it's up, the structure represents the β anomer. Due to the equilibrium present, one anomer rapidly converts to the other.

The common anomers have either a five- or a six-membered ring that contains an oxygen atom. A six-membered ring is a derivative of pyran; therefore, the monosaccharide is a pyranose. A five-membered ring is a derivative of furan, which makes the monosaccharide a furanose. (Figure 16-3 shows both a pyranose and a furanose.) The two anomers of glucose (refer to Figure 16-2) are α-D-glucopyranose and β-D-glucopyranose. Even though the Haworth projections indicate a "flat" ring, the actual structure of a pyranose ring is the chair conformer (see Figure 16-4).

Figure 16-5 shows one way to represent the equilibrium present during mutarotation. The unequal double arrows indicate that the dominant species in each case is the ring form.

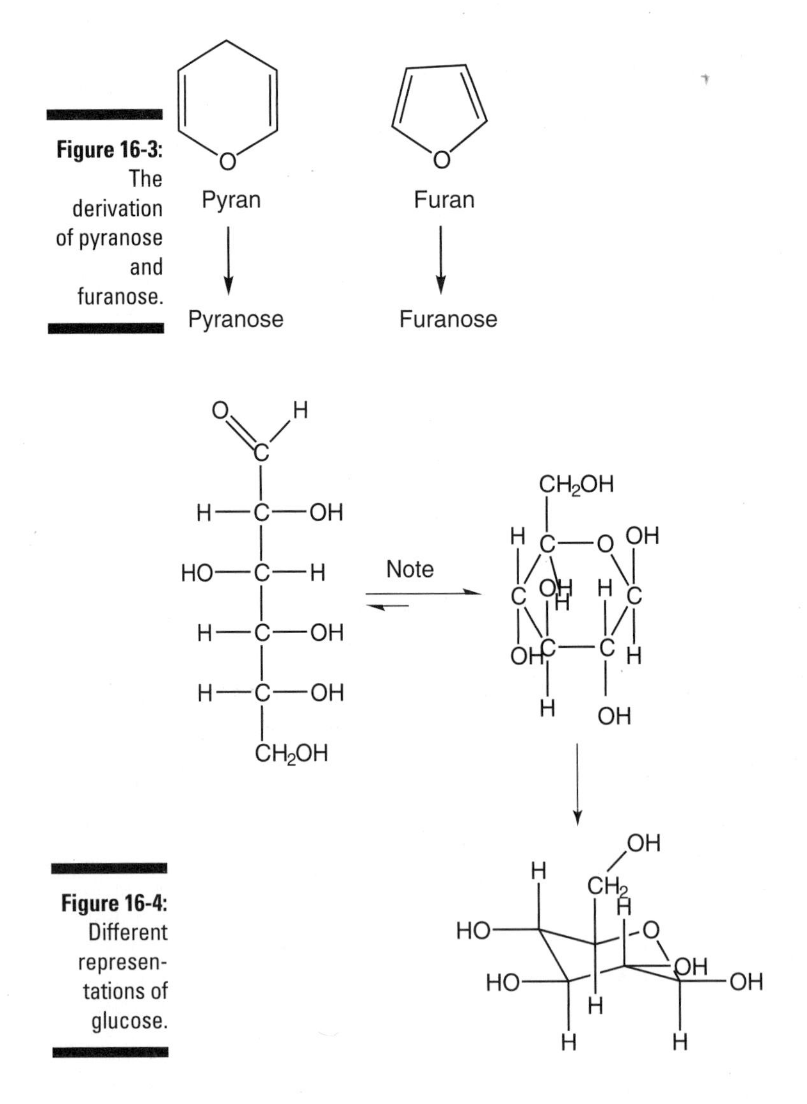

Figure 16-3: The derivation of pyranose and furanose.

Figure 16-4: Different representations of glucose.

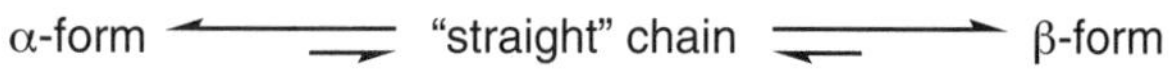

α-form ⇌ "straight" chain ⇌ β-form

Figure 16-5: Representation of the mutarotation process.

Due to mutarotation, a solution of pure α or pure β will change to a mixture. In the case of glucose, the mutarotation gives 36 percent α, 64 percent β, and negligible "straight" chain. The unequal distribution of the two anomers is due to the fact that the -OH on the anomeric carbon of the β form is equatorial, which for a chair conformer is more stable. The -OH on the anomeric carbon in the α anomer is axial, which means this anomer is slightly less stable.

Glycoside formation

The presence of a glycoside, which involves the formation of an acetal (see Figure 16-6) or a hemiacetal, can block mutarotation. Glycosides are different from the original carbohydrates in that they can't undergo mutarotation because the ring is "locked" (a locked ring can't reopen).

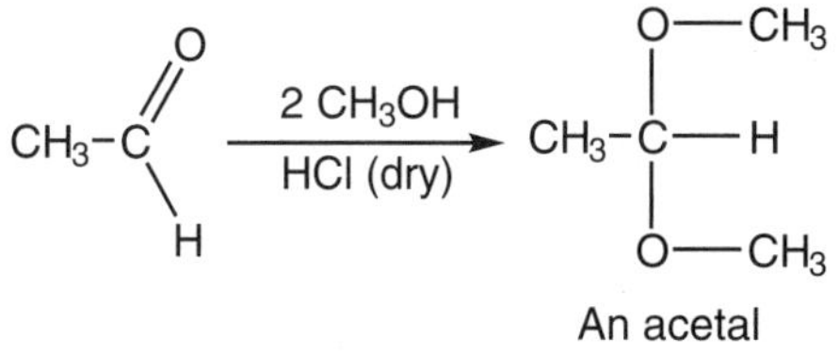

$$CH_3-CHO \xrightarrow[\text{HCl (dry)}]{2\ CH_3OH} CH_3-CH(O-CH_3)_2$$

An acetal

Figure 16-6: The formation of an acetal.

In the case of glucose, the acetal is a glucoside (see Figure 16-7). In general, the simple process shown in Figure 16-7 may form either an acetal or a hemiacetal to give a glycoside.

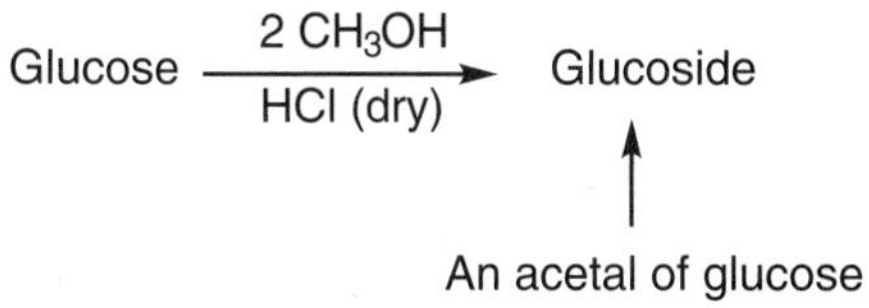

$$\text{Glucose} \xrightarrow[\text{HCl (dry)}]{2\ CH_3OH} \text{Glucoside}$$

↑ An acetal of glucose

Figure 16-7: The formation of a glucoside.

Following are some additional facts about glycosides to file away in your memory bank:

- Glycosides aren't susceptible to simple oxidation via Fehling's or Tollen's test (we explain these tests in greater detail later in this chapter).
- Glycosides don't form osazones (see the later "Osazone formation" section for more on these).
- A glycoside can undergo hydrolysis in an acid but not in a base.

Examining the many reactions of monosaccharides

The presence of alcohol and, in some cases, an aldehyde group makes monosaccharides susceptible to oxidation, whereas the presence of a carbonyl group makes monosaccharides susceptible to reduction. Because monosaccharides are the fundamental carbohydrate, you need to know what happens in the many reactions in which they're involved. The following sections are here to help you out with that. Welcome to the nitty-gritty of monosaccharide oxidation and reduction!

Oxidation of monosaccharides

Oxidizing the aldehyde group present in aldoses is easy; oxidizing the carbonyl group in a ketose is far more difficult. The susceptibility (or lack thereof) to simple oxidation is a useful method of distinguishing between aldoses and ketoses. The next sections explore the various types of monosaccharide oxidation reactions that can occur.

Oxidation with a metal ion

Carbohydrates such as aldoses that undergo oxidation with metal ions are referred to as *reducing sugars.* Both copper(II) ions and silver ions are capable of oxidizing aldoses. Oxidation by copper(II) ions is the basis for Fehling's test and Benedict's test, whereas oxidation by silver ions is the key to Tollen's test. (***Note:*** These tests work for any sugar with a hemiacetal, but they don't work on acetals or ketals.)

The oxidation of an aldose by copper(II) ions results in the reduction of the copper to copper(I) oxide, which has a formula of Cu_2O. A positive test has the bright blue copper(II) solution precipitating brick-red copper(I) oxide. The result? An aldehyde group becomes a carboxylic acid.

The oxidation of an aldose or ketose by silver ions results in the reduction of the silver ion to silver metal. The silver is normally present as $Ag^+(NH_3)_2OH^-$. A positive test has a coating of silver metal precipitating on the sides of the container as a mirror.

Tollen's test may result in the simple oxidation to a carboxylic acid, or it may cause fragmentation of the carbon backbone similar to the oxidation reaction seen with periodic acid (see the later related section). Figure 16-8 shows the general reaction that occurs when you conduct Tollen's test.

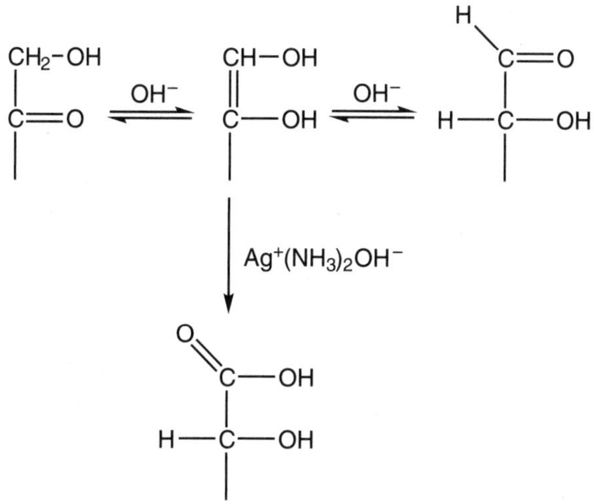

Figure 16-8: Tollen's test.

Oxidation with bromine and water

The oxidation of an aldose (not a ketose) with bromine and water results in an aldonic acid. An example of this reaction is shown in Figure 16-9.

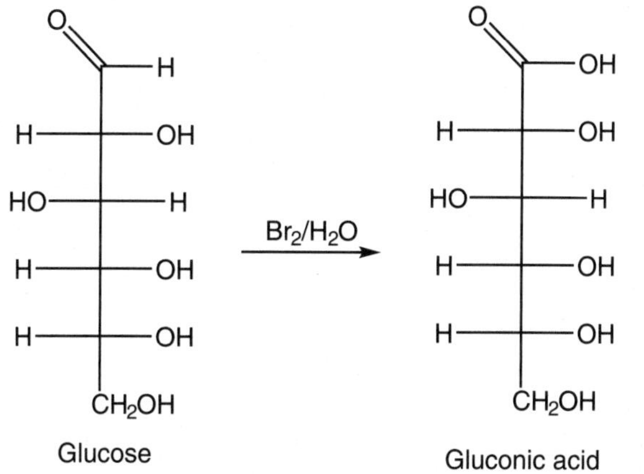

Figure 16-9: The conversion of an aldose (glucose) to an aldonic acid (gluconic acid).

Oxidation with nitric acid

Nitric acid oxidizes not only the aldehyde in an aldose but also the alcohol of the highest-numbered carbon atom. Figure 16-10 shows the oxidation of glucose by nitric acid.

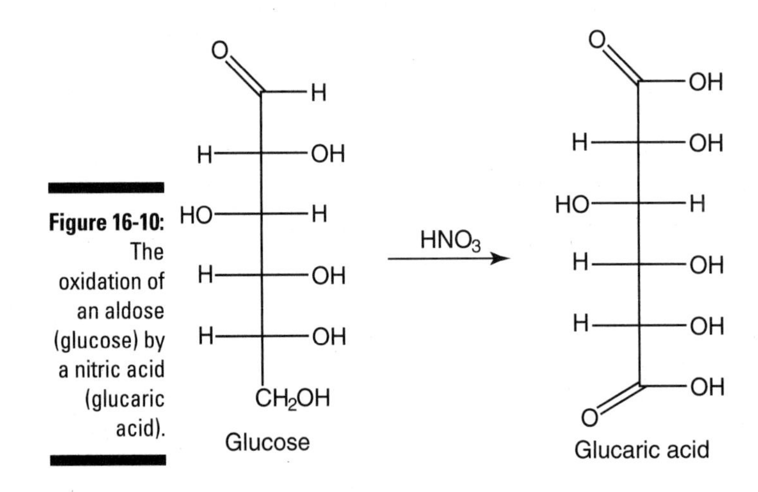

Figure 16-10: The oxidation of an aldose (glucose) by a nitric acid (glucaric acid).

Oxidation with periodic acid

Periodic acid is a stronger oxidizing agent than the others mentioned earlier in this chapter. It causes oxidative *cleavage* (fragmentation) of the carbon backbone. Normally a silver ion is present to facilitate the reaction through the precipitation of silver iodate ($AgIO_3$).

Figure 16-11 shows the three general types of oxidative cleavage that occur in carbohydrates, and Figure 16-12 shows the two additional possibilities that are less reactive. Look to Figure 16-13 for an example of oxidative cleavage of an aldotriose to produce two formic acid molecules and a formaldehyde molecule.

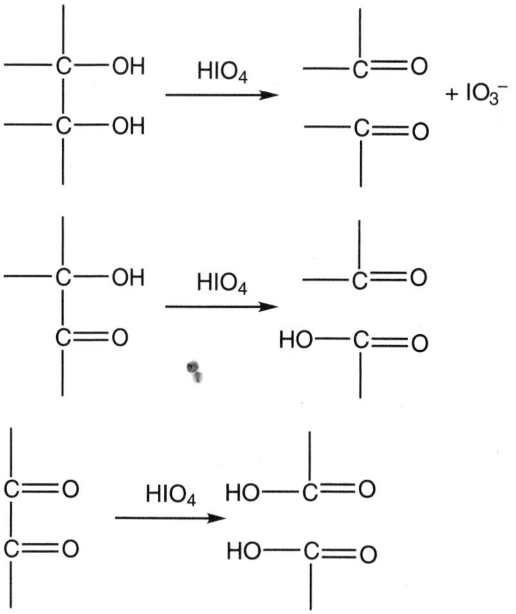

$$-\overset{|}{\underset{|}{C}}-OH,\ -\overset{|}{\underset{|}{C}}-OH \xrightarrow{HIO_4} -\overset{|}{C}{=}O,\ -\underset{|}{C}{=}O + IO_3^-$$

$$-\overset{|}{\underset{|}{C}}-OH,\ \underset{|}{C}{=}O \xrightarrow{HIO_4} -\overset{|}{C}{=}O,\ HO-\underset{|}{C}{=}O$$

$$\overset{|}{C}{=}O,\ \underset{|}{C}{=}O \xrightarrow{HIO_4} HO-\overset{|}{C}{=}O,\ HO-\underset{|}{C}{=}O$$

Figure 16-11: Oxidative cleavage in carbohydrates.

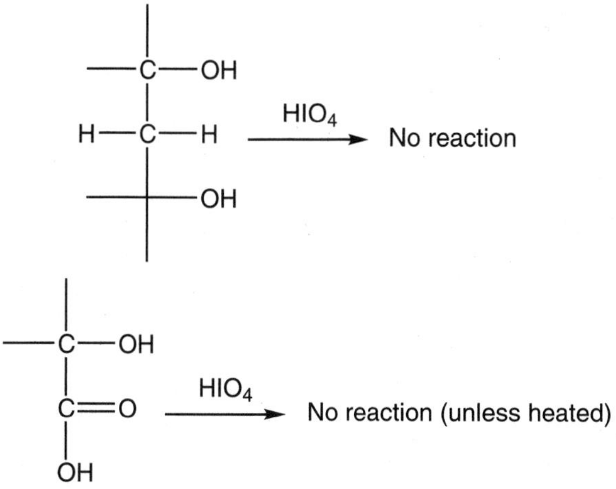

$$-\overset{|}{\underset{|}{C}}-OH,\ H-\overset{|}{\underset{|}{C}}-H,\ -\overset{|}{\underset{|}{C}}-OH \xrightarrow{HIO_4} \text{No reaction}$$

$$-\overset{|}{\underset{|}{C}}-OH,\ \underset{|}{C}{=}O,\ OH \xrightarrow{HIO_4} \text{No reaction (unless heated)}$$

Figure 16-12: Less reactive types of oxidative cleavage.

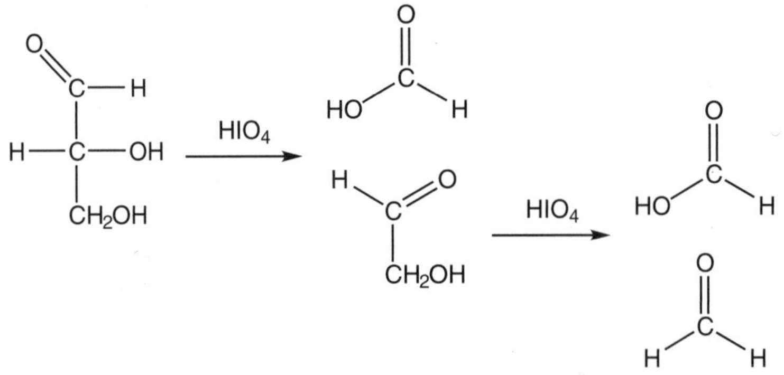

Figure 16-13: Oxidative cleavage of an aldotriose to two formic acid molecules and a formaldehyde molecule.

Reduction of monosaccharides

Reduction of a ketose yields a secondary alcohol, and reduction of an aldose yields a primary alcohol (called an *alditol*). A possible reducing agent is hydrogenation in the presence of a catalyst, such as platinum; another reducing agent is sodium borohydride ($NaBH_4$) followed by hydrolysis. Figure 16-14 illustrates the formation of an alditol.

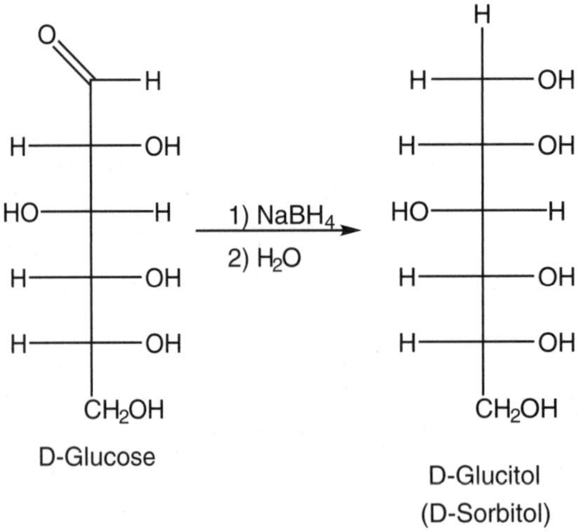

Figure 16-14: Reducing an aldose to form an alditol.

Osazone formation

The carbonyl group (and adjacent alcohol) oxidizes with excess phenyl hydrazine ($PhNHNH_2$) to form an osazone (see Figure 16-15). Osazone formation is very important in determining the relationship between various monosaccharides. For example, both D-glucose and D-mannose produce the same osazone, so they're epimers. *Epimers* differ by only one chiral center, which osazone formation destroys.

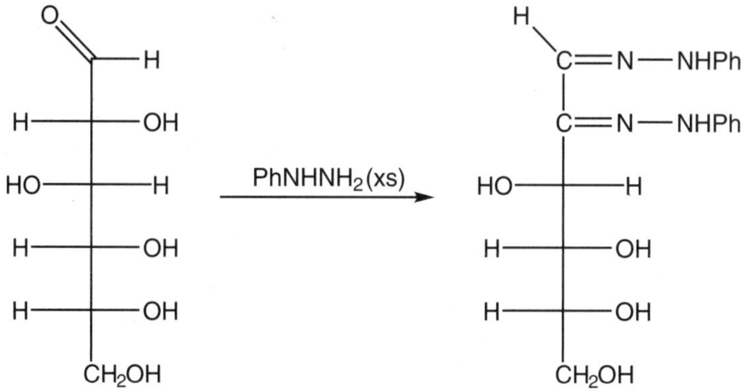

Figure 16-15: The formation of an osazone.

Synthesizing and degrading monosaccharides

To synthesize or degrade a monosaccharide is to convert one monosaccharide into another one. Historically, these processes were important in understanding the relationships between various monosaccharides. The next sections give you some insight into specific synthesis and degradation processes.

Kiliani-Fischer synthesis

Kiliani-Fischer synthesis is a means of lengthening the carbon backbone of a carbohydrate. The process begins with the reaction of hydrogen cyanide (HCN) with an aldehyde to produce a cyanohydrin. Treatment of the cyanohydrin with barium hydroxide followed by acidification yields an aldose with an additional carbon atom, as shown in Figure 16-16. The formation of the cyanohydrin creates a new chiral center as a racemic mixture.

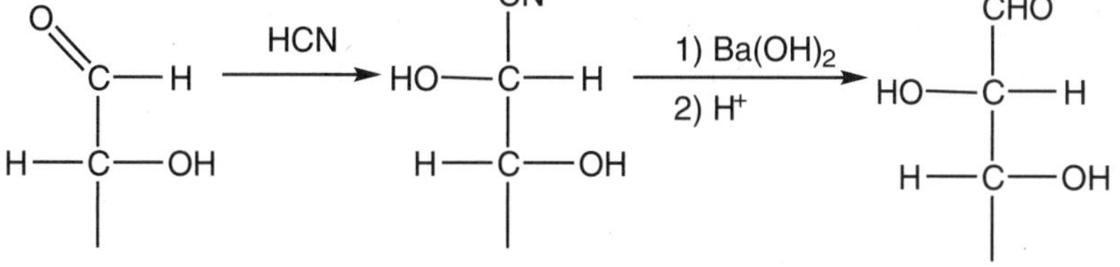

Figure 16-16: The bases for Kiliani-Fischer synthesis.

Figure 16-17 shows a specific example of Kiliani-Fischer synthesis.

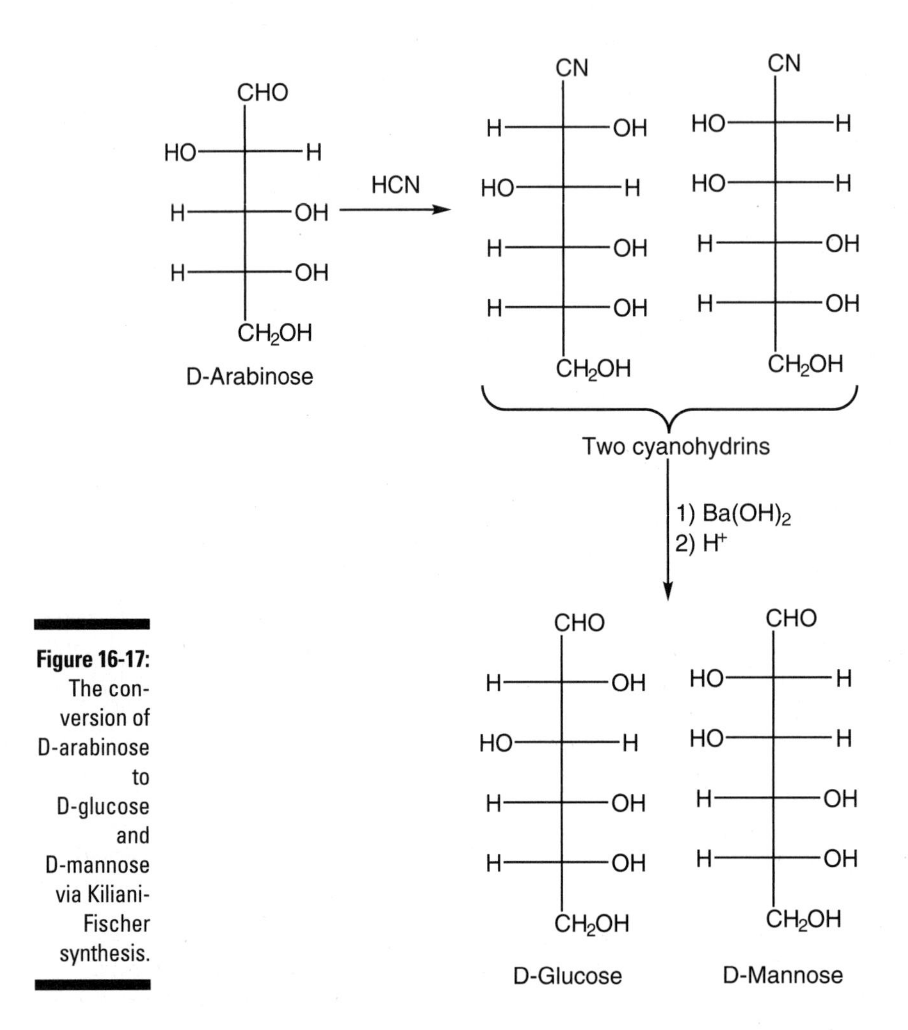

Figure 16-17: The conversion of D-arabinose to D-glucose and D-mannose via Kiliani-Fischer synthesis.

Ruff degradation

Ruff degradation shortens the carbon backbone by one carbon atom. The process begins by oxidizing the aldehyde group to a carboxylic acid (called a *glyconic acid*). Treatment of the resultant acid with hydrogen peroxide (H_2O_2) in the presence of iron(III) ions (Fe^{3+}) results in further oxidation, converting the carboxylic acid to carbon dioxide and generating an aldose with one less carbon atom. Figure 16-18 illustrates the Ruff degradation of glucose.

CHO
H—C—OH
HO—C—H
H—C—OH
H—C—OH
CH_2OH
D-Glucose

Br_2/H_2O →

COOH
H—C—OH
HO—C—H
H—C—OH
H—C—OH
CH_2OH
A glyconic acid (gluconic acid)

H_2O_2, Fe^{3+} →

CHO
HO—C—H
H—C—OH
H—C—OH
CH_2OH
D-Arabinose

+ CO_2

Figure 16-18: The Ruff degradation of glucose.

Meeting the (D-)aldose family

Aldose sugars make up a large part of the carbohydrate family, but the ones that are really worth knowing are part of the D-family. The simplest of these D-sugars is the triose glyceraldehyde. From there you have 2 tetroses, 4 pentoses, and 8 hexoses. Each of these aldose sugars has an enantiomer.

Figure 16-19 shows the structures of all members of the D-aldose family. The circle (head) represents the aldehyde, and the lines to the left or right indicate the orientation of the -OH on each of the chiral carbon atoms. The bottom carbon atom is an achiral -CH_2OH. (Note that the mirror image of each aldose in Figure 16-19 is the L-enantiomer.)

Figure 16-20 indicates the pattern of -OH placement for the bottom row of Figure 16-19, and Figure 16-21 reveals the overall pattern presented in Figure 16-19. Note that (+) indicates -OH is to the right and (–) indicates it's to the left.

The following mnemonic gives the bottom row in Figure 16-19, from right to left: ALL ALTruists GLadly MAke GUmbo In GALlon TAnks.

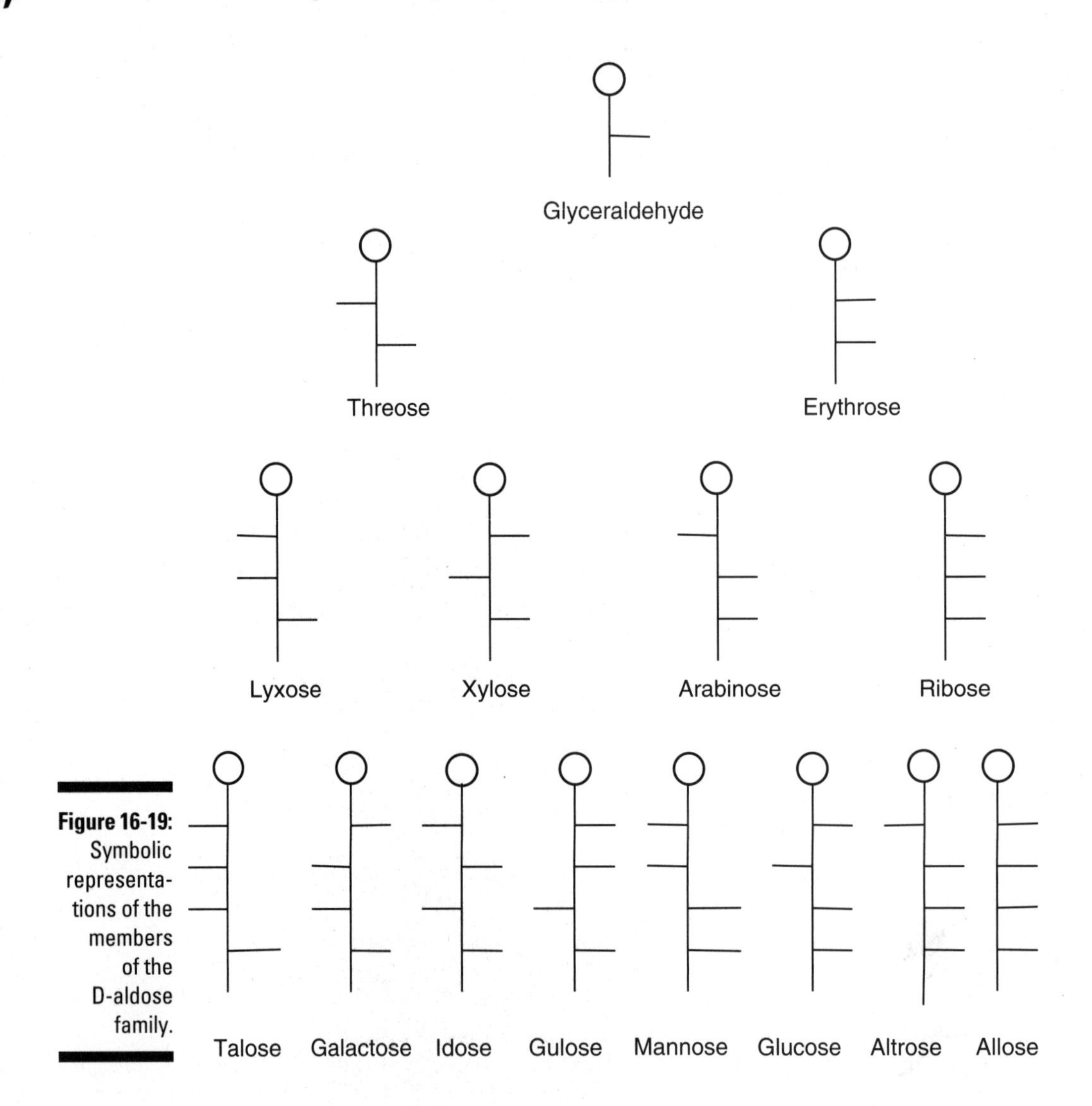

Figure 16-19: Symbolic representations of the members of the D-aldose family.

Figure 16-20: The relative positions of the -OH groups in the bottom row of Figure 16-19.

-	+	-	+	-	+	-	+
-	-	+	+	-	-	+	+
-	-	-	-	+	+	+	+
+	+	+	+	+	+	+	+
T	G	I	G	M	G	A	A

Figure 16-21: The overall pattern in Figure 16-19.

			G				
	T				E		
L		X		A		R	
T	G	I	G	M	G	A	A

Checking out a few disaccharides

Disaccharides are molecules that break apart into two monosaccharides during hydrolysis. Examples include sucrose, maltose, and cellobiose. We cover all three of these disaccharides in the next sections.

The properties of disaccharides aren't a simple combination of the properties of the two monosaccharides present. For example, the disaccharide lactose (milk sugar) is a reducing sugar containing a β-D-glucose linked to a D-galactose molecule.

Sucrose

Sucrose (pictured in Figure 16-22) is a nonreducing sugar containing a glucose joined to a fructose. Hydrolysis of sucrose gives a 50-50 mixture of the two monosaccharides as invert sugars. (The name *invert sugar* derives from the fact that there's an inversion of optical activity upon hydrolysis.) Sucrose is dextrorotatory (+), as is glucose; fructose, however, is levorotatory (–). The sum of the optical activities of the monosaccharides is levorotatory because fructose more than compensates for the activity of glucose.

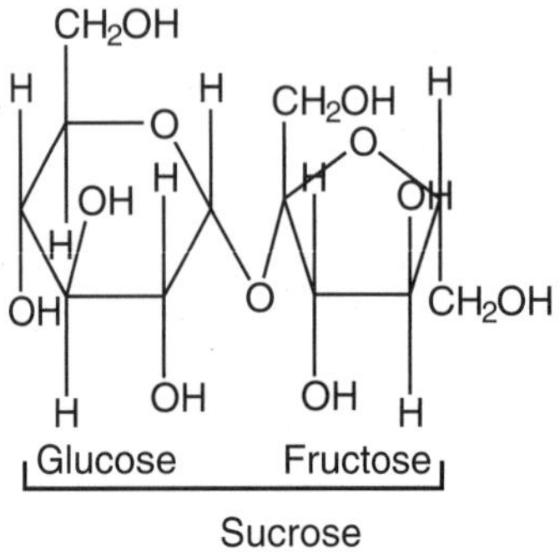

CH_2OH H H CH_2OH H O O OH H H OH H O CH_2OH H OH OH H

Glucose Fructose

Sucrose

Figure 16-22: The structure of sucrose.

Maltose

Maltose is a disaccharide made of two glucose molecules. The linkage involves the α anomer of the left glucose (see Figure 16-23). Maltose is a reducing sugar that's sometimes derived from starch α (see the later related section). The left ring in Figure 16-23 is locked and nonreducing; however, the right ring is reducing and can undergo mutarotation.

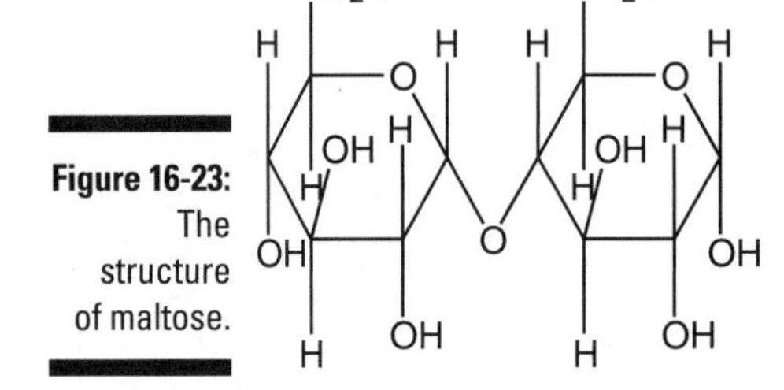

Figure 16-23: The structure of maltose.

Cellobiose

The hydrolysis of cellulose, a polysaccharide, sometimes yields the disaccharide *cellobiose*. When cellobiose, a reducing sugar, is hydrolyzed, two glucose molecules result. Unlike maltose, which we describe in the preceding section, the linkage involves the β anomer of the left glucose (see Figure 16-24).

Looking at some polysaccharides

Polysaccharides are polymers of monosaccharides (usually glucose). Examples include starch (which is 20 percent amylose and 80 percent amylopectin), glycogen (animal starch), and cellulose. We fill you in on all three of these polysaccharides in the following sections.

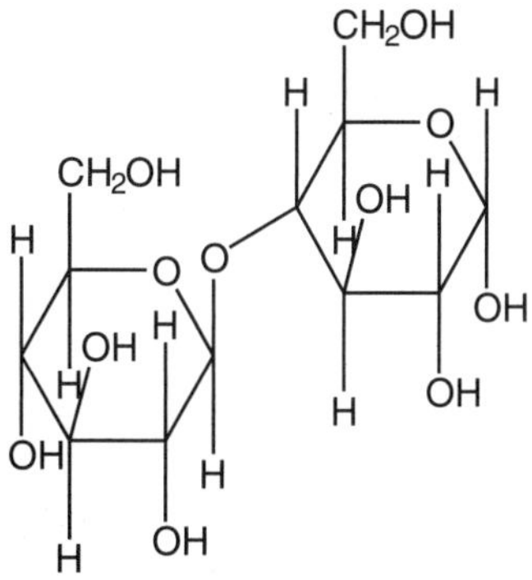

Figure 16-24: The structure of cellobiose.

Starch

Starch molecules contain multiple α anomers of glucose joined by a linkage between the anomeric carbon (#1) of one ring and the fourth carbon (#4) of the next ring. This linkage is therefore known as an α 1 → 4 linkage.

Starch has two components:

- **Amylose:** Amylose is soluble in hot water and nearly insoluble in cold water. The chain of glucose units curl into a helical structure with six glucose molecules per loop. This form of starch interacts with iodine (I_2) to produce an intense dark blue color. In general, amylose has a simple chain structure with each glucose attached to two others. The molecular weight of amylose ranges from 150,000 to 600,000 g/mole, which indicates that a single molecule of amylose contains anywhere from 800 to 3,500 joined glucose molecules. During digestion, the enzyme β-amylase attacks amylose at the nonreducing end and hydrolyzes the chain into maltose units. (The enzyme β-amylase is specific for α 1 → 4 linkages.)
- **Amylopectin:** Amylopectin is similar to amylose except that the glucose chain has branches. These branches involve linkages at the -CH_2OH position (#6), which makes them α 1 → 6 linkages. Amylopectin is water-soluble; it also interacts with iodine to form a reddish-purple complex. Typically, amylopectin is ten times the size of an amylose molecule. Digestion requires β-amylase (1 → 4 linkages) and a second enzyme to remove the branches (as in the 1 → 6 linkages).

Glycogen

Glycogen, animal starch, is similar to amylopectin, but it features more branching and tends to have a higher molecular weight. Glycogen occurs in the liver and muscle tissue. It interacts with iodine to produce a red color.

Cellulose

Cellulose constitutes about 50 percent of wood and about 90 percent of cotton. The molecular weight of this polymer is in the 1 to 2 million range. Cellulose molecules contain the multiple β anomers of glucose joined by a linkage between the anomeric carbon (#1) of one ring and the fourth carbon (#4) of the next ring. This is a β 1 → 4 linkage, which is why enzymes specific for α 1 → 4 linkages (and consequently humans) can't digest cellulose. The chain is typically unbranched, and hydrogen bonding between the chains lowers cellulose's solubility in water.

Treatment of cellulose with a mixture of sodium hydroxide (NaOH) and carbon disulfide (CS_2) yields cellulose xanthate. The action of acid on cellulose xanthate produces rayon or cellophane.

Discovering nitrogen-containing sugars

It may not seem like it, but you can replace an alcohol group (-OH) with an amine group ($-NH_2$) to create a nitrogen-containing sugar. Examples of nitrogen-containing sugars include glycosylamines (see Figure 16-25) and amino sugars (see Figure 16-26). In glycosylamines, the amine group replaces the alcohol group on the anomeric carbon. In amino sugars, the amino group replaces an alcohol group on a carbon atom that isn't anomeric.

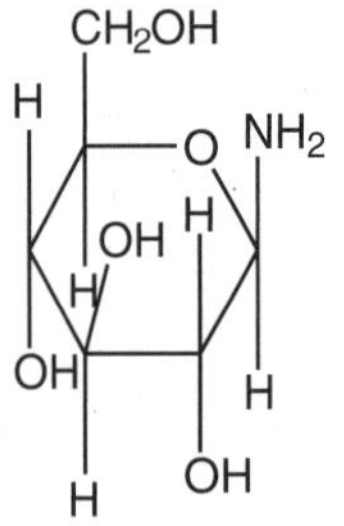

Figure 16-25: A glycosylamine.

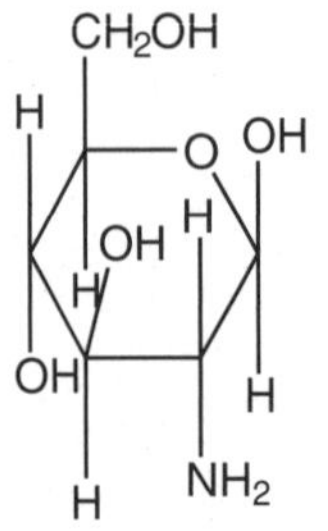

Figure 16-26: An amino sugar.

Lipids: Storing Energy Now So You Can Study Longer Later

Lipids are a diverse group of biologically important compounds that are linked only by their similar solubility in nonpolar or slightly polar solvents. Only a few lipids make an appearance in your average Organic Chemistry II course. Fats and fatty acids are highlighted, but steroids, prostaglandins, phospholipids, and other lipids generally aren't (check out *Biochemistry For Dummies,* written by us and published by Wiley, if you want to know more about any of these lipids).

Pondering the properties of fats

A fat is a solid (at room temperature) triester of glycerin (glycerol) with fatty acids (long-chain carboxylic acids). An oil is similar to a fat; however, an oil is a liquid at room temperature. The general structure of fats and oils appears in Figure 16-27.

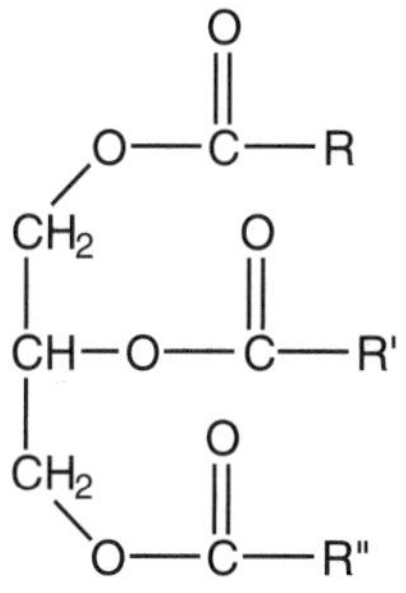

Figure 16-27: The general structure of fats and oils.

The melting point of a fat or oil depends on the size of the R groups and the level of unsaturation. The smaller the R groups and/or the greater the unsaturation, the lower the melting point is. The types of bonds present also help lower the melting point. The long chains of the R groups pack tightly if no double bonds exist, but if double bonds *are* present, they create "kinks" in the chain. Chains with kinks don't pack together well, thereby lowering the melting point (the better the chains pack together, the higher the melting point is).

Naturally occurring examples have fatty acids with an even number of carbon atoms, usually 10 to 20 without any branches. Double bonds are only present as the *cis* isomer. At least one double bond is present in an unsaturated fat; multiple double bonds are present in a polyunsaturated fat.

You can hydrogenate some of the double bonds in any polyunsaturated fat. Hydrogenation, whether complete or partial, results in an increase in the melting point, a fact that makes it possible to convert a liquid oil to a solid fat. (Note that partial hydrogenation may convert some of the *cis* double bonds into *trans* double bonds, thereby producing a trans fat.)

Soaping up with saponification

The *saponification* (the base-catalyzed hydrolysis of an ester) of fats has been important since ancient times. This process frees the glycerin and releases the fatty acids as carboxylate ions. The carboxylate ions, along with sodium or potassium ions from the base, create a soap (refer to Figure 16-28).

Figure 16-28: The general process for producing a soap.

$$\text{Fat} \xrightarrow[\text{or KOH}]{\text{NaOH}} \text{Soap}$$

Sodium stearate (shown in Figure 16-29) is an example of a type of soap. The carboxylate ion in the soap has a hydrophilic (or *ionic*) end and a hydrophobic (or *organic*) end. The presence of the hydrophilic (head) and hydrophobic (tail) ends is the key to the behavior of soap. The hydrophilic end is water-soluble, whereas the hydrophobic end is soluble in nonpolar materials such as oil (but not the glycerin-containing form).

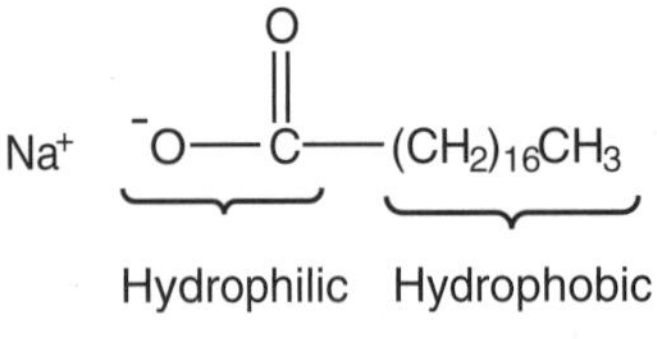

Figure 16-29: The structure of sodium stearate.

$$Na^+ \; \underbrace{{}^-O-\overset{\overset{\displaystyle O}{\|}}{C}}_{\text{Hydrophilic}}-\underbrace{(CH_2)_{16}CH_3}_{\text{Hydrophobic}}$$

Soap works because the nonpolar end dissolves in dirt (oil), leaving the polar end outside the dirt. This combination is known as a micelle (see Figure 16-30). To the surrounding water molecules, the micelle appears as a very large ion. These "ions" are water-soluble and repel each other due to their like charges, a behavior that causes them to remain separated. Metal ions in hard water ($Ca^{2+}/Mg^{2+}/Fe^{2+}$) cause a precipitate to form because they react with carboxylate ions to form an insoluble material (also known as the soap scum hanging around the bathtub or shower).

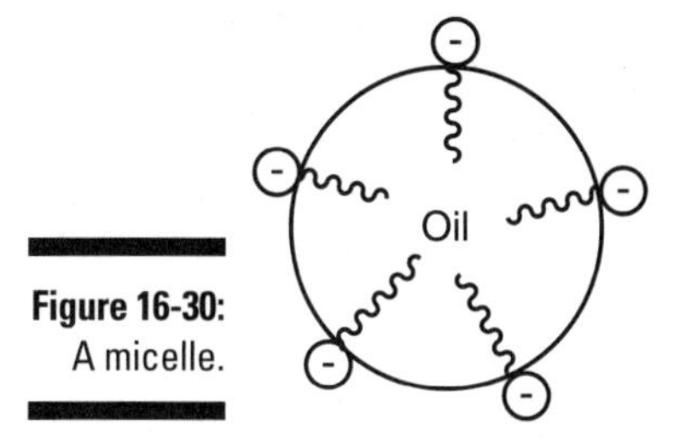

Figure 16-30: A micelle.

Detergents are like soaps, but they're derivatives of either phosphoric acid or sulfuric acid (see Figure 16-31). They've gained popularity because they don't precipitate in hard water.

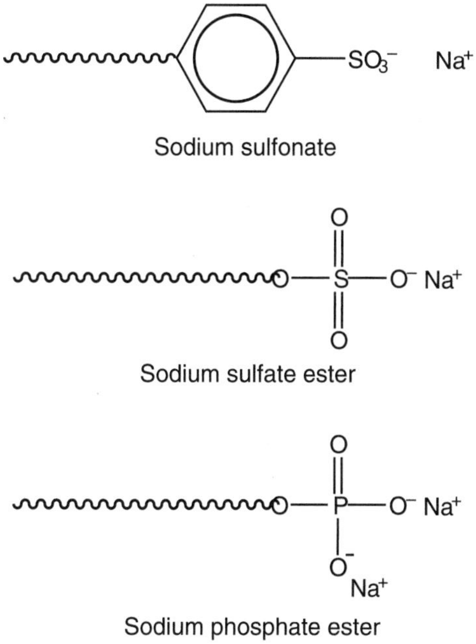

SO_3^- Na^+

Sodium sulfonate

O—S—O^- Na^+

Sodium sulfate ester

O—P—O^- Na^+

O^- Na^+

Sodium phosphate ester

Figure 16-31: Three generic detergents: a sulfonate, a sulfate ester, and a phosphate ester.

Bulking Up on Amino Acids and Proteins

Amino acids are the building blocks of proteins, which (along with polysaccharides and nucleic acids) are important biological polymers. Much of the behavior of proteins is better described in a biochemistry text. However, many of the properties of amino acids are as much a part of organic chemistry as they are a part of biochemistry.

The next sections familiarize you with the basic chemistry of amino acids, including their physical properties. They also walk you through the many ways in which amino acids can be created (synthesized).

Introducing amino acids

Amino acids contain an amino group and a carboxylic acid group. The biologically important amino acids are the α-amino acids (see Figure 16-32). In the α-amino acids, the amino group and the carboxylic acid group are attached to the same carbon atom. Note that the α carbon is chiral unless R is a hydrogen atom. The naturally occurring α-amino acids are the L enantiomers.

The difference between α-amino acids is the identity of the R group also attached to the α-carbon.

$$H_2N-\overset{H}{\underset{R}{C^{\alpha}}}-C(=O)OH$$

Figure 16-32: The general structure of α-amino acids.

Amino acids link to form amides and create a polyamide. In biochemistry, the amide gets replaced with a peptide, making the proteins polypeptides. The sequence of the amino acids dictates the primary structure of the protein.

Perusing the physical properties of amino acids

When proteins undergo hydrolysis, you wind up with 22 α amino acids, 20 of which are "regular" amino acids and 2 of which are derived amino acids. Amino acids are *amphoteric* (they possess the characteristics of both acids and bases and can react as either) because both acidic and basic groups are present. An internal acid-base reaction produces a dipolar ion known as a zwitterion (you can see the general structure of one in Figure 16-33).

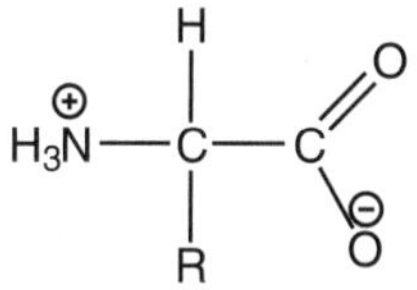

Figure 16-33: The general structure of a zwitterion.

The simplest amino acid is glycine (check out a glycine zwitterion in Figure 16-34). In glycine, the R group is a hydrogen atom. The ammonium portion of the zwitterion is acidic, and the carboxylate portion is basic. The presence of the charges in the zwitterions makes amino acids ionic solids (meaning they're nonvolatile). The ionic character makes the amino acids soluble in water but not in nonpolar solvents.

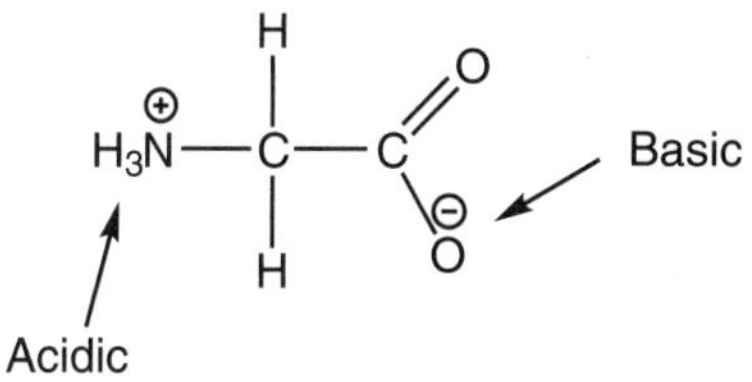

Figure 16-34: A glycine zwitterion.

The acidic and basic groups have K_a and K_b values, respectively. In general, the K_a values are approximately 10^{-9}–10^{-12} (pK_a = 9–12), and the K_b values are approximately 10^{-11}–10^{-12} (pK_b = 11–12). Glycine has a pK_a equal to 9.6 and a pK_b equal to 11.7. Therefore, the addition of an acid to a glycine zwitterion leads to protonation of the carboxylate end, and the addition of a base leads to deprotonation of the ammonium end (see Figure 16-35).

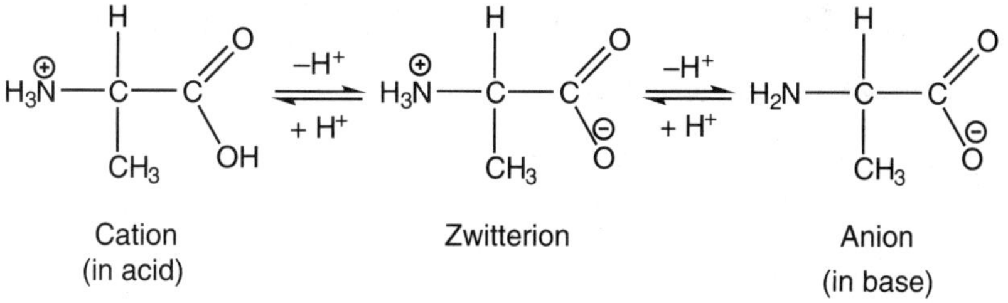

Figure 16-35: The action of an acid and a base upon the zwitterion of alanine ($R = -CH_3$).

The zwitterion form predominates at the isoelectric point (pI). Figure 16-36 shows where the pI values come from.

Figure 16-36: Determining the isoelectric point (pI).

$$pI = \frac{pK_{carb} + pK_{amine}}{2}$$

The cation form predominates at a pH < pI, whereas the anion form predominates at a pH > pI. In general, pK_{carb} equals pK_1 and pK_{amine} equals pK_2. However, complications may occur because the R group may be acidic or basic.

The conjugate base of every acid has a pK_b that's related to the pK_a as $pK_a + pK_b = 14.0$ or $K_aK_b = K_w = 1.0 \times 10^{-14}$.

Studying the synthesis of amino acids

You can synthesize amino acids in a number of ways, although the presence of both an acidic and a basic group may cause some complications. Amino acid synthesis serves as the application of previously introduced reactions, and most synthetic methods produce a racemic mixture. The sections that follow give you a look at the various ways in which amino acid synthesis can occur.

From α-halogenated acids

When the α-carbon of a carboxylic acid is halogenated, an amino group replaces the halogen in the halogenated acid with excess ammonia. Figure 16-37 illustrates this reaction.

From potassium phthalimide

When potassium phthalimide is added as part of a six-step synthesis of an amino acid, the resulting amino acid is present as a racemic mixture. Figure 16-38 illustrates multistep synthesis involving potassium phthalimide.

Study reaction sequences such as Figure 16-39 to note how the scheme develops and how you can change it to obtain a different product.

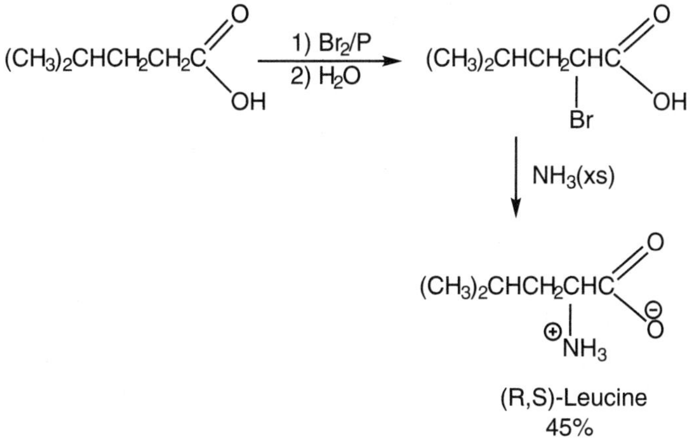

$(CH_3)_2CHCH_2CH_2COOH$ → 1) Br_2/P, 2) H_2O → $(CH_3)_2CHCH_2CH(Br)COOH$ → NH_3(xs) → $(CH_3)_2CHCH_2CH(NH_3^+)COO^-$

(R,S)-Leucine
45%

Figure 16-37: The synthesis of an amino acid from an α-halogenated acid.

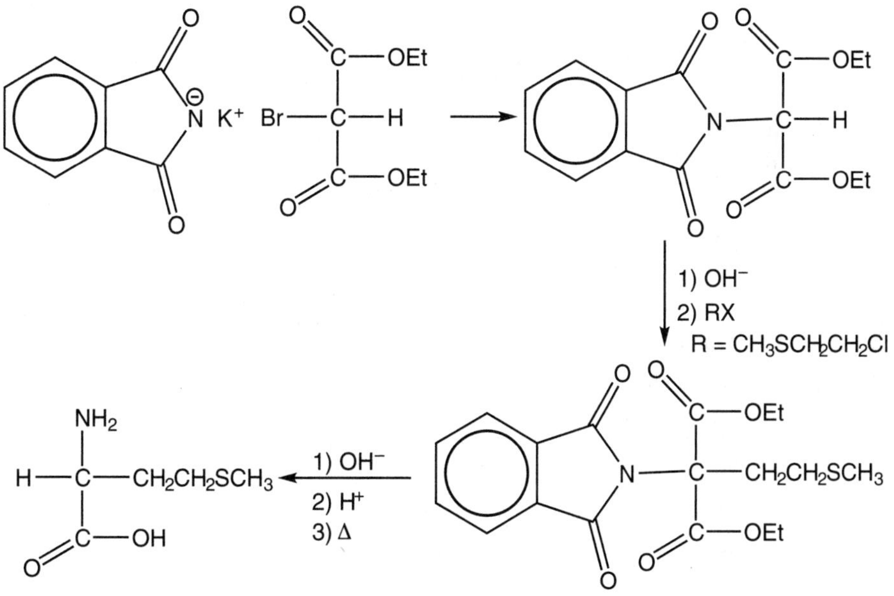

Potassium phthalimide (N^- K^+) + Br—CH(C(=O)OEt)(C(=O)OEt) → N-phthalimido—CH(C(=O)OEt)(C(=O)OEt)

→ 1) OH^-, 2) RX, R = $CH_3SCH_2CH_2Cl$ → N-phthalimido—C($CH_2CH_2SCH_3$)(C(=O)OEt)(C(=O)OEt)

→ 1) OH^-, 2) H^+, 3) Δ → H—C(NH_2)($CH_2CH_2SCH_3$)—C(=O)—OH

Figure 16-38: The synthesis of an amino acid ((±)-methionine) using potassium phthalimide.

From amido malonic esters

The multistep synthesis shown in Figures 16-39 and 16-40 leads to the synthesis of an amino acid. (Note that the diethyl acetamidomalonate shown in Figure 16-40 is the product of the reaction in Figure 16-39.) The example in the figures is a natural amino acid; however, you can synthesize other (not natural) amino acids with this procedure.

Sequences such as those shown in Figures 16-40 and 16-41 allow you to review many of the different reactions you encountered previously. If you don't recall the exact details of one of the reactions, go back and review the reaction.

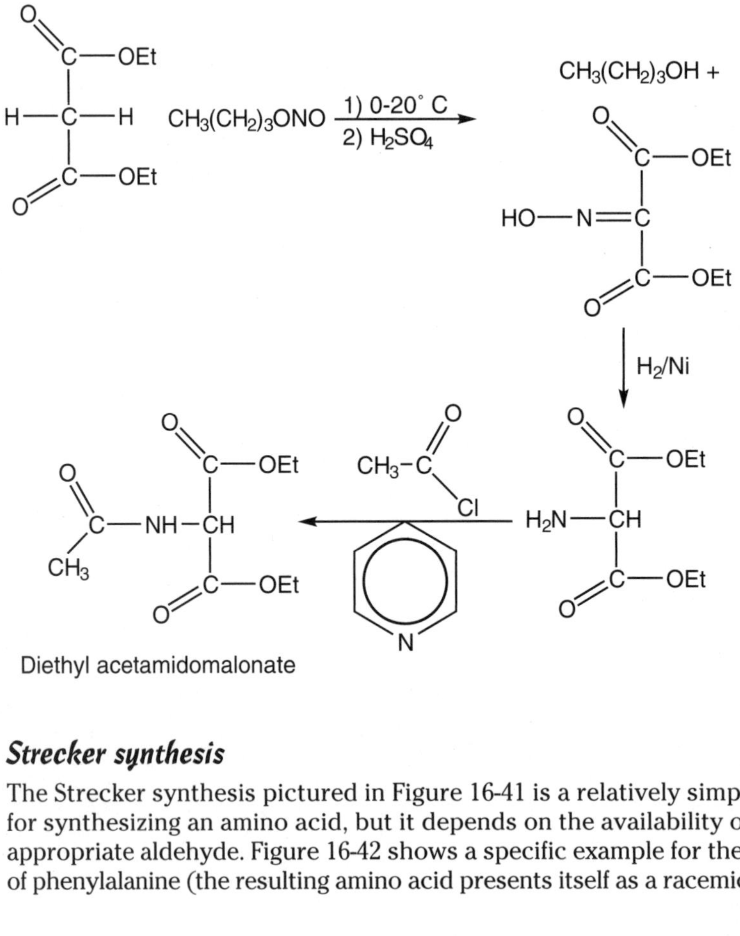

Diethyl acetamidomalonate

Figure 16-39: The beginning of the synthesis of an amino acid via an amido malonic ester.

Strecker synthesis

The Strecker synthesis pictured in Figure 16-41 is a relatively simple method for synthesizing an amino acid, but it depends on the availability of the appropriate aldehyde. Figure 16-42 shows a specific example for the synthesis of phenylalanine (the resulting amino acid presents itself as a racemic mixture).

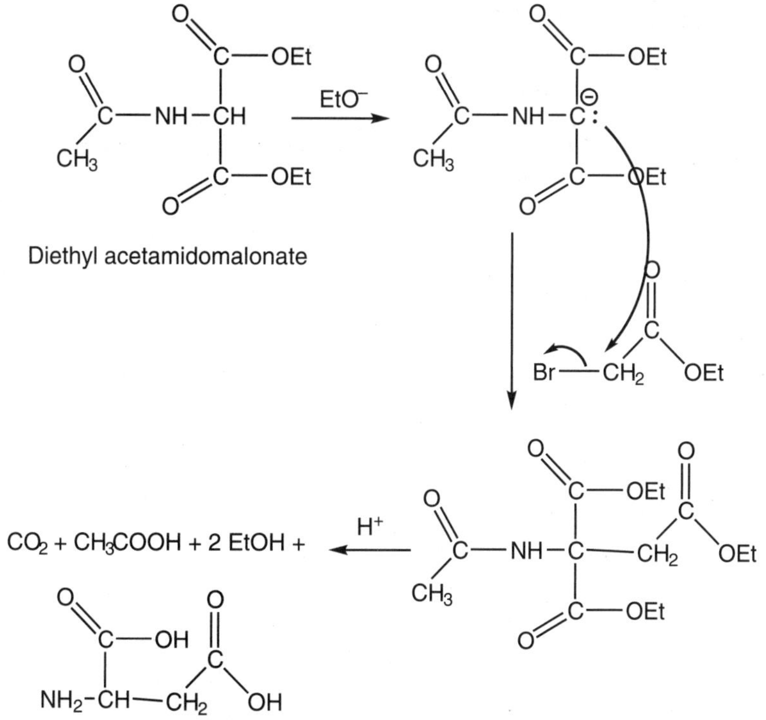
EtO⁻
Diethyl acetamidomalonate
OEt
NH
CH₃
Br
CH₂
H⁺
CO₂ + CH₃COOH + 2 EtOH +
NH₂-CH—CH₂
OH

Figure 16-40: The end of the synthesis of an amino acid via an amido malonic ester.

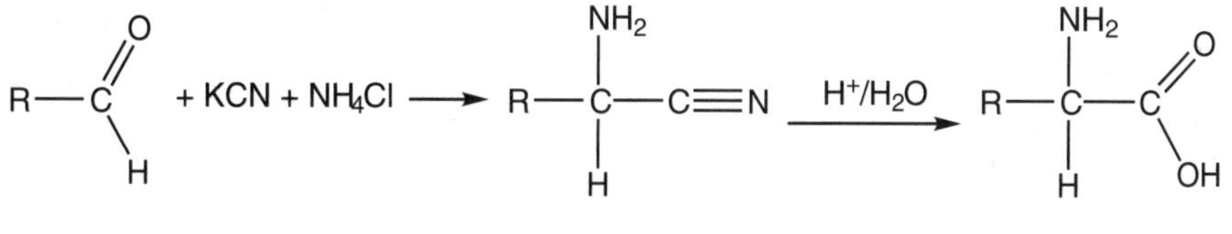
+ KCN + NH₄Cl
H⁺/H₂O
NH₂
R
H
OH

Figure 16-41: The general process of Strecker synthesis.

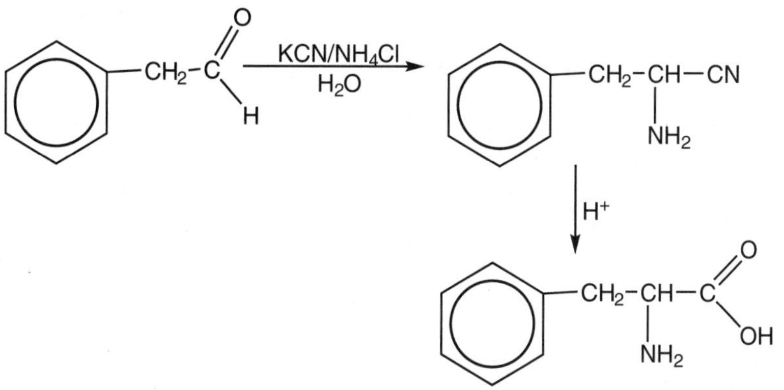
KCN/NH₄Cl
H₂O
H⁺
CH₂-CH—CN
NH₂
CH₂-CH—C
OH

Figure 16-42: The synthesis of (±)-phenylalanine via the Strecker synthesis.

Reductive amination

Reductive amination, shown in Figure 16-43, is a biologically important method of synthesizing amino acids. The reduction is caused by NADH in the presence of ammonia (NADH, nicotinamide adenine dinucleotide, is a biological reducing agent). The mechanism for this reaction and the other biologically important procedure for synthesizing amino acids (transamination) are more biochemistry than organic chemistry, so we don't touch on them here. You can check out *Biochemistry For Dummies,* written by us and published by Wiley, if you want to know more.

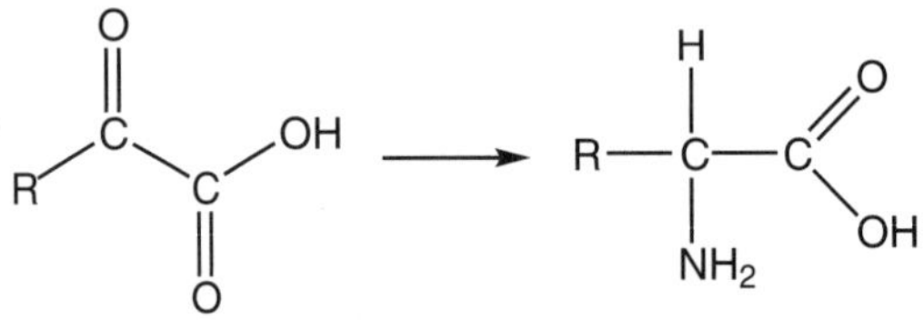

Figure 16-43: The general process of reductive amination.

Resolution of (±) amino acids

The resolution of (±) amino acids isn't really a synthetic method, but it's certainly useful in the production of a particular amino acid from a racemic mixture. In the resolution of (±) amino acids, an enzyme (a biological catalyst) interacts with only one enantiomer. (Why, you ask? Because enzymes are stereoselective.) The enzyme leaves one enantiomer unchanged and modifies the other into a different compound, which makes it possible to separate the enantiomer from the other compound by a number of techniques. After the enantiomer has been separated, all that's left is to reverse the process induced by the enzyme.

Part V
Pulling It All Together

"Okay—now that the paramedic is here with the defibrillator and smelling salts, prepare to learn about covalent bonds."

In this part . . .

In this part are a couple of chapters designed to help you pull together all those concepts you have been studying. In the first chapter we look at some strategies that you can use when designing a synthesis of a particular organic compound. You know that type of problem: "Starting with a piece of coal and a glass of water, synthesize DNA, naming all intermediates." Okay, maybe not quite that involved, but close. Then we tackle those dreaded roadmaps, walking you through a number of examples and giving you some tips to ease the way.

Realize that if you're using this book during a course, you're almost done when you hit this section. Hang in there; there's a light at the end of the tunnel (and we all hope it's not a train).

Chapter 17

Overview of Synthesis Strategies

In This Chapter

- Determining strategies for synthesis problems
- Figuring out a one-step synthesis
- Considering multistep synthesis for more complicated problems
- Solving sample problems with retrosynthetic and synthetic analysis

Synthesis is actually the reverse of predicting products. In synthesis, you have the product of a reaction and you must predict the reaction sequence necessary to form the product. While you may have hints as to the identity of the starting material, in most cases you need to predict the starting material, the reactants, and possibly the reaction conditions. You encounter two general types of synthesis questions — *one-step synthesis* and *multistep synthesis*. As the name implies, a one-step synthesis problem requires one "simple" answer. A multistep synthesis involves more than one reaction, and more than one answer may be correct. On an organic chemistry exam, one-step synthesis questions usually focus on the most recent reactions you have studied, whereas multistep synthesis questions usually involve a recent reaction in one step and one or more other reactions from any point in Organic Chemistry I or II in other steps.

When predicting products you should ask yourself three basic questions, and the same questions are also important when considering synthesis problems:

- What kind of reaction?
- What is the regiochemistry of the reaction (the chemical environment of the reactive site on this molecule for this particular reaction)?
- What is the stereochemistry of the reaction?

However, you need to apply these questions in a different way for synthesis problems than for predicting problems.

The answers to all three questions are found in the mechanism.

If the question starts with a ketone, what do you do? When predicting products, you must consider every reaction you know that begins with a ketone. When doing a synthesis problem, you must consider every reaction that produces a ketone.

On an organic chemistry exam, if the instructor takes time to draw the specific stereochemistry of a molecule, it's a hint that you should carefully consider the stereochemistry of any reactions you use.

To do any synthesis problem, you must know your reactions backwards and forwards. This takes time and effort. To learn them thoroughly, you need to practice over and over. During an exam, you don't have time to work out every possibility, so you *must* know the reactions.

Working with One-Step Synthesis

In a one-step synthesis problem, you know that only one reaction is necessary to form a particular product. The key is the functional group in that product. For example, if the functional group is an aldehyde, the reaction must be one of the limited number of reactions that you know form an aldehyde. Other information may limit your choice of reactions. For example, the problem may ask you to begin with an alcohol with the same number of carbon atoms, or you may have an alkene with a greater number of carbon atoms.

Next, you need to consider regiochemistry. Do any of the reactions you know give a product with the correct regiochemistry? This may allow you to narrow your choices.

Finally, you must consider the stereochemistry. Do any of the reactions you know give you the correct stereochemistry? Again, this may further limit your choices.

In some situations, more than one reaction can give you the correct product. Don't overanalyze the problem in an effort to choose between multiple correct answers, just pick one and go with it.

Tackling Multistep Synthesis

In most cases, a multistep synthesis is not a simple string of reaction steps. You need to look ahead (and behind). For example, you may have a perfectly good reaction that forms a *cis*-alkene, but that's the wrong answer because two steps later, you need a *trans*-alkene.

If you don't know your reactions, you can't possibly solve a multistep synthesis problem. Knowing the reactions includes the reactants, products, conditions, regiochemistry, and stereochemistry. Flashcards are a useful means of learning the reactions. On the front side of the card, write the reactants and conditions, and on the reverse side, write the products, regiochemistry, and stereochemistry. Learn the cards in one direction first (identifying what's on the back based on what you see on the front), and then learn them in reverse (knowing what's on the front when you look at the back). You must know the reactions backwards and forwards. (Shuffle the deck often.)

Many times, multistep synthesis problems start with an alkane. An efficient way of adding a functional group to that alkane is to brominate (or chlorinate) the alkane through a free-radical reaction.

A limited number of reactions can add or remove carbon atoms. For this reason, comparing the carbon backbones of the materials involved in the reaction is useful. These reactions require the presence or absence of certain functional groups.

Practicing Retrosynthetic and Synthetic Analysis

Retrosynthetic analysis is a method for tackling synthesis problems, especially multistep synthesis problems. The application of this technique involves working the problem backwards, starting at the final product and ending up with the initial reactants.

One of the great frustrations you may encounter during an organic chemistry exam is expecting to know the answer to a retrosynthetic analysis question immediately upon first reading the problem. Save yourself this frustration and accept the fact that you need to read and reread the problem multiple times.

When applying retrosynthetic analysis to a multistep synthesis problem, you must work backwards. If you become lost, as a last resort you may want to look at the forward reactions. However, the forward process often goes off on a tangent or leads to a cul-de-sac.

After finishing a multistep synthesis problem, spend some time working on another problem or task. Then come back and check each step in both the forward and the reverse direction. You should pay particular attention to both the regiochemistry and the stereochemistry of each step. In addition, if one of the steps involves a molecule with more than one functional group, make sure the reaction only alters the desired functional group.

To survive multistep synthesis questions on organic chemistry exams, you must practice, practice, and practice.

The rest of this chapter is dedicated to practicing retrosynthetic and synthetic problems. The following examples utilize some of the reactions that you have seen throughout this book.

Example 1

Figure 17-1 presents a typical retrosynthetic analysis problem. (The presence of a β-dicarbonyl compound indicates that the formation of a carbanion through the loss of a hydrogen ion from the α-carbon will probably be important; however, don't let this distract you from the task at hand.) The problem asks you to prepare a ketone. Your first question should be, "How can I prepare a ketone?" One answer to this question is to decarboxylate a β-dicarbonyl compound. For the compound in this problem, the reaction shown in Figure 17-2 works.

Resist the temptation to look ahead on this problem. Try to follow the steps and make your own predictions on what to do. Your choice may be as good as or better than the one presented here.

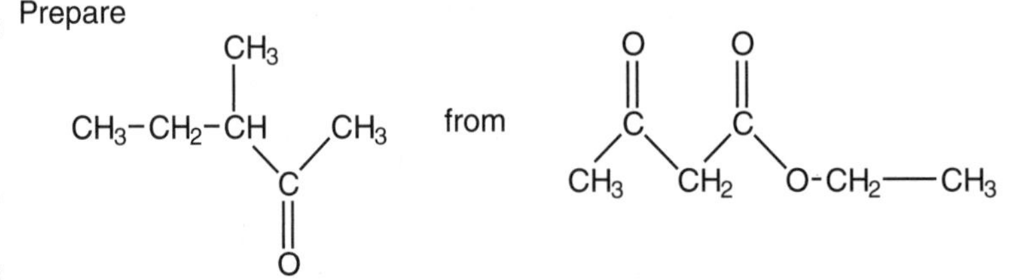

Figure 17-1: The first retrosynthetic analysis problem.

It may help to redraw the one or more compounds in a retrosynthesis problem so the general shape of the reactants and products are similar.

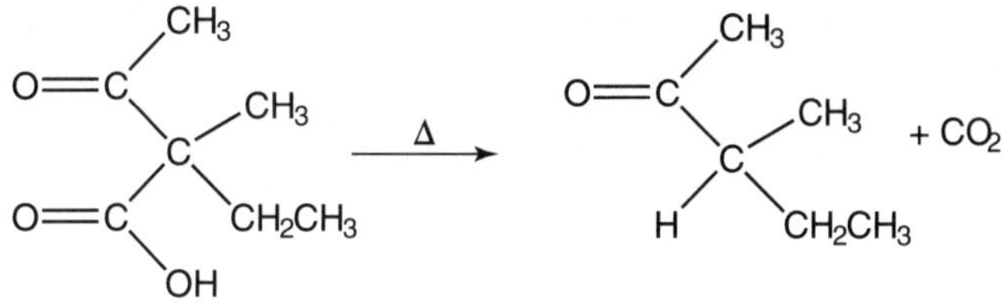

Figure 17-2: One step in the solution to the problem in Figure 17-1.

The next step is to determine what reaction produces the reactant in Figure 17-2. The carboxylic acid can be produced through the hydrolysis of an ester. You can now propose the reaction required to produce the particular acid. (The other compound in Figure 17-1 is an ethyl ester, so this process probably involves an ethyl ester.) Figure 17-3 adds the hydrolysis step to the solution to this problem.

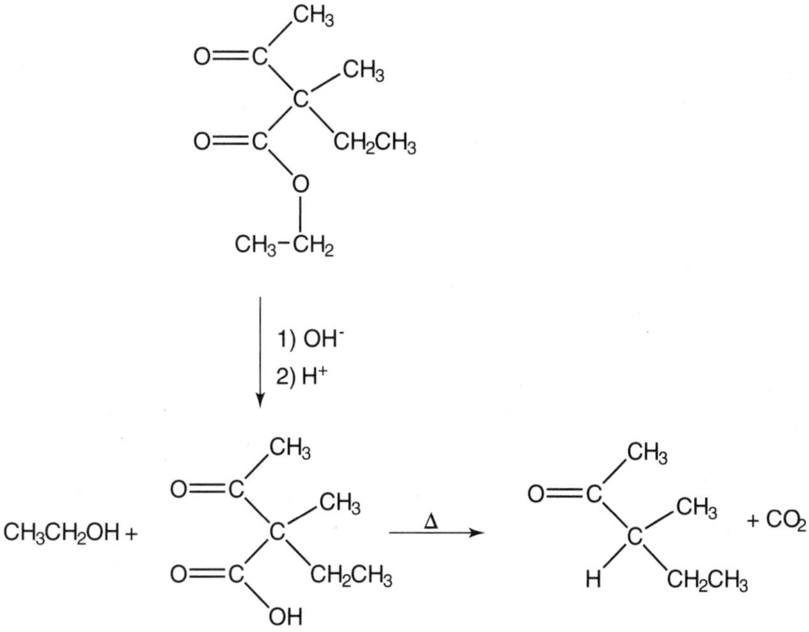

Figure 17-3: Adding the hydrolysis step.

Next, you need to determine how to produce the ester in Figure 17-3. (This is a β-dicarbonyl compound like the starting material. This relationship may be important.) One way to produce this ester is shown in Figure 17-4. This step requires you to start with a carbanion, which can form through the reaction of a strong base with a β-dicarbonyl compound. This step is in Figure 17-5. We use the *t*-butoxide ion as the base, but other bases are acceptable.

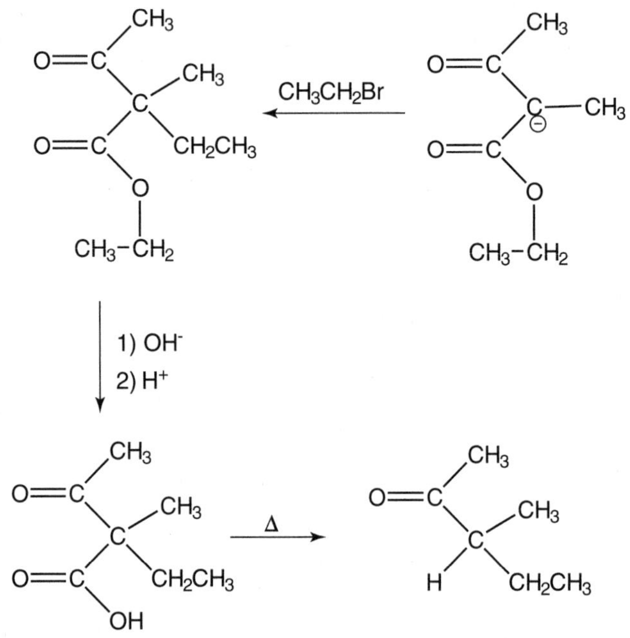

CH_3CH_2Br

1) OH^-
2) H^+

Δ

Figure 17-4: Adding CH_3CH_2OH + another step.

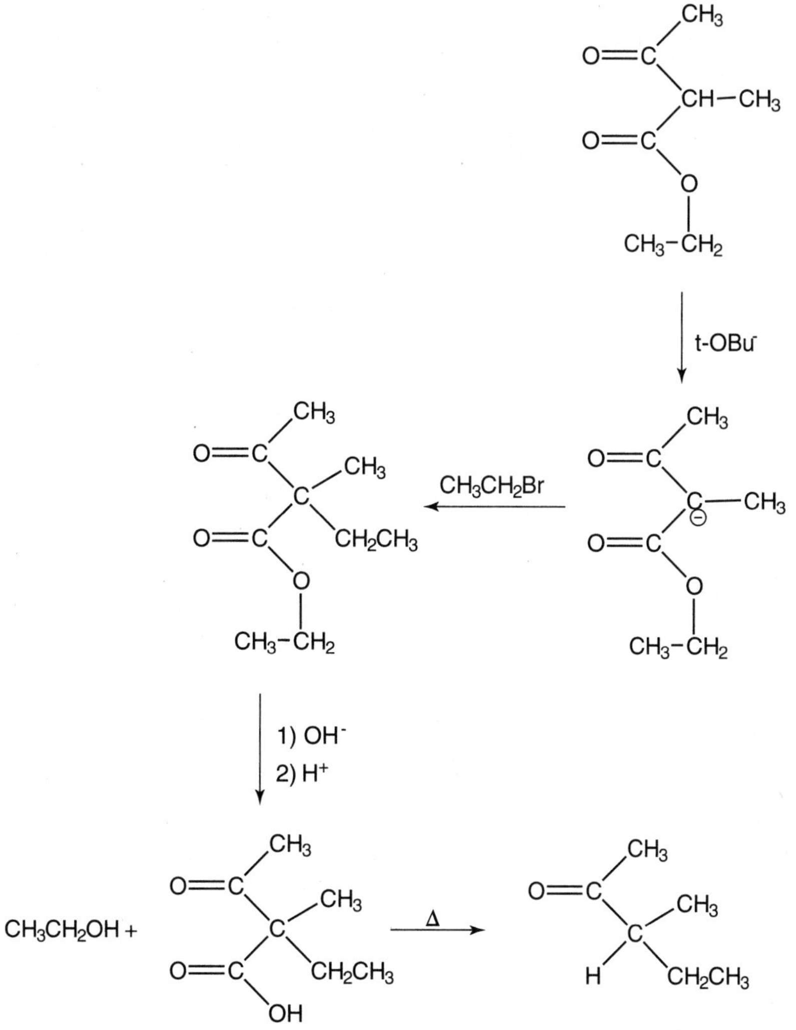

t-OBu⁻ → CH_3CH_2Br ← 1) OH^- 2) H^+ ↓ Δ →

Figure 17-5: Adding the reaction with the CH_3CH_2OH + *t*-butoxide ion.

Now, how do you prepare the starting material in Figure 17-5? The answer is to use a methyl halide such as bromomethane. This step appears in Figure 17-6.

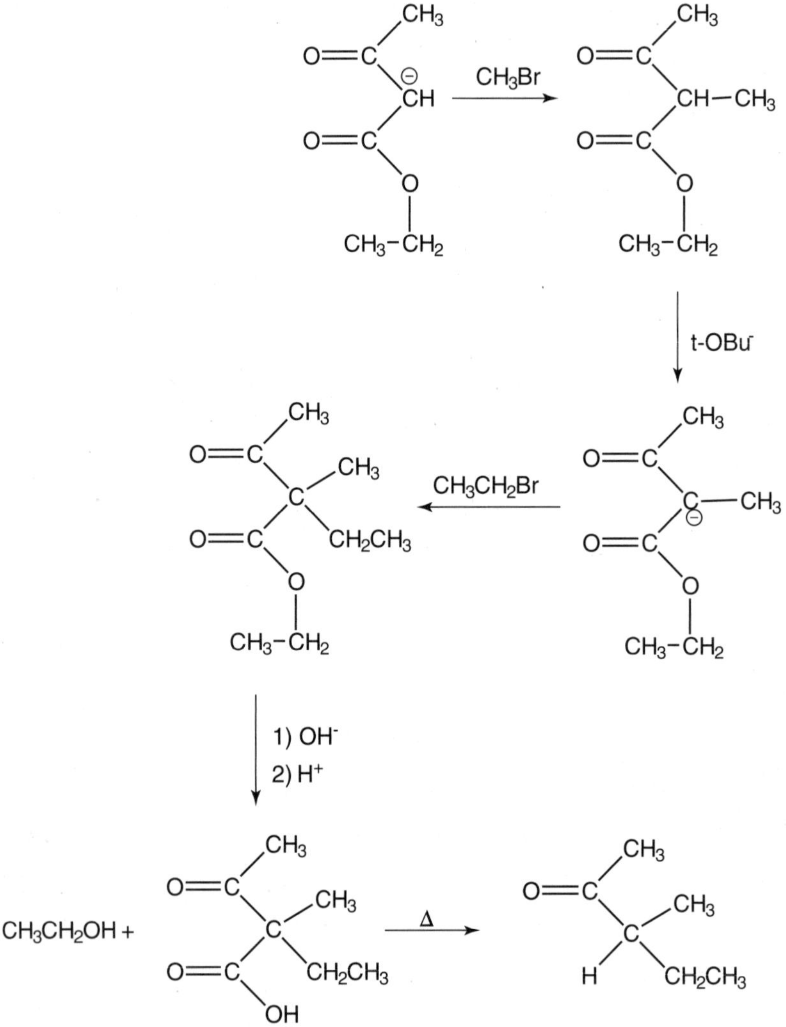

Figure 17-6: Adding the reaction with bromomethane.

You can link to the starting material from Figure 17-1 with one more step. You need to convert the starting material to the carbanion at the top of Figure 17-6. The complete answer is in Figure 17-7.

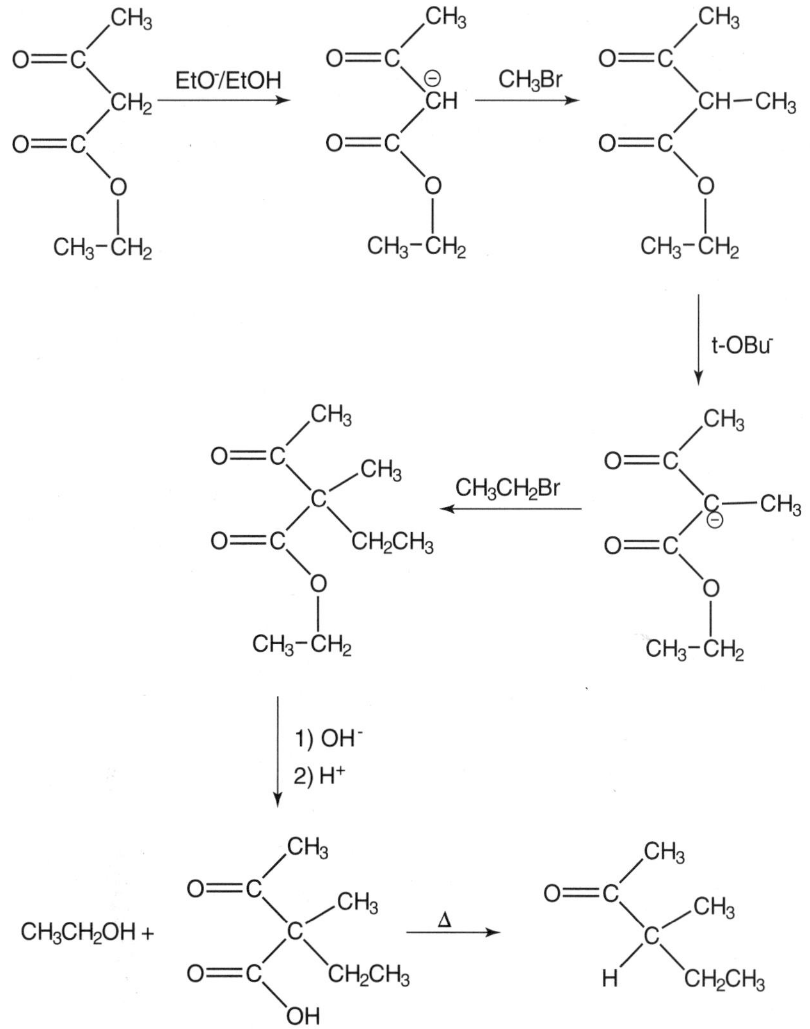

Figure 17-7: The solution to the problem presented in Figure 17-1.

Example 2

Figure 17-8 presents a synthetic problem and the answer is given in Figure 17-9. Try to solve this problem before looking at the answer. You may use any inorganic reagents and any organic reagent containing four carbons or less.

Synthesize

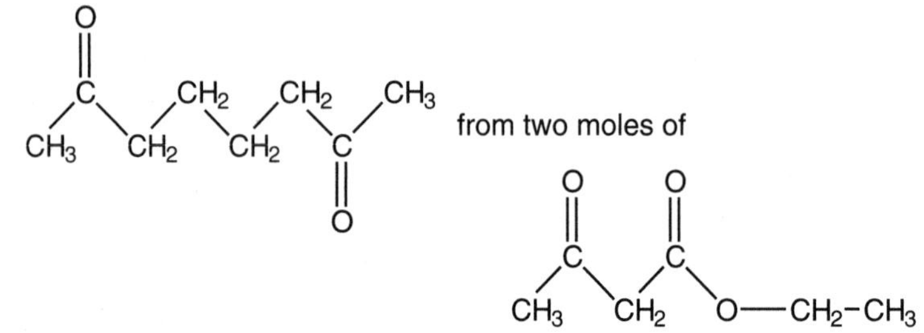

$CH_3C(=O)CH_2CH_2CH_2CH_2C(=O)CH_3$

from two moles of

$CH_3C(=O)CH_2C(=O)O{-}CH_2{-}CH_3$

Figure 17-8: The second analysis problem.

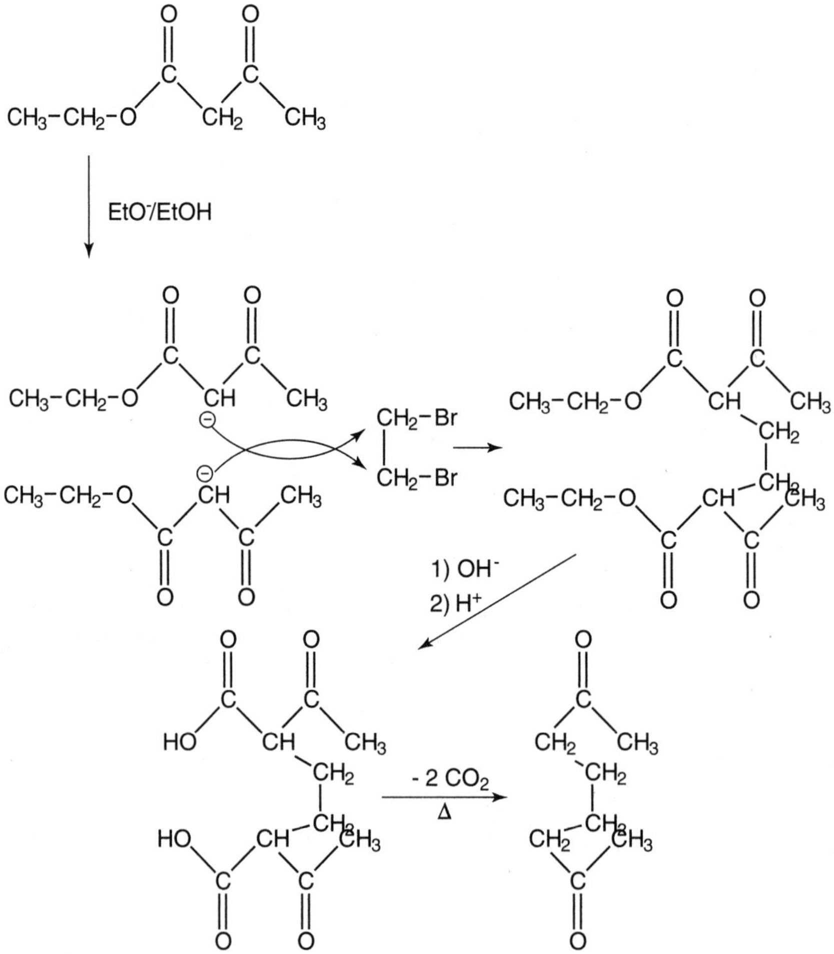

$CH_3{-}CH_2{-}OC(=O)CH_2C(=O)CH_3$

$EtO^-/EtOH$

$CH_3{-}CH_2{-}OC(=O)CH^{\ominus}C(=O)CH_3$ (two moles) + $CH_2{-}Br$ / $CH_2{-}Br$ →

1) OH^-
2) H^+

- 2 CO_2
Δ

Figure 17-9: The solution to the synthetic analysis presented in Figure 17-8.

Example 3

Another retrosynthetic analysis problem is in Figure 17-10. In this problem, you may use any inorganic reagents and any organic reagent containing four carbons or less.

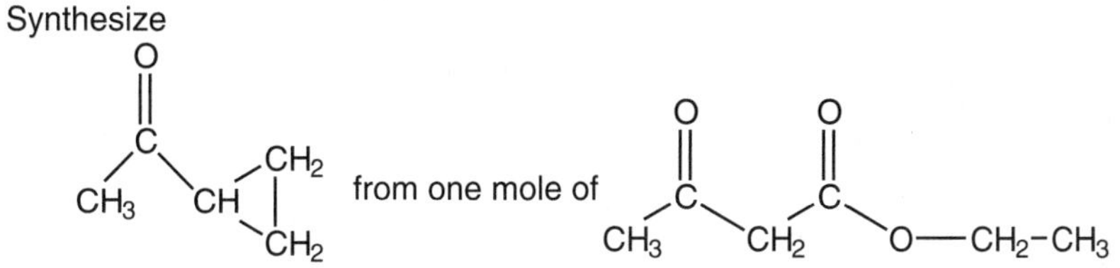

Synthesize

$$CH_3-C(=O)-CH(CH_2CH_2) \text{ from one mole of } CH_3-C(=O)-CH_2-C(=O)-O-CH_2-CH_3$$

Figure 17-10: The third analysis problem.

As seen previously in a variety of reactions throughout this book, the desired product can be formed by decarboxylation. Figure 17-11 illustrates this decarboxylation step.

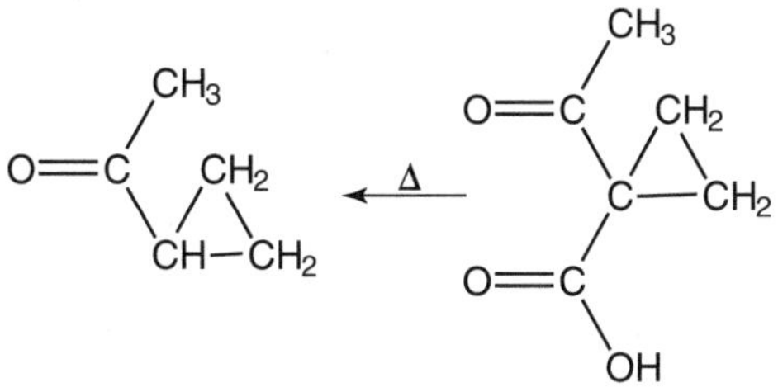

$$CH_3-C(=O)-CH(CH_2CH_2) \xleftarrow{\Delta} CH_3-C(=O)-C(CH_2CH_2)-C(=O)-OH$$

Figure 17-11: Decarboxylation to produce the desired product shown in Figure 17-10.

You now need to prepare the cyclopropane starting material. Figure 17-12 shows one way of preparing the cyclopropane starting material.

$$CH_3-C(=O)-CH(CH_2CH_2) \xleftarrow{\Delta} CH_3-C(=O)-C(CH_2CH_2)-C(=O)-OH \xleftarrow[2)\ H^+]{1)\ OH^-} CH_3-C(=O)-C(CH_2CH_2)-C(=O)-O-CH_2-CH_3$$

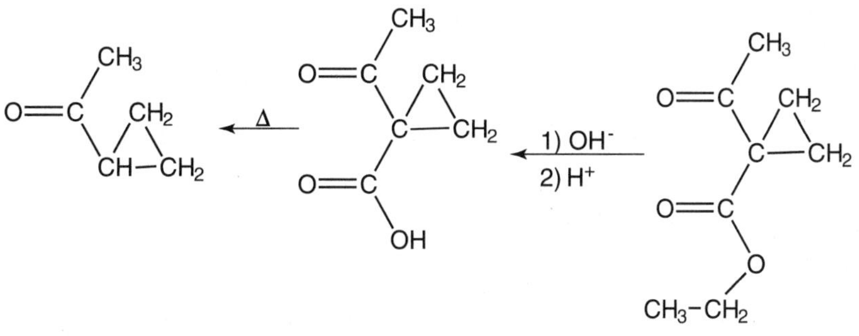

Figure 17-12: Preparation of the cyclopropane product.

Two steps, shown in Figure 17-13, are required to prepare the cyclopropane ring. Even though not normally required in a retrosynthetic analysis problem, Figure 17-13 also shows a partial mechanism.

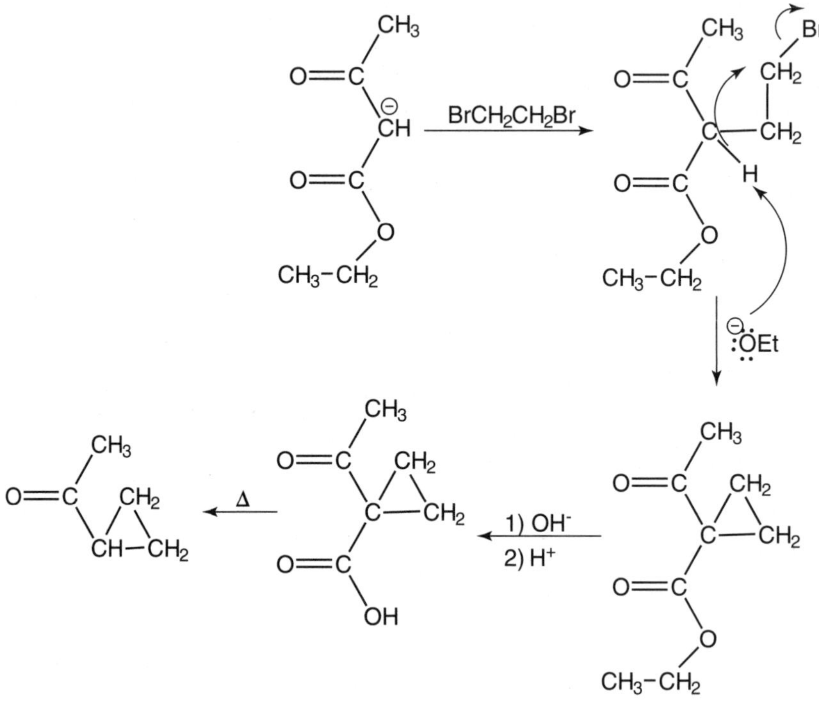

Figure 17-13: The formation of the cyclopropane ring.

Finally, you need to form the carbanion at the beginning of Figure 17-13. This step completes this retrosynthetic analysis problem. The complete solution is in Figure 17-14.

Example 4

Figure 17-15 presents a fourth problem, another retrosynthetic analysis, and Figure 17-16 shows one possible solution. You may use any inorganic reagents and any organic reagent containing four carbons or less.

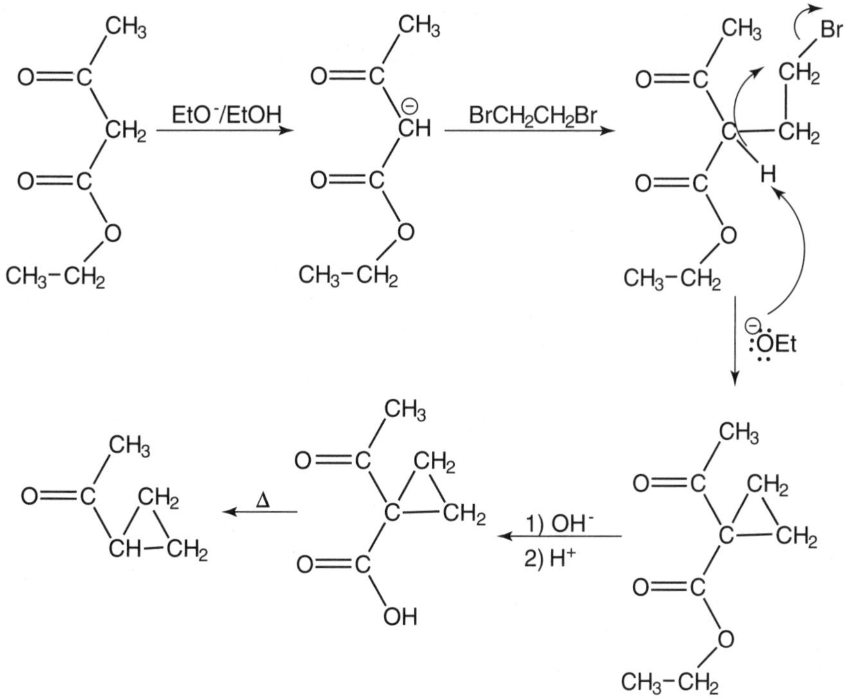

Figure 17-14: The complete solution to the problem presented in Figure 17-10.

Synthesize

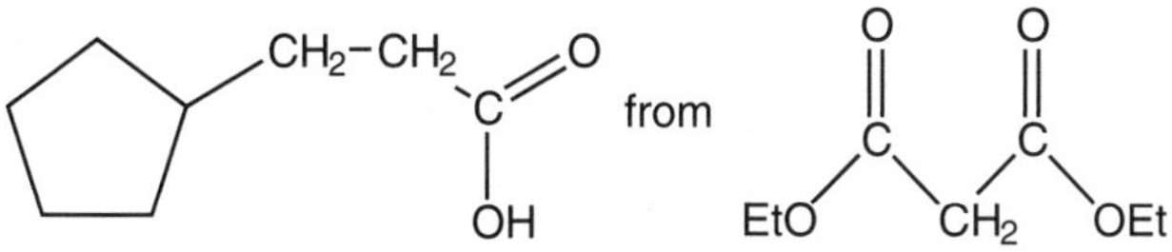

Figure 17-15: The fourth analysis problem.

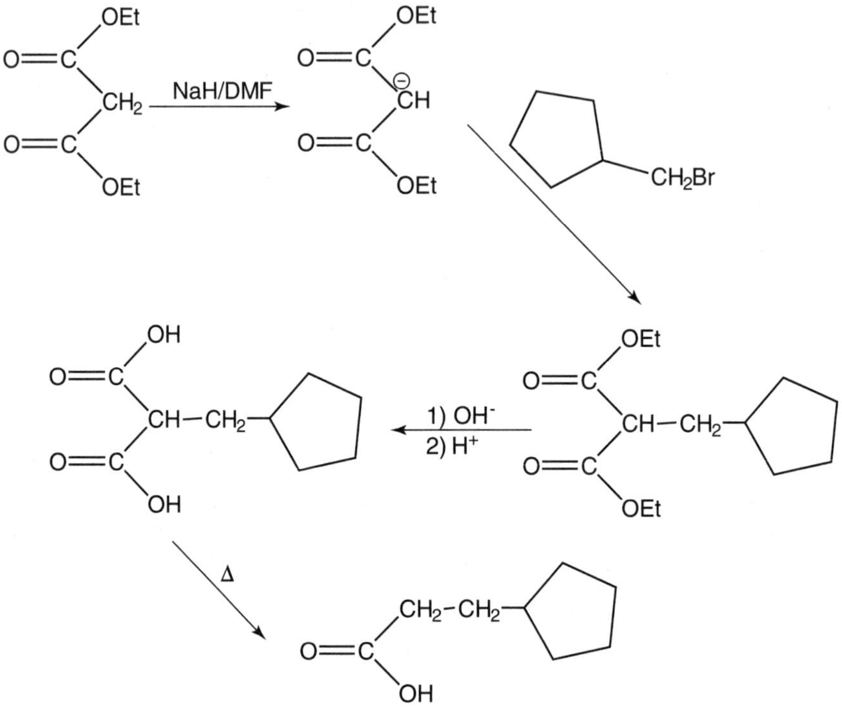

Figure 17-16: The solution to the problem presented in Figure 17-15.

Example 5

Figure 17-17 presents one last retrosynthetic analysis problem, with its solution in Figure 17-18. You may use any inorganic reagents and any organic reagent containing four carbons or less.

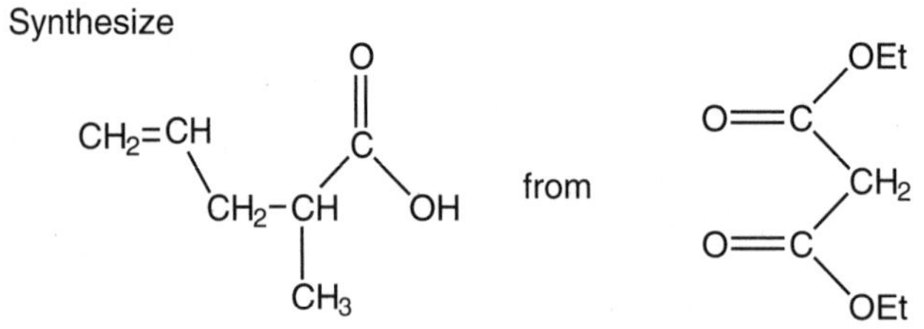

Figure 17-17: The last retrosynthetic analysis problem.

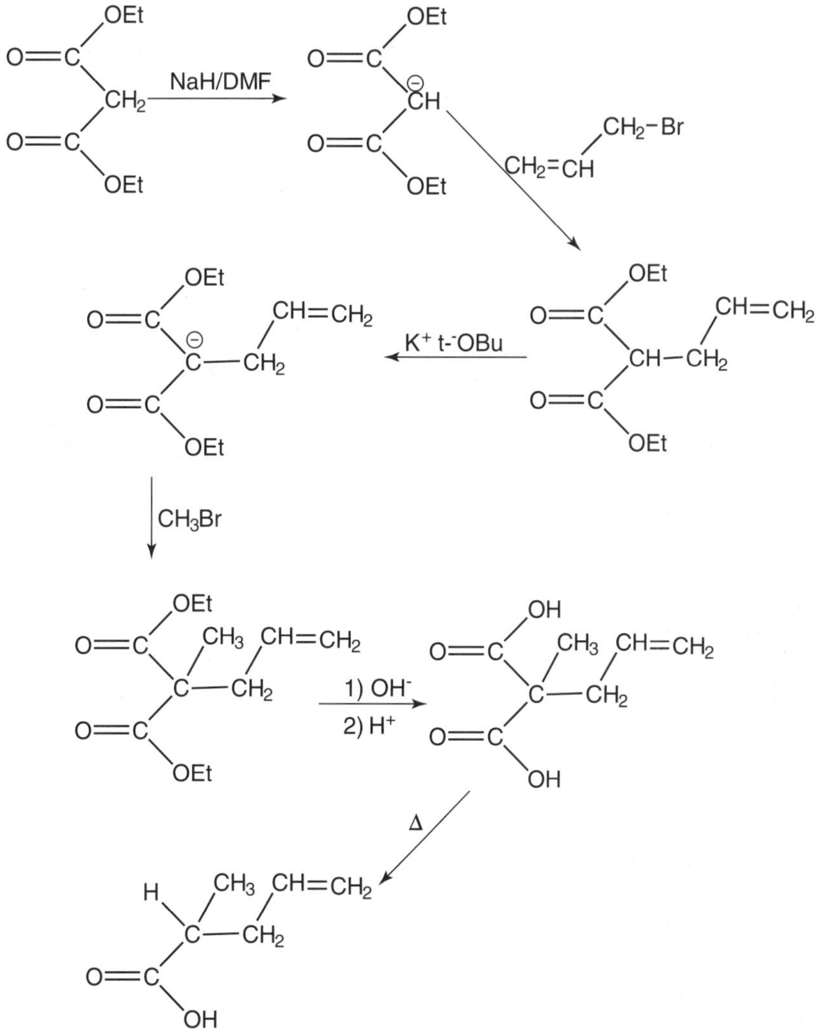

Figure 17-18: The solution to the problem presented in Figure 17-17.

Chapter 18

Roadmaps and Predicting Products

In This Chapter

- Mastering roadmaps
- Finding out how to predict products

Creating roadmaps and predicting the products of a reaction are two challenges found on many Organic Chemistry II exams. In Chapter 16 we give you some techniques on how to handle roadmaps. In this chapter we apply some of those techniques to show you how to build a roadmap without totally freaking out. Then we give you some tips in predicting reaction products.

Preparing with Roadmap Basics

Many Organic Chemistry II exams contain problems known as roadmaps that present you with a collection of facts you use to deduce the identities of a number of compounds.

Over the years, we have seen numerous students throw up their hands when faced with a roadmap problem. Indeed, many students simply skip the roadmap problems on their exams. That's a good way to lose a significant number of points and unnecessarily lower your grade.

If you know your reactions and the other rules, roadmaps aren't as difficult as they seem. The secret is to tackle the problem in small pieces, first by reading the problem and making a few notes; then continuing with the exam. Come back to the problem later and make a few more notes. Then go to some other part of the exam. Continue cycling from the roadmap to other questions on the exam until you have sufficient notes to attempt to solve the roadmap. If the roadmap defies solution at this point, return to the cycling procedure until you're ready to make another attempt or until that's the only part left on the exam.

Practicing Roadmap Problems

Now you get to take a look at number of roadmap problems. Some instructors include spectra on one or more of the compounds in the problem. We limit the amount of spectral data so you can focus on the approach to the solution of roadmap problems.

Look over the question before you look at its solution. See if you can develop a plan of attack on your own. Take your time.

Problem one

Compounds *A* and *B* are both hydrocarbons of molecular weight 98. They're both optically active but have different specific rotations. Ozonolysis of *A* gives formaldehyde (CH_2O) and a ketone; ozonolysis of *B* produces formaldehyde and another aldehyde. Treating either *A* or *B* with hydrogen/catalyst yields *C*, whose molecular weight is 100. *C* is an optically active compound. Give structures for *A, B,* and *C.*

Solution one

On scratch paper make a table consisting of three rows or columns labeled *A, B,* and *C* respectively. Each time through the problem, try to add some information to this table. For example, on your first pass you may find:

A	***B***	***C***
C_xH_y	C_xH_y	C_xH_{y+2}

The formulas for *A* and *B* reflect the fact that these are hydrocarbons, while the formula for compound *C* includes the information that catalytic hydrogenation yields an increase of two mass units (corresponding to two hydrogen atoms). In addition, the hydrogenation reaction indicates that *A* and *B* are alkenes.

On the second pass through, you may wish to estimate the values of *x* and *y*. If the molecular weight is 98, then the maximum value for *x* is 98 ÷ 12 g per atom of carbon = 8 (only whole numbers of atoms are possible). A value of 8 doesn't leave much room for hydrogen (2 hydrogen atoms bring the value to 98). If *x* is 7, then *y* needs to be 14, while if *x* is 6, then *y* needs to be 26. (For an alkane, if *x* is 6, the maximum *y* is 14; therefore, *x* cannot be 6 or smaller.) This limits *x* to 7, which requires *y* to be 14. This changes the table to:

A	***B***	***C***
C_7H_{14}	C_7H_{14}	C_7H_{16}

The formulas of *A* and *B* correspond to C_nH_{2n}, which means these are simple alkenes with no rings present. The formula for *C* (C_nH_{2n+2}) is that of an alkane (also with no rings).

Since *A* and *B* both undergo hydrogenation to form *C*, they must have the same backbone structure (which is the same as that of *C*). At this point, you may wish to sketch skeleton arrangements for seven carbon atoms. There are ten possibilities, but only two of the possible arrangements are optically active. These two arrangements are shown in Figure 18-1. The starred carbons are chiral carbons.

Figure 18-1: Optically active possibilities for seven carbon atoms (labeled carbon atoms are chiral).

```
C—C—C*—C—C—C              C
    |                     |
    C            C—C—C*—C—C
                   |
                   C
```

Formaldehyde and a ketone can only be produced from ozonolysis (of *A*) in a limited number of positions, which are indicated in Figure 18-2. The x marks the possible position of the double bond in the structures. Two of the three choices in Figure 18-2 would eliminate the chiral center. Thus the double bond in *A* must be the only one that doesn't involve the chiral carbon atom. This gives you not only the structure for *A*, but also the structure for *C* (the hydrogenated version of *A*). Compound *B* has the same skeleton but has only one place for the double bond if ozonolysis gives formaldehyde and an aldehyde (between the last two carbon atoms on the right side of the structure). Figure 18-3 gives the structures of *A*, *B*, and *C*.

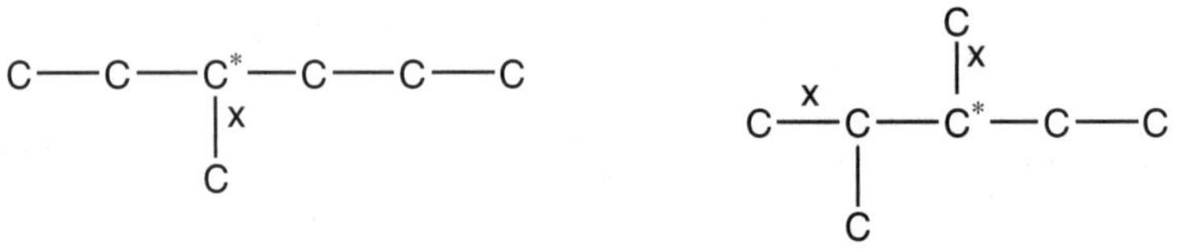

Figure 18-2: The possible positions (x) for the double bond in compound *A.*

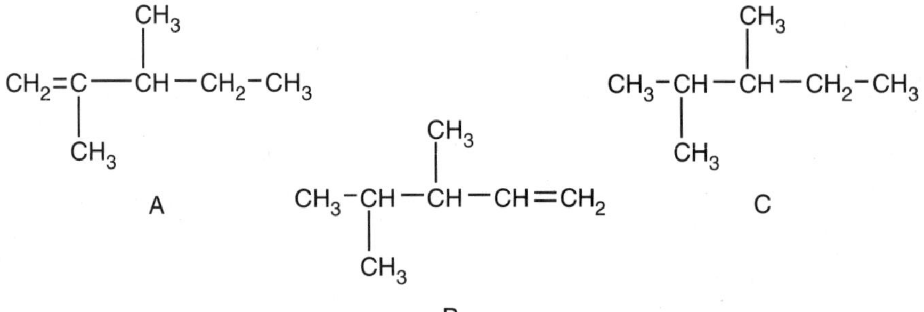

Figure 18-3: The structures of *A, B,* and *C.*

You should now work backwards from your answer to make sure these structures account for the facts in the problem.

Problem two

Unknown compound *D,* which has the formula $C_{10}H_{10}$, readily decolorizes a bromine in carbon tetrachloride solution. When *D* is allowed to react with hydrogen and a catalyst, *E* is obtained. The M^+ for *E* has an m/e value of 134. When *E* is heated with a mixture of nitric and sulfuric acid, only one product is produced. That compound has an M^+ of 179. Heating either *D* or *E* in hot chromic acid produces a solid *F,* $C_8H_6O_4$, whose melting point is in excess of 300 degrees Celsius. Give structures for *D, E,* and *F.*

Solution two

Your table begins with three columns.

D	*E*	*F*
$C_{10}H_{10}$	C_xH_y	$C_8H_6O_4$

Again, the solution involves several passes through the problem, gathering data. Don't worry if you gather this data in a different sequence than the one shown. You just need a collection of facts, and the order of acquisition is irrelevant.

- Compound *D* has m/e equal to 130, so the compound (at 134) has increased by four hydrogen atoms ($C_{10}H_{14}$). You now know the formulas for all three unknowns. You also know that *D* has either a carbon-carbon triple bond or two carbon-carbon double bonds. The reaction with bromine in carbon tetrachloride confirms this.
- The degree of unsaturation for compound *D* is six. The conversion of *D* to *E* accounts for two of the six. The remaining four (since they don't react) indicate an aromatic ring. (An insufficient number of carbon atoms are present to account for this degree of unsaturation by rings alone.)

  See *Organic Chemistry I For Dummies* by Arthur Winter (Wiley) if you're not sure how to determine the degree of unsaturation.
- The conditions leading to the formation of *F* indicate a disubstituted aromatic ring. In addition, the reaction conditions and formula indicate that *F* is a dicarboxylic acid. Disubstitution may be ortho, meta, or para.
- The reaction of compound *D* or *E* with a mixture of nitric and sulfuric acid nitrates the ring. The odd M^+ indicates an odd number of nitrogen atoms and the mass increase (179 – 134 = 45) indicates the product is mononitro.
- The two substituents in *D* contain four carbon atoms. These could be a methyl plus a C_3 or two C_2 groups.
- If *D* is ortho and if one group is methyl and the other group is C_3, then four mononitroproducts are given. If *D* is meta, then it has four mononitroproducts. If *D* is para, it has two mononitroproducts.
- If *D* is ortho and if both are C_2, then two mononitroproducts are given. If *D* is meta, three mononitroproducts are given. If *D* is para, then it has one mononitroproduct.
- The problem states that only one nitration product forms. Therefore, *D* (and by implication *E* and *F*) is para substituted. In addition, both C_2 branches must be identical, which means each has one degree of unsaturation (two alkenes and not one alkyne). The branches are $-CH=CH_2$. Hydrogenation converts the branches to $-CH_2-CH_3$. Heating with chromic acid eliminates the carbon atom furthest from the ring and converts the remaining carbon atoms to carboxylic acid groups ($-CO_2H$).

The structures of *D, E,* and *F* are found in Figure 18-4.

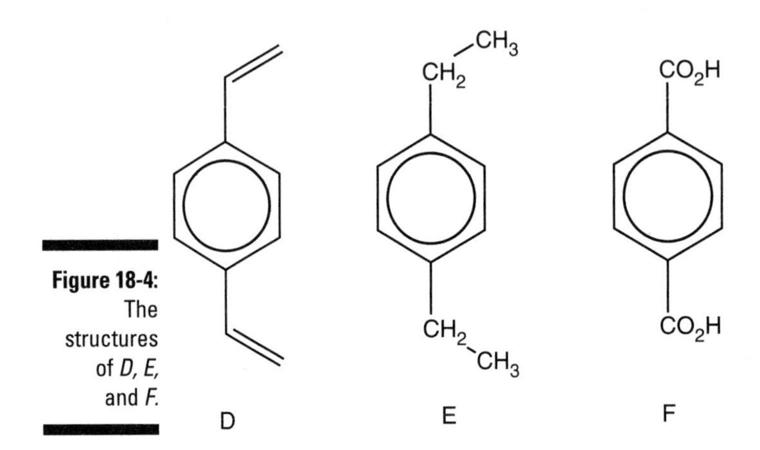

Figure 18-4: The structures of *D, E,* and *F.*

Problem three

G and *H* are isomeric compounds of formula $C_{12}H_{17}Br$. Both give an immediate precipitate upon reaction with alcoholic silver nitrate. They both react with NaOEt/EtOH to give *I,* a hydrocarbon with an M^+ m/e value of 160. *I* gives five singlets in the NMR, the one at δ7 being rather broad. *I* decolorizes bromine in carbon tetrachloride and adds one mole of HBr to give *G.* If peroxides are present, *H* is obtained. Ozonolysis of *I* produces *J* and *K,* both of which have a strong band at 1700 cm^{-1} in the IR. *K* is identical to the product obtained when methylacetylene is allowed to react with mercuric sulfate in aqueous acid. *J* gives two singlets and two doublets in the NMR. The low field doublets are in a 2:3 ratio to the upfield singlets. Nitration of *J* gives only two products, both of which have M^+ = 179 and an M + 1 intensity of 10.37. One of the two products strongly predominates over the other. Give the structures for *G* through *J.*

Solution three

As usual, begin with a table.

G	*H*	*I*	*J*	*K*
$C_{12}H_{17}Br$	$C_{12}H_{17}Br$	C_xH_y	?	?
m/e 241	m/e 241	m/e 160		

You need to collect the following facts (in any order).

- The signal at δ7 in the NMR of *I* indicates the presence of an aromatic ring (C_6). In addition, *I* has four other nonequivalent hydrogens that are not on a carbon atom that has hydrogen atoms on the adjacent carbon atom. *J* has four NMR signals; therefore, the hydrogen atoms are in four different environments. (Worry about the ratios later.)
- The strong band at 1700 cm^{-1} in the IR of *J* and *K* indicates the presence of a carbonyl group.
- *G* and *H* both give an immediate precipitate upon reaction with alcoholic silver nitrate. This means that the bromine is probably attached to a tertiary carbon.
- The reaction with NaOEt/EtOH is a dehydrohalogenation, which indicates that *I* is an alkene. Since both *G* and *H* give *I,* they must have similar regiochemistry. The m/e value (160) corresponds to the loss of HBr.
- The reaction of *I* with bromine in carbon tetrachloride confirms the prediction that *I* is an alkene. The reaction with HBr means that *G* is the Markovnikov addition product, and the reaction in the presence of peroxides makes *H* the anti-Markovnikov product.
- The ozonolysis of *I* along with the IR data indicate that *J* and *K* are aldehydes, ketones, or one of each.
- Methylacetylene is $CH_3\text{-}C{\equiv}CH$.
- The M^+ and M + 1 intensity for *J* indicates that it's a mononitro product. Since it has only two products, J must be para. The predominance of one product over the other means that one of the groups is a better director than the other.

The previous eight deductions give the foundation for solving the problem.

The easiest structure to deduce is that of *K*. The reaction of methylacetylene with mercuric sulfate in aqueous acid gives acetone. Therefore, *K* is acetone. (One down — four to go. Hooray for partial credit.) The loss of acetone means you have lost C_3H_6 from *I* and added an oxygen.

You now have all the formulas (and one structure):

G	***H***	***I***	***J***	***K***
$C_{12}H_{17}Br$	$C_{12}H_{17}Br$	$C_{12}H_{16}$	$C_9H_{10}O$	C_3H_6O
m/e 241	m/e 241	m/e 160	m/e 134	m/e 58

The presence of an aromatic ring in *I* indicates that all five compounds are aromatic. *J* has three carbon atoms that aren't part of the ring (and others that are). Para substitution indicates the presence of two branches — a C_1 and a C_2. The 2:3 ratios indicate methyl groups (two of them). If the C_1 is methyl, then the C_2 must contain the carbonyl, which may either be on the carbon next to the ring or on the carbon further away. If the carbonyl is next to the ring, you have your second methyl group (the carbon further from the ring). You now have the structure of *J*. See Figure 18-5 for the structures of *J* and *K*.

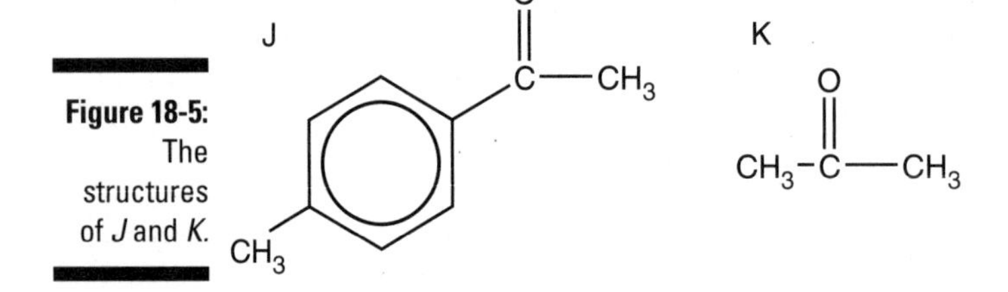

Figure 18-5: The structures of *J* and *K*.

Rearranging *J* and *K* leads us to *I* (see Figure 18-6).

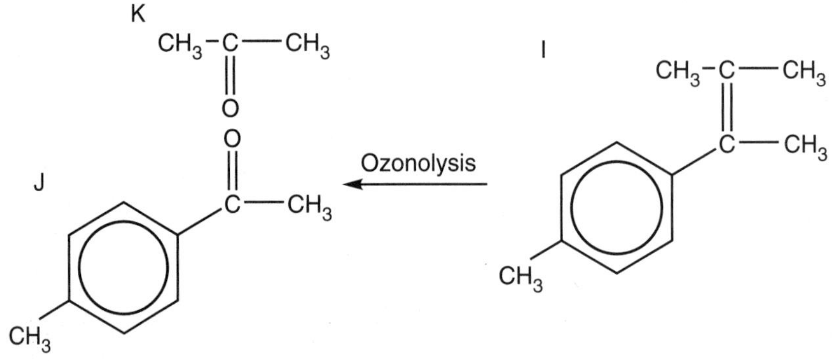

Figure 18-6: Finding *I*.

Adding HBr to *I* gives *G* and *H* as seen in Figure 18-7.

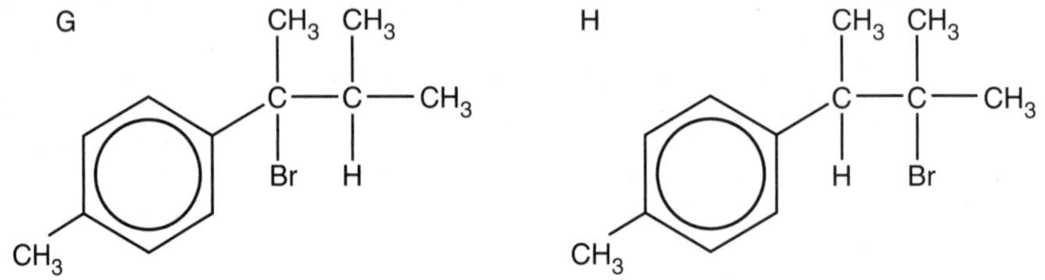

Figure 18-7: The structures of *G* and *H*.

Predicting Products

Being able to predict the products of a particular reaction is very important. It's easy to do when you only have one or two reactions to consider, but as you progress through an organic chemistry course, the number of reactions increases dramatically. Therefore, you need to keep some generalities in mind when considering a reaction: the type of reaction, the regiochemistry of the reaction, and the stereochemistry of the reaction.

In analyzing the type of reaction, you should pay close attention to both the starting material and any added reagents. In some cases, the reaction conditions are also important. In general, the reactions you studied most recently are important; however, organic chemistry exams are notorious for including any reaction since the beginning of Organic Chemistry I. In all reactions, the key is the functional group. Each functional group has a limited number of options, and you need to be aware of what these are. As a last resort, when dealing with a reaction that you don't recognize, you may want to start a mechanism. For example, most reactions done in acidic media begin with the protonation of an electron-rich site such as a nitrogen, oxygen, or carbon-carbon double bond.

Information about all three considerations (type, regiochemistry, and stereochemistry) appears in the mechanism.

Don't be distracted by large, complicated molecules. Focus on the functional group. The remainder of the molecule usually goes along for the ride without any changes.

The regiochemistry is important. For example, does the reaction follow Markovnikov's rule or is the reaction anti-Markovnikov? Or is the substituent a meta-director or ortho-para-director? You learned rules such as this for a reason; make sure you continually apply them.

Finally, the stereochemistry is important. This may be as simple as worrying about the inversion of configuration that occurs during an S_N2 mechanism. In other cases, however, it may be important to remember if the reaction is by syn-addition or anti-addition. Again, you should know these rules and always apply them.

We know that keeping all these rules and facts straight seems like a tremendous task, but the secret is practice, practice, and more practice. Eventually reactions and mechanisms will become natural. You'll develop a chemical intuition that will save you time and effort. In fact, you may actually start enjoying the challenge of roadmaps. (Okay, we know we went a little far with the last one, but occasionally it really does happen!)

Part VI

The Part of Tens

The 5th Wave By Rich Tennant

"So what if you have a PhD in organic chemistry? I used to have my own circus act."

In this part . . .

We know that your major goal in taking Organic II is to gain expertise in the specific Organic II topics, but we also know that you want to maximize your course grade. In these two chapters we offer some suggestions that we hope will maximize the results of your study. The first chapter is a tongue-in-cheek collection of ten suggestions to help ensure you don't pass. No; to be serious, these suggestions offer some guidance about general study techniques. The second chapter offers some definite ideas that will help you increase your individual exam score. All you need in addition to these tips is old-fashioned hard work.

Finally, as a bonus, we include an appendix full of named reactions to help you study.

Chapter 19

Ten Surefire Ways to Fail Organic Chemistry II

In This Chapter

- Discovering how to avoid the pitfalls of studying organic chemistry
- Finding out about effective study strategies
- Understanding how to use every resource available

You probably bought this book to help you succeed in your Organic II chemistry course. This chapter title caught your eye, and you thought, "I don't want to fail, I want to pass." Well, we decided to tackle the study habits and techniques part of the course in a different fashion than most other lists of advice, and these "tips" are based on what we've heard our student say throughout our years of teaching. So take the advice here in the good natured, tongue-in-cheek manner it was intended.

Simply Read and Memorize Concepts

A lot is sometimes said about organic chemistry simply being a memorization course. Yes, you need to do a lot of memorization, but applying the concepts requires much more than simple memorization. This is especially true of the second semester of organic chemistry. You will not only be applying the new concepts that you are learning, but also the concepts from the first semester course. It's a lot like learning French or any other foreign language. You need to know (memorize) the vocabulary and rules of grammar, but in order to be able to converse in that language, you must be able to put all that vocabulary together in the proper way (grammar) to convey an idea. The same is true of organic chemistry.

Don't Bother Working the Homework Problems and Exercises

The key to passing organic chemistry is work, work, work. Your brain needs repetition to form those neural pathways, so you must work and rework the homework exercises and exercises. And don't be sloppy — if you leave out formal charges or ionic charges or don't check the number of bonds for each carbon, you'll do the same thing on exams where those mistakes cost you points. You may want to start or join a study group if you're the type of person that benefits from discussing exercises with other students. And be prepared to wear down a lot of pencils and use a lot of paper. Don't try to conserve paper by writing and drawing so small that you can't see everything.

Don't Buy a Model Kit

Organic chemistry is a very visual subject, but most of us have difficulty visualizing a structure in three dimensions. Making a 3-D model of the structure allows you to detect finer points of conformation, steric hindrance, and so on. Making a model also helps you find those carbons with incorrect numbers of bonds.

Don't Worry About Falling Behind

One surefire way to flunk organic is to get behind. Organic is not one of those classes that you can ignore and then cram the night before a test. It's best done in small chunks, so plan on studying organic six or seven nights a week. Don't try to jam it all into two or three study sessions a week.

How long should you study each time? Good question, but no pat answer. The college-course rule of thumb that says you should spend two to three hours studying for every hour you're in class, but for organic, it's probably more than three for every class hour. However, you should know from your performance in Organic Chemistry I if this is reasonable for you or if you need even more. And remember, the quality of your studying is far more important than quantity.

Don't Bother Learning Reactions

Organic chemistry, especially Organic II, is all about the reactions and their mechanisms. If you don't know the reactions, you can't pass. You need to know the reactions by name, the reactants, the products, and the conditions, along with the reaction mechanism. Practice, practice, practice. Use flash cards. Quiz each other in your study groups. Write, write, write. Group reactions by product, by type, and so on. Do them forward and backward. Know those reactions!

If Your Textbook Confuses You, Don't Bother with Additional Resources

Sometimes organic textbooks aren't the easiest books to read, but no one says you can't use multiple sources. You already invested in this *For Dummies* guide, so use it. If you're having a difficult time with a particular concept, search the Internet and other organic chemistry textbooks until you find an author who explains in a way that makes sense to you. Use several sources and compare. Yes, all of that takes time, but it's worth it in the long run.

Don't Bother Reading the Chapter before Attending Class

Going into a lecture "cold" isn't a good idea. Okay, so maybe you don't need to read the entire chapter, but at least take 30 minutes or an hour to read the part your instructor will be covering that day. Don't try to work any exercises; just try to familiarize yourself with general concepts. Pay particular attention to vocabulary because it will help you get the most from the lecture. In other words, set the stage before you walk into class. Remember the six Ps — prior preparation prevents pretty poor performance.

Attend Class Only When You Feel Like It

Class attendance has a direct correlation to class grade. If you don't want to pass, don't come to class. Most people find that they can't really "get" organic just by reading the textbook. You need to watch your instructor draw structures, push electrons around (organic teachers are such bullies), and so on. You'll benefit from the in-class discussions, pick up some tricks of the trade, and maybe discover the logic of organic chemistry. Besides, your instructor may give a hint as to what's going to be on the next exam.

Don't Bother Taking Notes — Just Listen (When You Aren't Sleeping or Texting)

We said you should go to class; now we're saying that you should pay attention and take notes. A crucial survival skill for Organic Chemistry II: Learn to listen to your instructor, watch the board or screen, and take notes all at the same time. Have extra pencils handy (possibly with different colors). If you drop the one you are using, just grab another one. If you stop to pick up the dropped pencil, you may get so far behind that you end up dropping the course (especially if you're taking Organic II during a summer term!). Your lecture notes will probably be somewhat messy, so recopy them ASAP before you forget what they mean. The act of recopying notes within 24 hours of taking them is a great reinforcement of the material, and you gotta form those neural pathways. And don't even think about texting in organic class — unless you plan to repeat the class!

Don't Bother Asking Questions

If you're puzzled about something in class, ask about it either during class or during your instructor's office hours. Most professors teach because they like it, so they're happy to answer your questions. But if you go to your instructor for help, avoid saying "I just don't know anything," because what your instructor hears is that you haven't invested the necessary study time to know what you don't know. Go in with a list of specific questions, and go through them one by one. You have to learn the material yourself; the instructor is your guide. Don't worry about appearing to be stupid. The only stupid question is one that you don't ask. Better to ask questions than to give proof of your ignorance on the next test.

Chapter 20

More than Ten Ways to Increase Your Score on an Organic Chemistry Exam

In This Chapter

- Mastering some study strategies in preparing for an organic chemistry exam
- Avoiding the mistakes commonly made on exams
- Remembering that a carbon atom has four bonds

In this chapter we focus on those activities and techniques that will help increase your score on an organic exam. Many of these can also be used with other courses also. Many student "freeze" on an exam due to lack of self-confidence, but you can build up a confident nature by making sure that you know the material and by having a positive attitude. Think of the exam as an opponent you're about to battle and be eager to win (that is, to show your prof what you know). Who knows, you may actually grow to like Organic Chemistry. Sounds unlikely, we know, but stranger things have happened.

Don't Cram the Night before a Test

An organic chemistry test is not a test you can cram for the night before. Or even a few days before. (You probably found that out in Organic Chemistry I.) You need to study all along, a minimum of six days a week. Don't try to create review sheets just before an exam, either; make them all throughout the semester as you study the material. Then when exam times rolls around, you already have your sheets made and can start reviewing them. And don't underestimate the power of positive self-talk. Even when you get stuck in your studying, *never* say you can't understand the material, or that you're going to fail an exam or the course, and so on. Keep telling yourself that you can do it (I can pass the exam! I'm going to make an A in the course!) and pretty soon you'll believe it, and with hard work it can come true.

Try Doing the Problem Sets and Practice Tests Twice

When you're doing your problem sets or practice tests, work them twice. The first time you may have to refer to your notes or book. Make sure you understand and master the material that gave you trouble. Then work the problems again, but the second time try not to use your notes, and work them in a random order. Sometimes the context of the material gives you clues that may not be present on an exam, so you don't want to rely on them. Working problems in a random order is an especially powerful study habit when gearing up for the final exam.

Study the Mistakes You Made on Previous Exams

Use your exams to help fix mistakes in your knowledge and reasoning. Fully correct your errors on old exams as soon as possible. If your teacher thought that material was important enough to put on an exam, you will be seeing it again on another exam or the final. That corrected exam becomes part of your study/review material. Face it — every exam in organic chemistry is cumulative. Learn from the mistakes you made and don't make them again. (Don't worry, you'll have the opportunity to make brand-new ones!)

Know Precisely Where, Why, and How the Electrons Are Moving

When writing and studying reactions, pay attention to the mechanism, especially where, why, and how the electrons are moving. This is true whether you're doing homework problems or making your practice sheets. Be sure to use the right types of arrows — double arrows for equilibrium, single arrows for reactions, curved arrows or curved half arrows for electron movement, and so on. Also, don't try to combine too many mechanistic steps, especially on an exam. Take it one step at a time and your results will be clearer and easier to grade (and this is a *very* good thing). Keep in mind, though, that you may not be asked for the mechanism on an exam, just the reaction. In that case, only write the reaction. You can get yourself into trouble by volunteering extra information.

Relax and Get Enough Sleep before the Exam

You really have to be able to think when taking an organic chemistry exam. If you try pulling an all-nighter, you won't be able to think. So treat it the way athletes do a big match or game: Relax and get to bed early. Get up in plenty of time to have breakfast before your exam. Use positive self-talk. You may find that you want to isolate yourself from the other students before an exam so that you don't get into a question and answer dialogue and panic yourself.

Think Before You Write

Before you start to answer a test question, stop and think. Write down the assumptions that are pertinent to the question/problem on your scratch paper. Make a few notes and maybe even sketch out the reaction/answer before you start answering the question on the exam paper. Then work steadily and carefully on the answer. Make sure that everything you write down is clear and reasonable.

Don't write down unnecessary information on the exam. It takes time, distracts your professor as she grades the answer, and gives her more things to count incorrect. Pay attention to the question and fully answer it — no more, no less.

Include Formal Charges in Your Structures When Appropriate

Using formal charges sometimes allows you to pick out the most appropriate or likely structure, so use them when trying to decide between different Lewis structures. Certainly show them on exams when asked. If you *really* feel they're necessary, show them even if they weren't requested, but be absolutely sure they're correct. You don't want to volunteer unnecessary information that could be wrong.

And be sure to write your structures neatly on the exam, because if your instructor can't follow what you have done, you're going to lose points. If you get into the habit of drawing your structures neatly on your homework, you will do the same on an exam.

Check That You Haven't Lost Any Carbon Atoms

Losing carbon atoms (not accounting for all your carbon atoms) is easy to do when writing a multistep mechanism. Being sure to account for all your carbons is especially important in those reactions where you might be losing carbon dioxide or another small carbon-containing molecule. Even though many times your instructor won't require a balanced chemical equation, he will be upset if you lose carbons.

Include E/Z, R/S, cis/trans Prefixes in Naming Organic Structures

Forgetting to use prefixes is a common mistake that students make in the midst of an exam. Make sure you indicate the appropriate stereochemistry in your nomenclature, especially if your instructor takes time to indicate the specific stereochemistry of a compound. Again, get into the habit of doing this when you're working on homework exercises.

Think of Spectroscopy, Especially NMR, As a Puzzle

Spectroscopic data can be very useful on an exam, but think of it as individual pieces of a puzzle. Write down each absorption and assign a structural value to it, and then step back and look at the overall picture and try to see how all those individual pieces fit together. This is an especially valuable tip when faced with NMR and IR data since many times this data is far more detailed than UV-vis spectra or mass spectra.

Make Sure That Each Carbon Atom Has Four Bonds

We believe that more points have been lost on organic exams due to this one mistake more any other. Be sure that every carbon *atom* has four bonds. (This might not be true for ions.) Leaving off a bond, commonly to a hydrogen atom, is like waving a red flag in front of the grader. She gets that small smile, shakes her head, and down comes the red pen. Remember: Carbon makes four bonds!

Appendix

Named Reactions

Lots of chemical reactions take place in the diverse world of organic chemistry. Some are named after people; some are named after reactants or products. If you want to pass your Organic Chemistry II class, you need to know the following named reactions:

- **Acetoacetic Ester Synthesis:** The formation of a substituted acetone through the base-catalyzed alkylation or arylation of a β-keto ester.
- **Aldol Cyclization:** An internal aldol condensation.
- **Aldol Reaction:** The formation of an aldol (β-hydroxy carbonyl compound) through the catalyzed condensation of an enol/enolate with a carbonyl compound.
- **Cannizzaro Reaction:** The formation of an acid and an alcohol through the base-catalyzed disproportionation of an aliphatic or aromatic aldehyde with no α-hydrogen atoms.
- **Claisen Condensation:** The formation of a β-keto ester through the base-catalyzed condensation of an ester containing an α-hydrogen.
- **Claisen-Schmidt Reaction:** The production of an α,β-unsaturated aldehyde or ketone from an aldehyde or ketone in the presence of strong base.
- **Cope Elimination:** The pyrolysis of an amine oxide to produce a hydroxylamine and an alkene.
- **Crossed Aldol Condensation:** An aldol condensation involving different carbonyl compounds.
- **Crossed Cannizzaro Reaction:** A Cannizzaro reaction involving two different aldehydes.
- **Crossed Claisen Condensation:** A Claisen condensation utilizing a mixture of two different esters.
- **Curtius Rearrangement:** Similar to a Hofmann degradation with an azide replacing the amide.
- **Dieckmann Condensation:** The intramolecular equivalent of a Claisen condensation where dicarboxylic acid ester undergoes base-catalyzed cyclization to form a β-keto ester.

- **Diels-Alder Reaction:** The reaction of an alkene (dienophile) with a conjugated diene to generate a six-membered ring.
- **Friedel-Crafts Reaction:** The Lewis acid–catalyzed (usually $AlCl_3$) alkylation or acylation of an aromatic compound.
- **Gabriel Synthesis:** The reaction of an alkyl halide with potassium phthalimide to form, after hydrolysis, a primary amine.
- **Grignard Reaction:** The reaction of an organomagnesium compound, typically with a carbonyl compound to produce an alcohol, although it may be used in other situations.
- **Hell-Volhard Zelinsky Reaction:** A method for forming α-halo acid.
- **Hofmann Elimination:** Converts an amine into an alkene.
- **Hofmann Rearrangement:** A useful means of converting an amide to an amine.
- **Hunsdiecker Reaction:** A free-radical reaction for the synthesis of an alkyl halide.
- **Knoevenagel Condensation:** A condensation of an aldehyde or ketone with a molecule containing an active methylene in the presence of an amine or ammonia.
- **Malonic Ester Synthesis:** Synthesis involving a malonic ester or a related compound with a strong base such as sodium ethoxide. The ultimate product is a substituted carboxylic acid.
- **Mannich Reaction:** The reaction of a compound with a reactive hydrogen with aldehydes (non–enol forming) and ammonia or a primary or secondary amine to form a Mannich base (aminomethylated compound).
- **Michael Addition (Condensation, Reaction):** The addition of a carbon nucleophile to an activated unsaturated system.
- **Reformatsky Reaction:** A reaction leading to formation of β-hydoxy esters, using an organozinc intermediate.
- **Robinson Annulation:** The addition of a methyl vinyl ketone (or derivative) to a cyclohexanone to form an α,β-unsaturated ketone containing a six-membered ring.
- **Sandmeyer Reaction:** A reaction utilizing a diazonium salt to produce an aryl halide. The process begins by converting an amine to a diazonium salt.
- **Schiemann Reaction:** A means of preparing aryl fluorides.
- **Stork Enamine Synthesis:** A reaction leading to the formation of an α-alkyl or α-carbonyl compound from an alkyl or aryl halide reacting with an enamine.

Index

• A •

• B •

• C •

• D •

• E •

• G •

• H •

• I •

• O •

• Q •

• R •

• S •

• T •

• U •

• V •

• W •

• Y •

• Z •

Green Business Practices
FOR
DUMMIES

Completely revised and updated! New best practices to seal the deal and win the offer
Job Interviews
FOR
DUMMIES
3rd Edition
Joyce Lain Kennedy
A Reference for the Rest of Us!

Now revised and updated! Your plain-English guide to maximizing returns
Stock Investing
FOR
DUMMIES
2nd Edition
Get savvy tips on today's best investment opportunities — and pitfalls
A Reference for the Rest of Us!
FREE eTips at dummies.com
Paul Mladjenovic

Making Everything Easier!
2nd Edition
Knitting
FOR
DUMMIES
Learn to:
Pam Allen
Tracy L. Barr
Shannon Okey

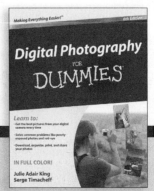

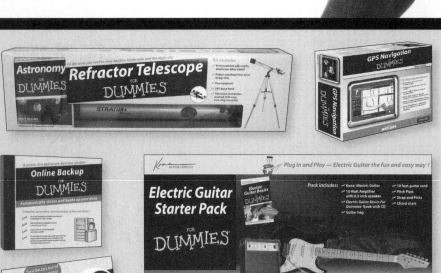